The International Society for Science & Religion was established in 2002 to promote education through the support of inter-disciplinary learning and research in the fields of science and religion. Our current membership of 140 comes from all major faith traditions and includes non-religious scholars. Inducted by nomination only, they are drawn from leading research institutions and academies worldwide. The Society embraces all perspectives that are supported by excellent scholarship.

In 2007, the Society began the process of creating a unique resource, *The ISSR Library*, a comprehensive reference and teaching tool for scholars, students, and interested lay readers. This collection spans the essential ideas and arguments that frame studies in science, religion, and the human spirit.

The Library has been selected through a rigorous process of peer review. Each constituent volume makes a substantial contribution to the field or stands as an important referent. These books exhibit the highest quality of scholarship and present distinct, influential viewpoints, some of which are controversial. While the many perspectives in these volumes are not individually endorsed by the ISSR, each reflects a facet of the field that is worthy of attention.

Accompanying the Library is *The ISSR Companion to Science and Religion*, a volume containing brief introductory essays on each of the Library's constituents. Users are encouraged to refer to the *Companion* or our website for an overview of the Library.

Physics and Cosmology
Scientific Perspectives
on the Problem of Natural Evil

Physics and Cosmology

Scientific Perspectives
on the Problem of Natural Evil

Nancey Murphy

Robert John Russell

William R. Stoeger, SJ

Editors

Vatican Observatory
Publications
Vatican City State

Center for Theology and
the Natural Sciences,
Berkeley, California

2007

Nancey Murphy is Professor of Christian Philosophy at Fuller Theological Seminary in Pasadena, California, USA.

Robert John Russell is Founder and Director of The Center for Theology and the Natural Sciences and Ian G. Barbour Professor of Theology and Science in Residence at The Graduate Theological Union in Berkeley, California, USA.

William R. Stoeger, SJ, is Staff Astrophysicist and Adjunct Associate Professor of Astronomy at the Vatican Observatory, Vatican Observatory Research Group, Steward Observatory, University of Arizona, Tucson, Arizona, USA.

Jointly published by the Vatican Observatory and
the Center for Theology and the Natural Sciences

Distributed (except in Italy and the Vatican City State) by
        The University of Notre Dame Press
        Notre Dame, Indiana 46556
        USA

Distributed in Italy and the Vatican City State by
        Liberia Editrice Vaticana
        V-00120 Citta del Vaticano
        Vatican City State

ISBN **978-88-209-7959-1**

ACKNOWLEDGMENTS

The editors wish to express their gratitude to the Vatican Observatory and the Center for Theology and the Natural Sciences for co-sponsoring this research. Particular appreciation goes to George Coyne, whose leadership at the Vatican Observatory made this and an earlier series of conferences possible. Thanks, also, to the staff at the Observatory and at CTNS for their work in organizing the conference, with special gratitude to Nate Hallanger for creating a website to facilitate interaction among participants prior to the conference.

Editing for this volume began with an initial circulation of proposed paper topics, and then of draft papers, before the conference in September 2005, and continued with interactions among the editors and authors after the conference. The editors express their gratitude to all of the participants for written responses to pre-conference papers and for their enthusiastic discussions during the conference.

Susan Carlson Wood, in Fuller Theological Seminary's School of Theology Faculty Publications Services, managed the copy editing, proof reading, and typesetting in expert fashion. The editors offer special thanks for her patient cooperation and expertise. We thank Fuller Seminary for its support of this phase of the project.

We also thank David Brewer, PhD student at Fuller Seminary, for assistance with preparing the index, Bonnie Johnston and other CTNS staff for preparing the cover design, and Francesco Rossi, on the staff at the Vatican Observatory, for overseeing printing and distribution of the book.

CONTENTS

CONTENTS

III. RESPONSES: THE GOD-WORLD RELATION

IV. CHANGING THE TERMS OF THE DEBATE

ABBREVIATIONS

CC   *Chaos and Complexity: Scientific Perspectives on Divine Action.* Edited by Robert John Russell, Nancey Murphy, and Arthur R. Peacocke. Vatican City State: Vatican Observatory; Berkeley, CA: Center for Theology and the Natural Sciences, 1996.

EMB   *Evolutionary and Molecular Biology: Scientific Perspectives on Divine Action.* Edited by Robert John Russell, William R. Stoeger, and Francisco J. Ayala. Vatican City State: Vatican Observatory; Berkeley, CA: Center for Theology and the Natural Sciences, 1998.

NP   *Neuroscience and the Person: Scientific Perspectives on Divine Action.* Edited by Robert John Russell, Nancey Murphy, Theo C. Meyering, and Michael A. Arbib. Vatican City State: Vatican Observatory; Berkeley, CA: Center for Theology and the Natural Sciences, 1999.

PPT   *Physics, Philosophy, and Theology: A Common Quest for Understanding.* Edited by Robert John Russell, William R. Stoeger, and George V. Coyne. Vatican City State: Vatican Observatory; Berkeley, CA: Center for Theology and the Natural Sciences, 1988.

QCLN   *Quantum Cosmology and the Laws of Nature: Scientific Perspectives on Divine Action.* Edited by Robert J. Russell, Nancey Murphy, and C. J. Isham. Vatican City State: Vatican Observatory, 1993.

QM   *Quantum Mechanics: Scientific Perspectives on Divine Action.* Edited by Robert John Russell, Philip Clayton, Kirk Wegter-McNelly, and John Polkinghorne. Vatican City State: Vatican Observatory; Berkeley, CA: Center for Theology and the Natural Sciences, 2001.

INTRODUCTION

Nancey Murphy

1 Background to the Volume

In September 2005 a group of philosophers, theologians, and scientists from four continents met at the papal villa in Castel Gandolfo, Italy, to consider one aspect of the problem of evil, so-called natural evil, in light of recent developments in physics and cosmology. Most of the essays in this volume were discussed and debated there, and then revised for publication.

This conference was one of many responses to a call from Pope John Paul II in 1979 for an interdisciplinary collaboration of scholars to seek a fruitful concord between science and faith. The Vatican Observatory, a century old astronomical research institution now located in Castel Gandolfo and at the University of Arizona, Tucson, has already responded in a number of ways. An initial conference was convened in 1988, and its proceedings published in *Physics, Philosophy, and Theology*. This was followed by a series of five conferences, cosponsored by the Center for Theology and the Natural Sciences in Berkeley, California, with the overall theme being "Scientific Perspectives on Divine Action." The present conference represents a continuation of that endeavor. This conference, cosponsored by the Center for Theology and the Natural Sciences in Berkeley, California, represents a continuation of that endeavor. The topic was chosen because of its importance to Christian theology *and* the fact of the relevance of contemporary science to its discussion. It has become traditional to distinguish among three different sorts of evil: moral, natural, and metaphysical (although these terms will be critically examined in some of the chapters below). Moral evil is what we usually think of evil, that is, human sin.[1] "Natural evil" is being used here to designate the suffering that comes to humans and other sentient creatures as a result not of the evil acts of others but by means of natural causes—tsunamis, earthquakes, and hurricanes being particularly salient examples as this introduction is being written. "Metaphysical evil" has been used to designate the sorts of weaknesses and limitations to which humans are subject.

A great deal of philosophical and theological attention has been devoted to the problem of moral evil throughout Christian history and long before. Indeed, some would take an attempt to account for the human predicament in its failures as one of the marks of a religious worldview. It is not clear that contemporary science has a great deal to add to these discussions. The most common Christian response is to argue that free will is a necessary prerequisite for the loving response that God desires of human creatures, and that freedom will inevitably be misused.[2]

---

[1] In the past, angelic sin would have been included as well.

[2] One point at which science and philosophy of science may have a say in this discussion is via their relevance to the issue of free will itself. Is it possible to reconcile freedom and moral responsibility with law-governed neurobiological processes? See Russell, Murphy, Meyering, and Arbib, *NP*; and also Murphy and Brown, *Did My Neurons Make Me Do It?*

However, natural science is quite relevant to the problem of natural evil (and metaphysical evil as well). Classically natural evil has been explained in terms of moral evil. There is more than one way to do this, but the most familiar theological account contends that natural evil and death entered creation through the catastrophic fall of original humanity (or perhaps anticipated by the prehistoric fall of rebellious angels). Natural evils on this account are both a divinely imposed punishment for human disobedience, and also a natural consequence in that human (or angelic) rebellion broke the chain of command from God to the human and then to the natural worlds, allowing God's originally perfect creation to manifest disruptions such as earthquakes and droughts.

In many respects both biblical criticism and science have called this defense into question. The Genesis story of a first set of parents is no longer considered to be historical. We no longer have the hierarchical cosmos in which such accounts were developed, and natural history shows that there must have been millions if not billions of years of death before humans entered the scene. How can these natural evils now be reconciled with the creator's perfect goodness? One kind of response is to show that the evils of the sorts we find in our world *must* be permitted or produced in the process of realizing the goods that God intends for creation. Science's contribution is to consider the law-like character of natural processes, recognizing that much suffering is simply the consequence of the regularities of nature. In addition, recent work on the "fine-tuning" of the constants and basic laws of physics suggests that the laws themselves had to be almost exactly as they are in order that life exist in the universe.

This is a bald statement of the problem to which this volume is addressed and a rough hint at the direction in which many of the authors look for resources. But note that much of the work herein will involve not only responses to but refinements or even rejections of the questions and resources as I have characterized them.

The earlier project on divine action was closely related to this one. The rationale for the focus on God's action in the natural world was again twofold. First, many of the other topics that arise in the dialogue between theology and science are integrally related to this issue. Second, this was a problem upon which scientific insights could be brought to bear. It was modern science that made divine action problematic, in a way that it had not been in earlier periods. Before the rise of modern science it was not difficult to reconcile divine action, whether of the general or special sort, with natural causation: in the vocabulary of medieval theologians such as Thomas Aquinas, God is the primary cause, and natural entities are secondary causes, much like a carpenter and the carpenter's tools. When modern scientists such as Isaac Newton began to emphasize the law-like character of natural processes, this perfect regularity in nature was seen as a testament to God's sovereignty: just as humans obey (imperfectly) God's moral law, so too nature obeys (perfectly) God's "natural laws." But as the concept of the laws of nature became a pillar of modern understanding, problems arose for traditional theological accounts. While it was easy to maintain a role for God as creator and sustainer of the whole natural order, it became difficult to understand *special* divine acts, that is, those that could not be seen as mediated by the regular workings of the laws of nature. Many theists rejected the concept of miracles as irrational intrusions in God's orderly creation. Much of the development of modern theology of a more liberal sort can be seen as an attempt to maintain Christian teaching about God's involvement in "salvation history" without resorting to accounts of supernatural intervention in the natural order.

Most of the participants in that series of conferences were convinced of the necessity of holding on to some concept of "special divine acts," yet they were reluctant to achieve this by taking the route of the more conservative theologians who insist on God's miraculous contravention of the laws of nature, a route that often leads to inattention to the results of science or, worse, to their outright rejection.

However, if the rise of modern science was the source of the problem of divine action, happily, science itself may well have contributed to its solution. The mechanistic conception of nature sponsored by Newtonian physics has given way to a much richer and more complex view of the world, the implications of which we are only beginning to grasp. The scientific topics chosen for the meetings were (1) quantum cosmology (combined with the topic of the "fine-tuning" of the laws of nature), (2) chaos theory, (3) molecular and evolutionary biology, (4) neuroscience, and (5) quantum mechanics.[3]

The problem of natural evil is related to the topic of divine action: To the extent that one finds theologically, philosophically, and scientifically plausible accounts of special divine action, the question becomes acute— why does God seem to do so little to prevent and alleviate the suffering of innocent victims of natural disasters, diseases, and so forth?

2 Overview of the Volume

The editors have divided this volume into four sections. The first includes history of the issue and a critical analysis of how the history has often been understood, followed by two chapters that provide typologies: one of types of suffering, the other of the various "shapes" of defenses. The second section comprises chapters that address the problem of suffering head-on, with resources from science, theology, and philosophy. The third section contains essays that address the issue by offering reformulations of typical understandings of the relation between God and the world. Finally, essays in the fourth section claim, in one way or another, that the question of the volume needs to be reframed.

---

[3] Papers from these conferences can be found in Robert J. Russell, Nancey Murphy, and C. J. Isham, eds., *Quantum Cosmology and the Laws of Nature,* hereafter *QCLN*; Robert John Russell, Nancey Murphy, and Arthur R. Peacocke, eds., *Chaos and Complexity,* hereafter *CC*; Robert John Russell, William R. Stoeger, and Francisco J. Ayala, eds., *Evolutionary and Molecular Biology*; hereafter *EMB*; Robert John Russell, Nancey Murphy, Theo C. Meyering, and Michael A. Arbib, eds., *Neuroscience and the Person,* hereafter *NP*; and Robert John Russell, Philip Clayton, Kirk Wegter-McNelly, and John Polkinghorne, eds., *Quantum Mechanics,* hereafter *QM*. For an evaluation of this project see Robert John Russell, William R. Stoeger, SJ, and Nancey Murphy, eds., *Scientific Perspectives on Divine Action: Problems and Progress* (Vatican City State: Vatican Observatory; Berkeley, CA: Center for Theology and the Natural Sciences, 2007).

2.1 History, Definitions, Typologies

This section contains four chapters. The first is **Niels Christian Hvidt**'s "The Historical Development of the Problem of Evil." Hvidt begins with a brief overview of the issue in the Christian tradition, but then widens his focus to other religions and to earlier sources. The problem of suffering has been said to be the touchstone of all religions. Max Weber's characterization of the problem in world religions has been influential. Weber argues that there are only three satisfactory types of response. One is the doctrine of karma, found in Hinduism and some of its offspring. A second is cosmic dualism, as found in Zoroastrianism and Manichaeism. The third is predestination, which he attributes to both Calvinism and Islam.

Christian responses to suffering draw upon both Hebrew Scriptures and Greek and Roman philosophy. There is no single approach to suffering in the Old Testament. Hvidt notes that there is a general difference between pre- and post-exilic thought. Pre-exilic writings tend to emphasize more that God is the source of all, and that therefore the LORD must be seen as sending both good fortune and calamity. The latter is understood to be largely deserved because of the people's unfaithfulness to the covenant. After the exile Hebraic thought bears the marks of association with dualistic cosmologies, and hence there is more of a tendency to countenance the effects of personified evil apart from God. The book of Job, however, suggests that all answers to the question of why the innocent suffer will be inadequate.

Hvidt claims that in the New Testament the "retaliatory theology" of the Old Testament is countered by six factors: recognition (as already in Job) of the problem of innocent suffering; the convictions that suffering can be fruitfully transformed, that it can be a means of testing, that people can suffer vicariously for others, that there is an inherent "economy of sin and suffering," and, finally, that God's love is always greater than his justice. The New Testament also emphasizes, as do many of the authors in this volume, that suffering can only be fully addressed in the context of eschatological hope.

Augustine has been perhaps the most influential writer on evil in the West. His writings incorporate several themes. One is the idea of evil as privation of good, taken over from Plotinus, but tempered by the contention that all of creation is essentially good—a universe containing all gradations of good is better than one with only the most noble of creatures. In addition there is the issue of human freedom, its misuse, and the idea that suffering is a just retribution for sin for *all* humans in that the "original sin" of the first parents is passed to the children.

While later theologians up to the Enlightenment followed closely in Augustine's footsteps, a major change in writings on evil and suffering occurred in the modern period. Hvidt takes the work of G. W. F. Leibniz as a turning point. Leibniz maintained Augustine's themes of evil as privation and of suffering as due to moral evil, and argued on logical grounds that this must be the best of all possible worlds. Leibniz coined the term "theodicy" to refer to arguments intending to show that God is not responsible for evil. Although Leibniz thus built heavily on the tradition in his responses to the problem of suffering and evil, his approach nevertheless represented a fundamental change: Prior to Leibniz, God was offended by the mess his creatures had made of his good creation through disobedience and sin, to the point that only God's son could solve the crisis. Particular features of the 1755 Lisbon earthquake, such as the fact that it occurred while most people were in church, resulted in widespread rejection of Leibnizian optimism. One might say that the burden of proof shifted from assuming God's existence and goodness and then trying to make sense of suffering, to questioning whether there can be a good God, given the world as we know it. So with Leibniz and the Enlightenment, the focus changed so that God became the one who needed justification for the evils in the world, with human reason as the judge regarding God's innocence or culpability.

Thus Hvidt's essay is structured in terms of "theodic analogies" prior to Leibniz and "a history of reception" of the problem after him. Later responses include Immanuel Kant's outright rejection of the adequacy of any theodicy, attempts as in G. W. F. Hegel's work to couch the problem in terms of development, and recent work such as that of John Hick intended to revive early Eastern accounts of God's purposes for suffering in human life.

One of **Terrence W. Tilley**'s main concerns in "The Problems of Theodicy: A Background Essay" is to call attention to the distinction between *theodicies* and *defenses* as different rhetorical practices. In constructing a defense, one is not attempting to warrant a claim that God exists or even to give a theory about God's intentions in allowing evil in the world, but merely to deflect a potential rebuttal of Christian belief, namely, the claim that the existence of evil in the world is *inconsistent* with God's goodness and power. Here one has only to show the possibility of reconciling God's goodness with the existence of evil.

In contrast, a theodicy is an attempt to *explain* the evil in the world. Because the practice of theodicy originated as a response to David Hume's and others' claims that the presence of evil in the world defeats design arguments, it is closely associated with the view that it is theists who bear the burden of proof. Thus, the evil needs to be explained as a step in an argument for the plausibility of God's existence.

Tilley attributes the first full-blown theodicy to William King, a contemporary of Leibniz, writing at the beginning of the eighteenth century. By the middle of the century, when design arguments were falling on hard times, theodicies came to predominate over defenses. Tilley notes that in Hick's very influential *Evil and the God of Love* something of the same shift can be seen between the first and second editions!

Tilley sets out a variety of criticisms of the practice of theodicy. One is the anthropocentrism of thinking that it is we humans who have to justify God. Another is that even if a theodicy succeeds, it only renders plausible a thin theism, and nothing like the Christian faith in a salvific, triune God. Third, theodicy is impractical. It creates an abstract concept of evil and contributes to evil itself by distancing and distracting its practitioners from the concrete, existential evils around them. It is a practice at home in the academy, and not in places of Christian ministry. In fact, the instrumental justification of evil can even deny evil and turn it into good. Finally, Tilley argues that theodicists tend to anachronism in attributing this sort of discourse to premodern writers.

So this essay stands as a warning to other authors in this volume not to contribute to the "evils of theodicy." He ends with a brief note about defenses. A defense rebuts the atheologian's claim that the existence of evil is logically incompatible with the existence of an omniscient, omnipotent, and omnibenevolent God. Tilley credits Alvin Plantinga with solution of this logical problem. Yet his solution requires the assumption of libertarian free will, ignores social evils, and requires the logical possibility that natural evil is demonically caused. This is the opening for Tilley's later chapter in which he will present a scientifically sensitive defense on the basis of the inevitably of evil in God's permitting genuine creativity to emerge in God's creation.

So far it has been taken for granted that we all know what constitutes suffering. **Wesley J. Wildman**, in his chapter "The Use and Meaning of the Word 'Suffering' in Relation to Nature," argues that while "natural evil" is used frequently in this volume, including in the title, "suffering" is a more useful category for framing the problem with which we are concerned here. In particular, "suffering" directly describes what most troubles us about so-called natural evil without biasing subsequent moral assessment of natural disasters, predation, and disease processes. Thus, he sets out to describe and classify the sorts of suffering that are found in nature, as well as to survey the causes of and natural responses to suffering.

Wildman considers a hierarchy in natural processes running from mere change through disruption of the integrity of complex systems, physical injury response, conscious pain, emotional distress, and finally existential anxiety. He recommends that "suffering" be used only for cases including injury response at the lower end and existential anxiety at the most extreme. However, he recognizes that it may be controversial to apply the term "injury response" to processes found in plants and in animals without central nervous systems. His justification is that even plants do *respond* to injury, for example by releasing chemicals that fight off predators or signal conspecifics. One of these is jasmonic acid, whose release is inhibited by aspirin, suggesting a possible evolutionary link with the development of pain responses.

Having considered the causes of suffering in nature, Wildman concludes that decay and dissolution are inevitable features of complex self-organizing systems and so suffering due to injury is universal in the biological realm. Death itself is not a form of suffering, though death in its psychological and social and natural connections is a profound cause of suffering. Physical pain is widespread, and emotional distress is universal among higher organisms. The highest form of suffering, existential anxiety, seems to be present only in human beings, as attested by the distinctively human phenomenon of suicide. Yet it may not be universal among human beings, owing to neurological deficits or trauma affecting some that prevent the requisite sensitivity from arising.

Wildman intends his survey of the forms and causes of suffering in nature to support his recommendation that "evil" should be limited to the realm of cognition, where suffering can be controlled or avoided and yet is caused cruelly or ignored selfishly. When "natural evil" is used in reference to lower-level causes of suffering such as earthquakes, its use should be seen as loosely analogical, *as if* there were an intentional agent deliberately using the earthquake to inflict suffering. He also aims to create a burden of demonstration for those making alternative proposals for the extent of suffering in nature, and substantively different recommendations for how to use the language of suffering.

Christopher Southgate and **Andrew Robinson** have titled their chapter "Categories of Good-Harm Analysis and Varieties of Theodicy: An Exploration of Strategies for Responding to the Problem of Evil." They point out that many responses to the problem rest on arguments to the effect that there are certain goods which are related to and thereby provide justification for the existence of the harm in question. As a way of understanding the basic structure of responses to the problem, they offer an analysis of the possible ways in which these goods and harms may be understood to be related. There are three ways in which such a good-harm analysis (GHA) may be formulated:

1. Property-consequence GHAs: a consequence of the existence of a good is the possibility of it causing harms. The classic instance of this is the free-will defense in respect of moral evil.
2. Developmental GHAs: the good is a goal which can only develop through a process which includes the possibility (or necessity) of harm. The most familiar version of this is John Hick's "Irenaean" theodicy that regards the world as a "vale of soul-making" in which virtue is learned through a process that involves suffering.
3. Constitutive GHAs: The existence of a good is inherently, constitutively inseparable from the experience of harm or suffering.

They then point out that each of these categories of good-harm analysis may have three types of "reference": human (the relevant goods and harms are restricted to humans); anthropocentric (the good accrues to humans, but the harm to a range of creatures); or biotic (both the goods and the harms may be experienced throughout the biosphere). The three possible categories and three possible references give rise to nine types of defense. Southgate and Robinson suggest that each logical type encompasses a family of varieties of defense; each shares the underlying logical structure of the type, but they may differ widely in actual content.

A neglected area of theodicy is consideration of "biophysical evil," harms to nonhuman creatures. Southgate and Robinson explore the application of their taxonomy to this area, comparing Southgate's own work on biophysical evil with that of Holmes Rolston, Jay McDaniel, and Ruth Page. Southgate and Robinson show that responses to biophysical evil tend implicitly to combine two or more types of theodicy. They suggest that, fruitful as such combinations can be, future work in this area would be clarified if the categories and references of the underlying good-harm analyses were made explicit.

The authors note that property-consequence and developmental arguments have tended to dominate the theoretical literature in theodicy, but observe that where the focus is on the seeming "overplus" of suffering in creation, or on the intensity of the suffering of the individual creature, constitutive positions come to the fore. Furthermore, whereas theoretical and practical responses are conventionally regarded as being logically distinct exercises, they argue that their taxonomy demonstrates how the conceptual and the existential aspects of the task may be understood to be in continuity. They conclude that there may be a constitutive instinct to the understanding of the relation of goods and harms within Christian theodicy, an intrinsic relationship between suffering and flourishing life. This would be in tune with the mysterious good represented by the passion and death of the Christ.

2.2 Scientific and Philosophical Responses

This section includes six chapters that are intended, in various ways, to address the problem of suffering at the hands of nature, the first employing more scientific resources and progressing toward more purely philosophical arguments.

In "Entropy, Emergence, and the Physical Roots of Natural Evil," **William R. Stoeger, SJ,** points out that the universe has gradually evolved from a very simple, hot, and homogeneous state to one of overwhelming richness and complexity. The course of this development is marked by the emergence of layer upon layer of intricately organized structures, capable of behavior impossible at lower levels. But it is also marked by the fragility, transience, and the inevitable dissolution and death of those complex systems during the course of evolution. What characteristics of nature are at the root of these pervasive contrary tendencies? Would a universe with the emergence of intricately organized, intimately and diversely interrelated entities be possible without transience and fragility?

Stoeger answers that entropy, both thermal and gravitational, as well as gravity itself, are deeply implicated both in emergence and in dissolution and decay. Any complex structure is composed of parts in very precisely ordered and specific functioning relationships. To effect those relationships and maintain them requires the availability of the components themselves and the energy and the proper environment within which to combine them. There is usually only one, or at most a few, ways of doing so such that the complex system works. But there are many more possible arrangements of the components such that the system will not work or will fall apart. Without a proper environment providing specific material and energy input, systems naturally tend toward one of those nonfunctioning states—their entropy or disorder tends to increase. This is merely to describe the second law of thermodynamics.

Gravity provides the overall context which supplies the fundamental macroscopic and microscopic differentiation that enables complex systems to emerge. In particular, it triggers the formation of stars, which manufacture the elements that are the building blocks of complex chemical and biological systems, as well as the low-entropy energy necessary for powering the processes of complexification on the nearby cool planets and moons associated with them. Gravitational entropy—disorder—increases as gravitational systems condense into more and more compact configurations, a black hole being the system of maximum gravitational entropy. This is the opposite of thermal entropy, disorder, and provides a reservoir of energy to enable the emergence and maintenance of complex systems elsewhere in the universe.

Without entropy, the susceptibility of systems to increasing disorder, with their attendant dissolution and decay, there would be little if any emergence of organization or complexity in our universe. This is simply because new, more complex systems must come from the rearrangement of less complex systems and from the energy they supply. This in turn directly requires the dissolution, both to provide their components and also to open ecological niches within which to prosper. Any universe which has its own internal integrity and dynamisms and is material, relational, interconnected, and evolving, and at the same time is open to new possibilities, predictable, and potentially personal, will have to be this way. It is only a static, complete, or thoroughly externally micromanaged universe that would be able to escape these requirements. Obviously such a boring universe would not be in keeping with the divine priority of inviting creation to deeper relationships, and ultimate communion.

The task of **Robert John Russell**'s chapter, "Physics, Cosmology, and the Challenge to Consequentialist Natural Theodicy," is to assess the possibility of constructing a robust "natural theodicy," that is, a theodicy whose focus is not on moral evil but on the suffering, disease, death, and extinction that characterize the evolution of life and the underlying physical processes which make natural evil at the biological level possible. Russell works with a consequentialist natural theodicy in which natural evils are an unintended consequence of God's choice to create life through the biological processes of evolution. While such theodicies are "reasonably robust" when considering evolutionary biology, the implications brought to theodicy by the physics underlying biology are severe, even if seldom made explicit in the literature. Hence Russell moves below biology to physics and back in time and space to the physical cosmology which describes our universe. Here we discover the fundamental laws of physics and the fundamental constants of nature that make our universe "fine-tuned" for the possibility of life and thus to what Russell calls "cosmic theodicy."

Russell concludes that a consequentialist cosmic theodicy faces two severe challenges. First, according to the doctrine of creation *ex nihilo* the physical laws and constants of nature are created, not given a priori. Hence the first challenge is stark: Why did God choose to create *this* universe with *these* laws and constants? Are there *no other possibilities open to God* to create a universe in which life could arise by natural processes and yet without the physical natural evils that characterize our universe?

The first challenge leads Russell to a prior scientific question: can contemporary physics help to explore the question of other possible universes in which life could arise? Here some of the research surrounding the anthropic principle, such as a multidomain mega-universe, might be helpful. The general implication of these theories is that the fine-tuning of the laws and constants of our universe for life is merely the random result of the existence of many universes embodying different laws of physics and different values of the constants. Perhaps God's choice is not at the level of the laws and constants of *our* universe but at some megalevel of universes, laws, and constants. Yet this in turn just reopens the question of cosmic theodicy at a higher level of abstraction: (1) Would slightly different values of the constants have produced a universe in which life evolved and in which the extent of natural evil was lessened? (2) Would completely different sets of these values have produced a universe in which life evolved and in which the extent of natural evil is lessened? If Russell can show to any reasonable extent that there are *no* other possible universes in which life could arise by natural processes but with significantly less suffering, then a consequentialist natural theodicy succeeds in the sense that God created our universe because these particular laws and constants are *uniquely* necessary for life. He explores these questions briefly as a way of pointing out directions for future research.

Even if a consequentialist cosmic theodicy at this level of abstraction can be supported there is a second and even deeper challenge: *is life worth the price of such extensive suffering?* This second challenge points us in a different direction and leads to a much different conclusion about the entire project of natural, versus moral, theodicy. This view is based on the fact that natural theodicy is framed within the doctrine of creation. But the Christian tradition's response to evil has always been to set the present world in a wider, eschatological context: the new creation proleptically begun in Christ and forming the ultimate future of our—and every—universe. It is only when the new creation is the starting point for reflecting on evil that Russell believes we can hope to give a response to its origin and meaning in this present, broken world. He closes, then, by turning to an eschatological framework and by hinting at ways in which it might offer a fruitful interaction with science, particularly with questions about the laws and constants of nature.

In "Science and the Problem of Evil: Suffering as a By-product of a Finely Tuned Cosmos," **Nancey Murphy** takes for granted the value of arguments such as Stoeger's and Russell's, detailing ways in which suffering appears to be a by-product of natural conditions necessary for human life. She attempts to sum up in a diagram the necessary conditions, and then show how the various evils can be seen to be unwanted but necessary consequences of these conditions. For example, the capacity to feel pain (or something equally noxious) is necessary to provide for the flexible responses of higher organisms to their environments without unnecessarily damaging themselves, but it is also the most direct cause of physical suffering.

The focus of Murphy's paper, however, is not to make these detailed arguments since other authors are better equipped to do so, but rather to consider the sorts of objections to which these "necessary by-product" arguments are subject. These objections take the form of the question: "Yes, but could God not have made the world better in x manner?" where x stands for questions of a variety of types and levels. She includes the following: (1) Could God not perform more special divine acts so as to alleviate more suffering? (2) If the fine-tuning requirements pertain to life as we know it, why not life of a different sort: (2.1) without pain? (2.2) without carnivory? (2.3) with different biology? (2.4) nonbiological? (2.5) nonmaterial? (3) What about a nonmaterial universe? (4) One of radically different material? (5) Why not begin with the world as envisioned eschatologically? (6) Finally, might it be the case that no world at all would be better than this one?

Other authors have addressed many of these questions, so again Murphy's goal is more focused. She first raises the question of whether the problem of suffering is about the world or a problem with how we speak about it. Suffering itself is a problem in the world; we must do as much as we can to combat it. Suffering becomes an *intellectual* problem only when we think and talk about it, and then only against the background of a number of things we hold to be true, such as the existence, goodness, and power of God. How much of the rest of our knowledge needs to be in place to get the problem of suffering off the ground? Murphy asks this in order to see whether some of the objections she considers are so contrary to fact as to be not only unanswerable but meaningless.

Questions 1 through 2.3 leave enough of our knowledge of the world intact so that we can have some sense of the state of affairs in question and evaluate its possibility (and preferability). Questions 2.4 through 4, however, call for a blank slate with regard to our general knowledge of reality. We believe that we can attach significance to these questions because they are in proper grammatical form and we know the meanings of the words. But if the world were as different as is proposed, then all of the referential and semantic relations would be different in ways that we cannot specify. We cannot make any more full-blown description of the state of affairs in question. Thus, she concludes that the sentences are in fact meaningless. She warns that the intellectual problem of suffering can easily grow beyond the boundaries of sense, and it is important to attempt to locate that boundary so that empty words do not create an unnecessary stumbling block to faith in a loving God.

In "The Lawfulness of Nature and the Problem of Evil," **Thomas F. Tracy** begins with the recognition that what we are now learning about the universe suggests that an astonishingly restrictive set of conditions must be satisfied if life is to arise at all. Further, these conditions necessitate that the history of life unfold by trial and error, with vast numbers of individuals being generated and destroyed. If we affirm that God is the source of the physical laws that characterize our universe, then we cannot avoid the question of why God chooses to create a world that has this terribly costly structure.

Philosophical discussions of this problem tend to focus on the question of whether we can generate a *morally justifying explanation* for God's permission of evil. Although this is a matter of importance to many theologians as well, the central concern of the Biblical religious traditions is to articulate a vision of *what God is doing* to address the presence of evil in our world. First and foremost, they are concerned with salvation, not explanation. This chapter seeks to integrate these sometimes disparate theoretical and practical approaches to the problem of evil by exploring the prospects for a theologically informed explanation, a "thick defense."

Any attempt to give a justifying explanation of God's permission of evils will need to address three key questions: (1) What is the good for the sake of which God permits evils? (2) What is the relation of evils to this good? (3) Is it morally justifiable to pursue this good in this way? Christianity typically has answered the first question by affirming that all finite things exist for the good of being-in-relation-to-God, and that for rational creatures this relationship takes the form of loving communion. We can then address the second question by thinking through the logically necessary conditions for the possibility of realizing this good. Modern discussions typically contend that among these necessary conditions is moral freedom. This makes it possible to explain moral evil as a product of responsible human choices, but it is less widely recognized that moral freedom also brings with it the inevitability of natural evils. Tracy argues that any world in which rational moral agency is possible must have a lawful, impersonal, and amoral structure. This is the "nomic condition" for the existence of finite persons, and it must be paired with a further, more restrictive, set of "anthropic conditions." The occurrence of natural suffering is an unavoidable by-product of meeting these conditions.

It is entirely possible to grant this point, however, and contend nonetheless that the world need not include *so many* evils. William Rowe has argued that an omnipotent and perfectly good being would prevent intense suffering whenever it could do so without loss of a greater good. Familiar instances of apparently pointless suffering provide sufficient evidence, therefore, that there is no such being. Wesley Wildman offers a similar "argument from neglect" in this volume. Tracy considers two ways of responding to this argument. The first approach (developed by Stephen Wykstra) contends that we have a radically incomplete grasp of the goods that God may bring about by permitting these evils. This reply, however, undercuts the possibility of providing even a partial explanation of God's permission of evils. The approach Tracy favors appeals instead to limits on our understanding of the conditions that must be met in order to realize familiar goods. We cannot very fully comprehend the preconditions and consequences of God acting to eliminate apparently pointless evils. This difficulty is deepened when we recognize that evils and goods are not always linked as means to ends; some evils must be permitted simply as a side effect of meeting the nomic and anthropic conditions. God has a purpose for permitting these evils, but the evils themselves have no purpose and will therefore appear to us to be preventable by God without loss. Indeed, we can allow that some that some of these evils may in fact be preventable by God, but only by substituting an evil that is equally bad.

This defeats the argument from excessive evils, but it does so by creating an epistemic stalemate. Just as the objector cannot make an evidential case that there are evils that God should have prevented, so the theist cannot provide sufficient evidence to show that there are no such evils. The project of explaining God's permission of evil must remain incomplete and hypothetical, and it will be most fruitful when it draws on the resources of the theological traditions that speak of what God is doing to overcome evil and that look ahead in hope to the eschatological completion of God's creative and redemptive work.

In "Divine Action and the 'Argument from Neglect'" **Philip Clayton** and **Stephen Knapp** respond to what they regard as the strongest reason to doubt the existence of an active, benevolent, and personal God: the occurrence of innocent suffering in cases where it seems that such a God could easily prevent it. Answering Wesley Wildman's "argument from neglect" (see below) they invoke the hypothesis that divine action is constrained by the conditions of a universe created, at least in part, as an environment in which finite rational agents could evolve. If God intervened whenever the operation of natural laws was about to produce an instance of innocent suffering, the universe would lack the predictability required for scientific inquiry and, more generally, for finite agents to acquire an understanding of their own existence.

Although this hypothesis may be a familiar one, it faces some serious objections. For example, it might seem to raise an obvious question: perhaps God cannot intervene in *every* case where we would expect a benevolent and sufficiently powerful being to do so; but why can God not intervene at least in some of the more spectacular cases, such as the Holocaust, or the Indian Ocean tsunami of 2004? To answer that question, Clayton and Knapp analyze what it means to suppose that God might intervene on some occasions but not others. They notice an important but unappreciated *moral* difference between the conditions of divine and human action. Because human beings lack the knowledge or power to respond in all cases where they might be morally obligated to do so, we forgive ourselves for succumbing to "compassion fatigue," or for choosing, however tragically, to rescue some victims at the expense of others. But God has no such excuse; if God intervenes *even once* to prevent or relieve a case of innocent suffering, God has no reason not to intervene in all other similar cases. Hence it is not clear how God might break the laws of nature in some cases without having to do so with a frequency that is incompatible with God's reasons for creating the universe in the first place. Clayton and Knapp call this the "not even once" principle.

Does this mean, as many have argued, that divine action is altogether excluded from our universe, so that a personal God becomes, in Wildman's phrase, "virtually irrelevant" to human affairs? The authors draw on emergence theory and recent work in the philosophy of mind to defend their claim that God could perform acts of communication with human beings without breaking natural laws. But even in an emergence-based theory, divine actions will still be subject to the "not even once" principle: even if God can communicate with human minds without breaking natural laws, it will not be morally permissible for God sometimes to warn us of impending disasters and other times not.

It should be noted that Clayton's and Knapp's approach to the problem of innocent suffering, and hence their answer to the "argument from neglect," is explicitly hypothetical. While it is not sufficient, they argue, merely to establish the logical possibility that God has a reason not to intervene, it is also not necessary to claim to know the precise reason why God does not intervene to prevent innocent suffering. What the theist needs—and what they claim to provide—is a plausible account of why divine intervention might not occur in cases where one would expect it.

Terrence W. Tilley's second chapter is titled "Toward a Creativity Defense of Belief in God in the Face of Evil." Rather than treating all evils as due either to individual sin or diabolical intervention, as does Plantinga's free will defense, Tilley offers a different account of both natural and social evils. Many so-called natural evils are not merely natural, but the result of a collision of natural forces with humanly created environments. Hence, he finds that many events labeled as natural evils are also social evils, the result of human projects. He thus proposes a creativity defense that he finds more in line with the religious concepts as actually employed in the Christian traditions than is the Plantingan free will defense. If God as the primary creative agent of the universe has made creatures in the divine image and likeness, then it is possible that God has made them secondary creative agents who share in the divine creativity. Some nonhuman natural evil, then, is simply inevitable in a universe that is itself creative. Much of the power of social evil or social sin, unthematized by Plantinga and others, is attributable to human creativity run amok. Social evils are the result of the corruption of human creativity in the creation of social structures analogous to the way individual sin is the corruption of human freedom; as such, social evils are a constituent of the "natural evils" that are "natural" disasters.

Tilley claims that a creativity defense (CD) is an alternative response to the logical problem of evil that does not require the possibility of libertarian free will or demonic power. Thus, a CD has the potential both to provide a defense of God's goodness in the face of manifold forms of evil beyond the hackneyed "sin and suffering" typically found in theodicies and defenses and to accept a traditional understanding of *poena et culpa*. Tilley thinks that he avoids crossing the line he has drawn between defenses and theodicies, but he believes that he nonetheless may have blurred that distinction in this chapter in that he seems to be explaining how evil can exist in God's world even if he attempts to avoid the theodicists' trick of explaining evils away—and thus getting humanity "off the hook" while appearing to get God "out of the dock."

2.3 Responses: The God-World Relation

Approaching from three directions, first science, then Christian theology, and finally a survey of philosophical and world-religious perspectives, each of the authors in this section maintains that the problem of suffering cannot be addressed adequately on the basis of more typical understandings of the relation between God and world.

In "Natural Evil in a Divinely Entangled World," **Kirk Wegter-McNelly** argues that theological accounts of natural evil need to consider not only the question of the relation between natural evil and the divine will but also the question of the relation between natural evil and the possible modes of God's presence in creation. To this end, he develops an account of God's relation to physical processes in terms of an analogy drawn from quantum physics. God, he argues, can be helpfully envisioned as "entangled" with the physical world. Thinking of the God-world relation in this way leads to the notion that God's relation to the physical world produces a kind of world in which God's presence is effectively obscured by the physical world's relative autonomy. The relevance of this analogy for a discussion of natural evil is that it produces a conceptual image of God's relation to physical processes in which the divine act of establishing and maintaining a relation with the processes of the physical world becomes the basis for their relative causal independence and thus the possibility of processes that harm life, that is, natural evil.

Wegter-McNelly criticizes views of evil that, to his mind, reduce or ignore its "evilness." He resists the call to provide a robust theological explanation of evil. He argues instead that theology's task is to defend in a more limited way against the claim that the existence of natural evil contradicts the presence of a loving, powerful, and active God—what he terms a defense of the "compossibility" of natural evil and divine action. He asks whether a by-products-type account of natural evil constitutes a Tilleyan theodicy or defense. He distinguishes between Nancey Murphy's maximalist use of the argument and Thomas Tracy's minimalist use of the same argument, concluding that the latter more successfully avoids the trap of theodicy. He portrays his own analogy for the God-world relation as a complement to Tracy's minimalist approach.

Along the way Wegter-McNelly provides a brief history of the concept of entanglement in the development of quantum theory, as well as a brief introduction to the basic mathematical concepts needed to appreciate the place of entanglement within the quantum formalism (using Dirac notation to discuss spin-½ systems). His quantum thought experiment yields an example of an empirically identifiable causal relation between two physical objects that manifests itself relationally but, remarkably, not in the individual behavior of the related objects. This kind of entangled relation then serves as the basis for his theological analogy of a God who relates to the physical world in such a fashion that the relation itself becomes the basis for the world's causal independence.

In "Why is God Doing This? Suffering, the Universe, and Christian Eschatology," **Denis Edwards** addresses the issue of natural evil from the perspective of a theology based on the Christ event. The intense suffering caused by the South Asian tsunami of December 26, 2004 raised a fundamental question for many people who had been brought up on the widespread notion of an interventionist God: Why is God doing this? This question is a particular example of a wider issue: How can theology respond to the fact that science sees the costs of emergence as built into the processes of the universe at the level of initial conditions, laws, and constants? This chapter attempts a partial response to these questions by proposing an alternative view of divine action, one based on Karl Rahner's theology, which is noninterventionist and radically eschatological.

This notion of divine action is built on the foundational concept of divine *self-bestowal*. The Christian gospel proclaims a God who gives God's self to creatures in the Word made flesh and in the Spirit poured out in grace. This divine self-revelation points to the meaning of creation itself: God creates in order to give God's self to creation as its final fulfillment. The story of the universe is part of a *larger* story, the story of divine self-bestowal. This divine self-bestowal is not only the future final fulfillment of creation; it is already at work as the most immanent thing in the universe and its creatures. This immanent creative love works in a particular way: it enables creatures not only to exist but to become what is new. This involves a second foundational concept: God gives to creation itself the capacity for *self-transcendence*. This capacity is intrinsic to creation, but comes from the creative act of God. Divine self-bestowal and creaturely self-transcendence characterize not only creation, grace, and incarnation, but also the final consummation of all things.

In this theology of divine action, dependence on God and creaturely integrity are directly related. The more creatures are grounded in God, the more they possess their own integrity. God's action enables creaturely freedom and autonomy to flourish. Divine power is not the power of a despot. The cross and resurrection redefine divine power as the transcendent capacity for love. It is a power-in-love that involves vulnerability to and waiting upon the freedom of humans and the integrity of natural processes. God's action is not an intervention from without, but the act of an immanent God. This one act of self-bestowing love (primary causality), an act that embraces both creation and salvation, finds expression in a variety of created, secondary causes, and these can be seen as objective and special divine acts. These include words, persons, and events.

This view of divine action provides an alternative to the concept of an interventionist God who arbitrarily brings suffering to some and healing to others. It sees God as working in and through natural processes in a way that respects the integrity of the processes and the freedom of human beings. It sees God as lovingly accompanying each creature in its life and its death. It sees the resurrection as a promise that creaturely suffering and death will be redeemed, as all things are taken up and find their fulfillment in God's self-bestowing love. It argues that while we do not fully understand why so much suffering is built into the universe, there is reason to entrust ourselves and our universe to the absolute mystery of self-bestowing love.

In "Incongruous Goodness, Perilous Beauty, Disconcerting Truth: Ultimate Reality and Suffering in Nature," **Wesley J. Wildman** seeks to bypass traditional theodicy arguments in search of another theological destination. He attempts to use the reality of suffering in nature as a source of pressure on theological ideas about the character of ultimate reality. Specifically, he sets up what he calls a reverent competition among three prominent theological approaches to God, and uses suffering in nature as a test for that competition.

The three views in this theological competition are determinate-entity theism, process theism, and ground-of-being theism. By the former, Wildman refers to theological views that portray God as a being with determinate characteristics, such as omnipotence, personality, will, intentions, plans, and powers to act in history and nature. By the latter, Wildman refers to theological views in which God is beyond the categories of existence and nonexistence and ultimately has no intentional, agential consciousness. Process theism proposes a God who is not creator but a creative force within the natural process seeking always to maximize value.

With the parties to the reverent competition introduced, Wildman evaluates them in terms of the way they handle various facets of suffering in nature. Though he confesses to some attraction toward all three of these ways of conceiving God in relation to the challenge of suffering in nature, he argues that the alternatives to ground-of-being theism face significant conceptual problems in mounting their response. These problems derive chiefly from their attempts to make of God a moral agent unambiguously good in a humanly recognizable way. This is most evident when determinate-entity theism has to face the "argument from neglect," which alleges that a loving determinate-entity deity should intervene in human affairs and natural processes to educate and guide, to punish and redirect, just as human parents do for their children. The absence or infrequency of such intervention is culpable neglect; this demonstrates that the determinate-entity idea of God is morally and metaphysically incoherent. Process theism must come to terms with a similar charge, namely, that God's perpetual striving to maximize value and ease suffering is fundamentally ineffective, and thus that God is too incompetent to be worthy of worship.

Ground-of-being theism can avoid the argument from neglect and the argument from incompetence but must face its own challenge, namely, that its vision of ultimate reality is morally incomprehensible to human beings. Wildman admits that rejecting a personal center of divine consciousness and activity is religiously indigestible to many people. The theologians among them will understandably continue the struggle on behalf of determinate-entity and process theisms, representing the interests of their religious constituencies. But he also points out that, for some others, ground-of-being theism is the very breath of spiritual life, the fecund source of both suffering and blessing, and the basis for the profound sacredness of nature, and even of suffering in nature. He then discusses the main strategies for defending determinate-entity and process theisms against the attacks of ground-of-being theism, arguing that "best of all possible worlds" arguments, "inevitable by-product" arguments, "kenosis" arguments, and "eschatological" arguments are finally ineffective.

The conclusion of the reverent competition is that ground-of-being theism has more to commend it than is usually recognized in contemporary theology and especially in the theology-science dialogue. And it sponsors a compelling interpretation of suffering in nature as neither evil nor a by-product or condition of the good, but rather part of the wellspring of divine creativity in nature.

2.4 Changing the Terms of the Debate

The volume concludes with two complementary chapters that argue, as did some of the preliminary chapters, for conceptual and linguistic changes in traditional ways of focusing debates on evil.

Brad J. Kallenberg's chapter, "The Descriptive Problem of Evil," proposes a metaphor for language as the cane in the hand of the blind person. The better one becomes at getting around with the cane, the more he or she is apt to forget the cane but *through* the cane perceive the objects scraped and tapped by the other end. A defective cane may distort the world perceived by the blind person. So too, defective use of language threatens to muddy our understanding of the things we talk about. When discussing something as difficult as natural evils, a frequently undetected defect in our language use is "overly attenuated description." Kallenberg sketches three ways in which descriptions of "natural evil" have been unwisely attenuated in contemporary conversations about theodicy. Confusion enters the conversation (1) when occasions of suffering are described in self-distancing ways, (2) when the narrative context is overly compressed, and (3) when the range of acceptable causal explanation excludes nonefficient forms of causality.

Kallenberg then describes ways in which the lexical shortcuts taken in discussion about "natural evils" can be corrected by expanding descriptions with respect to narrative space and narrative time (i.e., including relevant details both prior and subsequent to the tragedy in question). However, it remains to be seen whether conversants are willing to pay the cost involved. For, in order to talk most clearly about "natural evil," and thus understand the problem most deeply, those doing the talking must not only employ descriptions that are relevantly expanded with respect to narrative time and space. They must also attend to practice. If we take Ludwig Wittgenstein as a reliable guide for the workings of language, "*practice* gives the words their sense." Frankly, attention to practice and to the description of practice is out of step with the longstanding tendency of theodicists to aim at simple explanation. While Kallenberg grants that explanation *may* put an end to our questions (as explanations occasionally do), more often than not, what we are really after are not explanations per se, but rather means for telling the story of our lives in ways that encompass tragedy *satisfactorily*. What is achieved when descriptions are expanded in the ways he advocates is the emergence of a pattern that is invisible when description is compressed.

Kallenberg provides an example of how the juxtaposition of rich descriptions and appropriate action makes it possible for the trained eye to perceive a pattern that is a function of both components. It is precisely this pattern that succeeds as a satisfactory response to natural evils. Hugh of St. Victor understood the brokenness of human beings and their world as gesturing toward what human life is for. Humans alone are creatures capable of imitating divine redemption by their appropriate responses to the world's brokenness. Doing so is the pursuit of wisdom, and this pursuit is manifest in all aspects of human knowing, including the mechanical arts.

A practical corollary of Hugh's theological vision is found in a teenaged shepherd, Bénezet, in southern France, who believed he received a divine calling to build a bridge over the Rhone at Avignon. Eventually he enlisted enough help and the bridge was built. It takes several dense pages for Kallenberg to recount the benefits—in terms of human safety, politics, ecclesiology, and even theology—that the bridge provided, benefits of which the young Bénezet could not have dreamed. Thus, Kallenberg illustrates his points regarding the necessity of expanded description combined with practical response.

Don Howard's chapter, "Physics and Theodicy," provides a fitting conclusion after Kallenberg's in that Howard, too, disputes the distinguishability of the natural in "natural evils" and provides his own examples of the ways in which the story needs to be expanded both before and after the event in order for the event itself to be evaluated. Howard's first story is the drowning of Edith Babson, which led to her brother designating gravity as our enemy number one and endowing the Gravity Research Foundation. The purpose of the foundation was to seek a scientific means of shielding humans from gravity's ill effects. While that goal failed, the foundation did considerable good in promoting basic research in general relativity. Howard's point: naming gravity (or the laws of nature) as the cause of evil can be either a way of absolving ourselves from any guilt or responsibility, or it can be the first step in taking responsibility—responsibility to explore the science that might have in the past or may still in the future prevent suffering from "natural" causes. Current prime examples include the lack of any early warning system for tsunamis in the Indian Ocean, the failure to build levees in New Orleans suitable for category 5 hurricanes, and future probable effects of further global warming.

Along the way, Howard asks what we are blaming when we blame nature. The most common response is that it is the laws of nature—the connection with the problem of evil then being that these are the laws God designed. However, there are a number of questions to raise. One is whether nature should be understood in terms of laws, rather than, say, in terms of forces. Another is, why "blame" the laws rather than the initial conditions, particularly when thinking about the cosmos as a whole, which is unique? Third, there is the question of divine action: how to apportion "responsibility" among the laws, initial conditions, and ongoing special divine actions.

But, finally, the important question is how to apportion responsibility between nature and ourselves. This is increasingly important as scientific knowledge grows because we have increasing ability to perceive the role of human activity in having changed natural processes, increasing ability to predict the effects of our activity on the environment, and, therefore, increasing responsibility to focus scientific research on issues that are more likely to produce good and eliminate suffering.

3 Postscript

It is tempting to use editorial license to sum up the message of these fifteen chapters, but I shall resist. The reader will surely perceive a great deal of overlap, agreement, and complementarity. But there are also deep divides that a handy summary would too easily dismiss.

I shall content myself with the expression of a wish. It seemed to the organizers of the conference at which these chapters were discussed that too little scholarly attention had been given to the problem of suffering at the (metaphorical) hands of nature, and that this was a topic that could be aided by reflection from the perspective of current science. We are happy to have provided these fifteen chapters in the hope that it may stimulate more thinking and research on this topic that is so massive a stumbling block for sensitive souls in search of a loving God.

I. HISTORY, DEFINITIONS, TYPOLOGIES

Niels Christian Hvidt

Terrence W. Tilley

Wesley J. Wildman

Christopher Southgate and Andrew Robinson

THE HISTORICAL DEVELOPMENT
OF THE PROBLEM OF EVIL

Niels Christian Hvidt

1 Introduction

The problem of evil confronting faith in a God who abounds in goodness and power may well be the greatest challenge to theistic religions.[4] Some go so far as to consider it the problem that "brought older theism to the ground."[5] The problem of evil has been said to constitute "the core of present-day theological thought" since modern human beings cannot ask about God in their cultural surroundings without inquiring about evil.[6] Because of the ubiquitous presence of suffering in human history, especially after Auschwitz, theodicy has become an integral part of theology and the philosophy of religion, as it addresses one of the most existentially real problems of the most diverse faith traditions.[7]

Gottfried Wilhelm Leibniz was the first person to use the label "theodicy" for philosophical arguments arising from the problem of evil: first in his letter of 1697 to Queen Sophie Charlotte of Prussia; later he entitled his famous philosophical study on the problem from 1710 *Theodicy (Essays de Théodicée),* the only full-length work he saw fit to publish during his lifetime. But already in the first booklet he authored, *Philosopher's Confession,* at age 26 in 1672, and in subsequent writings he reflected on the problem of evil. The word "theodicy" is made up of the Greek word for God *(Theos)* and justice *(dike)* in reference to Romans 3:5–6: "But if our injustice serves to confirm the justice of God, what should we say? That God is unjust to inflict wrath on us? (I speak in a human way.) By no means! For then how could God judge the world?"

The purpose of theodicy is to propose reasons why God is not responsible for evil in the world.[8] Humans have obviously reflected upon this question long before Leibniz. There is rich historical material in the "tradition of the problem of theodicy" and we do well in tracing its historical roots.[9] Nevertheless, a number of scholars argue that the project of theodicy strictly speaking is primarily a concern of modern times, as its basic components are mainly issues of modernity: the problems of evil, the extent of divine power and goodness, human freedom, and the belief in the human being's rational powers.[10] As Terrence Tilley writes, "[t]he practice of constructing theodicies only became possible in the context of the Enlightenment."[11]

---

[4] Alston, "Religion, History of Philosophy of."

[5] Williams, *Revelation and Reconciliation,* 151.

[6] Vermeer, *Learning Theodicy,* 1; ref. to Peukert, *Wissenschaftstheorie, Handlungstheorie, fundamentale Theologie,* 335.

[7] Schumacher, *Theodizee,* 9.

[8] Ricœur, *Le mal,* 26.

[9] Walter Sparn, *Leiden, Erfahrung und Denken: Materialen zum Theodizeeproblem,* 139. Janssen argues that theodicy is a fundamental problem of modern times, reaching to the "roots of the concept of God." Janssen, *Gott, Freiheit, Leid,* 6.

[10] Janssen, *Gott, Freiheit, Leid,* 1–2.

[11] Tilley, *Evils of Theodicy,* 221.

Reflection on suffering has indeed been on the mind and lips of believers throughout the ages, but only in the Enlightenment does it become a fundamental and independent *concept,* with Leibniz as its "inventor."[12] Thus, we do well in speaking of "theodic analogies" in the traditions of theodicy prior to Leibniz and of "a history of reception" of the problem after him.[13]

St. Augustine, "the fountainhead of most Western Theodicies,"[14] has been followed and quoted most often in the considerations on the problem of evil, and we shall take a closer look at his exposition due to its historical impact.[15] In Augustine, reflections on evil appear in the framework of other philosophical and theological considerations, such as the goodness and perfection of God and the nature of fallen creation. It is embedded in the constant adoration and explication of God's greatness. Only in the Enlightenment do theodic reflections become detached from such a framework of adoration. Only here is the ultimate reference point shifted from God to humanity. Only here is God the one in the court of justice accused by the problem of evil with the possible outcome that people can reject belief in God if the theodic project does not yield enough evidence in favor of God's innocence. Only here, at least seen on the background of the medieval reflections on the problem of evil, is atheism considered a possible outcome if the theodic arguments fail.

Without true consensus as to the first articulation of the problem of evil and its true significance and role in the different philosophical and religious proposals,[16] theodicies have really only come across as "orientations."[17] John Bowker points out that this is partly due to the fact that there is really not *a* or *the* problem of evil.[18] Rather, religious reflections that have ensued from suffering have differed according to historical, cultural, and religious settings. However, there is great consensus on the importance of addressing the problem of suffering in relation to God. Already G. W. F. Hegel called theodicy "the ultimate goal and interest of philosophy."[19] Some scholars such as Susan Neiman have argued that the problem of evil lies at the very root of the developments of modern Western

---

[12] Janssen, *Gott, Freiheit, Leid,* 1; Marquard, *Abschied vom prinzipiellen,* 43; Oelmüller, *Die unbefriedigte Aufklärung,* 194. Schumacher is critical towards Janssen when he writes that the "root of theodicy" is to be found in modern times; he is more convinced than Janssen that the medieval discussion of God is related to the problem of evil in ways similar to the theodicies of modern times. He proposes Thomas Aquinas as a thinker who formulated full-fledged theodic arguments, see Schumacher, *Theodizee,* 27, 29.

[13] "Theodizee-Analogien," "Rezeptionsgeschichte"; see Geyer, "Theodizee VI – Philosophisch," 232.

[14] Inbody, *The Faith of the Christian Church,* 143.

[15] Another ancient Christian writer who made a "school" of reflection on the problem of evil is Irenaeus, who suggested that God allows (some) evil for the moral development of the human being. For the sake of brevity we shall present his view embedded in the reception of John Hick, who has become influential in modern times.

[16] Janssen, *Gott, Freiheit, Leid,* ix.

[17] Schumacher, *Theodizee,* 9.

[18] Bowker, *Problems of Suffering in Religions of the World.*

[19] "letzte Ziel und Interesse der Philosophie," Hegel, *Vorlesungen über die Geschichte der Philosophie,* 455; quoted in Janssen, *Gott, Freiheit, Leid,* 17.

philosophy since the sixteenth century and that its history should be re-oriented in terms of different philosophical reactions to the problem of evil, with Lisbon and Auschwitz as fundamental experiences that formed Western thinking: One of the book's "central claims" is then that "the problem of evil is the guiding force of modern thought."[20]

However, the project of theodicy is not merely one of apologetics aimed at securing the rationality of theistic faith. The question of why God allows so much evil is more than a *mind wrecking* issue—it has *faith wrecking* potential as well. Believers throughout the ages have lost their faith because of personal encounters with sickness and suffering which they believed God would have prevented, if he existed. Thus, theodicy is not merely an issue of academic theology or philosophy, not merely "an intellectual stumper," but as Robert Price observes, "theodicy quickly intrudes itself upon the agenda of even the simplest believer as soon as serious misfortune strikes. The technical jargon of theology may remain in the ivory tower, but the stricken believer must ask, 'How can God have let this happen to me?'"[21]

Some believers have not been tempted to lose their faith when confronted with evil as much as to change their view on God. A famous example is C. S. Lewis. In his own theodicy, *The Problem of Pain,* Lewis was confident that God could make good use of suffering and allowed it for various positive purposes: "We are good stuff gone bad, a defaced masterpiece, a rebellious child . . . rebels who need to lay down our arms. Pain and suffering are frequently the means by which we become motivated to finally surrender to God and to seek the cure of Christ. That's what we need most desperately. That's what will bring us the supreme joy of knowing Jesus. Any suffering, the great Christians from history will tell you, is worth that result."[22] Thus, for instance, he called pain "God's megaphone to rouse a deaf world."[23] However, years later, after the loss of Joy, Lewis's wife, he became less confident as to the intelligibility of suffering: "Not that I am in danger of ceasing to believe in God. The real danger is of coming to believe such dreadful things about Him. The conclusion I dread is not 'So there's no God after all,' but 'So this is what God's really like. Deceive yourself no longer.'"[24] Lewis continues:

> Your bid—for God or no God, for a good God or the Cosmic Sadist, for eternal life or nonentity—will not be serious if nothing much is staked on it. And you will never discover how serious it was until the stakes are raised horribly high. . . . Nothing will shake a man— or at any rate a man like me—out of his merely verbal thinking and his merely notional beliefs. He has to be knocked silly before he comes to his senses. Only torture will bring out the truth. Only under torture does he discover it himself.[25]

---

[20] Neiman, *Evil in Modern Thought,* 2–3.

[21] Price, "Illness Theodicies in the NT," 309.

[22] Ref. in Strobel, *The Case for a Creator,* 44.

[23] Lewis, *The Problem of Pain,* 93.

[24] Lewis, *A Grief Observed,* 6. The book formed the basis for the film *Shadowlands* starring Anthony Hopkins as Lewis and Debra Winger as Joy.

[25] Ibid., 37–38. The difference in Lewis's approach in the two works could also partly reflect different approaches to the issue, with *The Problem of Pain* being the

Just as theodicies were formed by the painful experiences of individuals, they often emerged as responses to collective experiences with evils. This was the case with "The Great Lisbon Earthquake" on November 1, 1755, a "watershed event in European history,"[26] with the horrors of Auschwitz—or more recently with the tsunami caused by the earthquake that hit deep under the Indian Ocean near Sumatra on December 24, 2004.[27] Sociological evidence furthermore suggests that life-threatening disease can shake trust and faith in the goodness and providence of God. Thus in one trial, some breast cancer patients "talked about their loss of faith in a loving and protecting God as a result of their ordeal with the illness and their perception that God did nothing to help them."[28] Diseases associated with religious and social stigmatization, especially AIDS/HIV, have proven particularly inductive to lack of faith, as a patient reported in one study: "Before I found out I was HIV positive, I believed in God, I believed in saints, and when I found out I was HIV positive, I lost hope, I lost faith, and I lost my spirit. I was a bad person. A gray person. I thought I was never going to get out of that stage."[29]

An important paradox complicates things further: Suffering can be both an *obstacle* and an *occasion* for faith ("*Anstoss* des Glaubens" in the double sense).[30] In spite of the attention to how the problem of evil has deterred people from faith in God, there is growing awareness that the sufferings people encounter have, for many, become the pathway to faith in God.[31] In fact, although it might be difficult to confirm empirically, evidence suggests that suffering has led a greater number of people to faith than the problem of evil has led people to atheism. Others simply report that their faith has helped them find meaning and comfort during apparently meaningless suffering, that is, that faith considerations have provided important perspectives on the "positive reinterpretation of disease."[32] The resources found in religious belief for coping with disease have been among the great rediscoveries of both the medical and social sciences.[33] The conviction that suffering holds important spiritual potential

---

more theoretical approach, and *A Grief Observed* the more practical approach.

[26] Kozak and James, "Historical Depictions of the 1755 Lisbon Earthquake."

[27] The 2004 tsunami led to a number of articles in major newspapers and magazines on how different religious leaders reflect on the problem of evil.

[28] Gall and Grant, "Spiritual Disposition and Understanding Illness," 520; ref to Gall and Cornblat, "Breast Cancer Survivors Give Voice."

[29] Quoted in Pargament et al., "Religion and HIV."

[30] Neuhaus, *Theodizee Abbruch oder Anstoss des Glaubens,* 13. Anstoss = "impulse"; *Anstoss nehmen an* = "take offence at."

[31] Dransart, *La maladie cherche à me guérir;* Echeverria, "The Gospel of Redemptive Suffering"; Ekstrom, "Suffering as Religious Experience." A number of articles from the social sciences investigate how spiritual meaning is structured in severe illness and how opportunities for meaning can arise from such disease. Viktor Frankl had already written that suffering informs spiritual meaning, which in turn ameliorates suffering, Frankl, *Man's Search for Meaning.*

[32] Bussing, Ostermann, and Matthiessen, "Role of Religion and Spirituality in Medical Patients."

[33] Abernethy et al., "Religious Coping and Depression among Spouses of People with Lung Cancer"; Ano and Vasconcelles, "Religious Coping and Psychological Adjustment to Stress"; Bosworth et al., "The Impact of Religious Practice and

has been the object of much reflection in Catholicism, with John Paul II's *Salvificis Doloris* as an important example.[34] At the same time there is growing awareness that pathological religious reflections, also called "religious struggle"[35] on the origin of suffering often is a direct cause for decreased health and quality of life.[36]

Sadly, religious struggle can emerge out of endeavors to rationalize suffering. Some evils are just so painful, especially severe depressions, that those who experience them would revolt if they were in any way to conceive of them as caused by God as means to anything whatsoever. Rather, those reflecting on the good that may come out of evils suffered in faith normally simply state that, although we may not be able to understand why things are as they are, we can nevertheless believe that God is able to turn suffering into something good, as Paul says: "We know that all things work together for good for those who love God, who are called according to his purpose" (Rom 8:28).

It is impossible to draw hermetically closed borders between theoretical and practical approaches to suffering. Especially the biblical reflections on evil contain evidently practical perspectives. Historically, many thinkers who worked on theoretical solutions to the problem of evil did so after having encountered suffering in their own lives. They became *theoretical theodicists* out of *practical experience*. Furthermore, many theodicists have pondered how their theodic endeavors can find practical applications. Both Terrence Tilley and Kenneth Surin argue that Augustine's and Irenaeus's reflections on suffering always were embedded in a practical framework that gave sense to their theoretical considerations,[37] as summarized by Michael Quirk: "The theories they invoked were legitimate tools in a *practical* response to a *lived* encounter with the evils of their day."[38] However, Tilley and Surin consider constructing theodicies "a destructive practice" when they are purely metaphysical as in Leibniz. They are not alone in their critique of trying to theorize on human sufferings, as Franz Rosenzweig, Herman Cohen, and Emmanuel Levinas have done the same.[39]

Religious Coping on Geriatric Depression"; Fabricatore, "Stress, Religion, and Mental Health"; Fabricatore et al., "Stress, Religion, and Mental Health"; Gall and Cornblat, "Breast Cancer Survivors Give Voice"; Koenig et al., "Religious Coping and Depression among Elderly, Hospitalized Medically Ill Men"; Koenig, Larson, and Larson, "Religion and Coping with Serious Medical Illness"; Pargament, et al., "God Help Me (I)"; Pargament, Koenig et al., "Religious Coping Methods as Predictors"; Phillips et al., "Self-Directing Religious Coping."

[34] See presentation in Echeverria, "The Gospel of Redemptive Suffering."

[35] Fitchett, "Religious Struggle"; Pargament, "God Help Me (II)"; Pargament, Koenig et al., " Religious Struggle as a Predictor."

[36] Butter, "Development of a Model for Clinical Assessment of Religious Coping"; Pargament, Koenig et al., "Religious Struggle as a Predictor"; Pargament et al., "Red Flags and Religious Coping."

[37] Especially Augustine's *Enchiridion* is "Bishop Augustine's response to a layperson's problem" (Tilley, *Evils of Theodicy*, 86, 118–19). Surin, "The Problem of Evil"; idem, *Theology and the Problem of Evil*, 10ff.; Tilley, *Evils of Theodicy*, 221.

[38] Quirk, Review of Surin.

[39] Kearns, "Suffering in Theory," 57.

At the same time, people in ministry have often addressed theoretical and metaphysical questions when helping those who encountered suffering in their lives and, as a result, also in their faith. As Quirk comments, even theoretical approaches to the problem of evil can be important on the practical level. The best pastoral care for some who are crushed by the question of why God allows her or him to suffer may be to provide reasons why God is not to blame. Thus, as Alister McGrath writes, it is important to keep in mind that theoretical theodicies "constitute a signifying practice, and that as theodicists they are producers of systems of signification. As such, the formulation of a response to the 'problem(s) of evil' (and suffering) is always 'interest-relative.'"[40] In other words, theoretical reflections on the problem of evil may be of comfort to those who suffer and is always potentially relevant to their need to make sense of often senselessly painful experiences.

2 A Problem for Most Forms of Religious Faith

The problems for faith ensuing from the evils in the world are felt in most theistic religions that believe in a personal and individual deity with unlimited or at least some power, goodness, and free will. Theodicy has even been said to be not confined to monotheistic religions, but to be "the touchstone of all religions." Many religious scholars hold polytheistic and polydemonistic forms of faith to be the exception to this dialectic between the problem of evil and faith. First, the deities of Greek or Roman mythology, for instance, were seen as too intimately linked with the experiences of humans to stand aloof as those who determined their lives from above or outside human experience. Second, they were believed to let curses and blessings descend on human beings in often whimsical or at least irrational ways. And third, they themselves were believed to be subject to fate and hence could not alone explain the ultimate origin of all there is in the world, both good and evil. Thus, religious reflections did not advance the rationality of human experience and hence could not serve to answer questions regarding the origin of evil and suffering.

Some researchers argue that it would make sense for polytheistic believers to ask about the origin of and reasons for suffering as long as some of the painful experiences of humans could be traced back to and explained by the workings of deities or demons. Others again insist that whether or not the origin of evils could be partly or fully explained, the question remains unanswered why trouble hits good and bad alike. Faith does not per se provide a vision of any just distribution of good and evil in this world.[41]

Many religious scholars have written on theodicy, but the most influential of them is Max Weber, who presented his views in *The Sociology of Religion*. He considered a vast spectrum of religions and their reflections on the problem of evil. In Weber's opinion only three systems of thought had answered it satisfactorily: (1) The notion of karma. (2) The decree of predestination of the *Deus absconditus,* which Weber identified primarily in Islam

---

[40] Surin, "The Problem of Evil," 196.

[41] Gerlitz, "Theodizee I – Religionsgeschichtlich," 210.

and Calvinism. (3) The dualism of Zoroastrianism, Gnosticism, and Manichaeism.[42]

2.1 Karma

The idea of karma, the entire cycle of cause and effect, may well constitute the most intimate link between moral and natural evil, and Weber did indeed consider it "the most radical solution of the problem of theodicy."[43] According to this idea, the world is a self-contained cosmos. All evils in our present lives, whether the result of evil actions others may have done to us or natural occurrences that we experience as evil such as disease and disaster, can be explained by the ethical merit or demerit of our actions in former incarnations. Each new reincarnation carries the guilt of the former lives and the resulting evils are distributed justly in the present life as in "a universal mechanism of retribution."[44] The only way to improve the karma of future reincarnations is by acting as virtuously as possible in this life.

According to the interpretation by a number of scholars, Hindu religions therewith render the problem of evil obsolete or at least consider it to be illusory or non-existent. Since God is not the author of the evils people experienced, God is not to blame. At least in the time of the Upanishads (400 to 500 CE), and in implicit form most probably already in Vedic texts, the idea of karma is identified as the response to the problem of evil.

The idea that the Hindu religions do not contain theodic considerations has been modified to some extent through the research of Wendy O'Flaherty.[45] In her reading, the notion of evil is different from the Biblical notion of sin: "Evil is not primarily what we do; it is what we do not wish to have done to us."[46] Thus, the Sanskrit notion of pa\pa\ coincides with the failings of nature. The evils that people experience in their lives are the result of moha or ma\ya\. In O'Flaherty's interpretation, it is God who brings about both, and the evils that humans experience in their present lives are not per se automatic outcomes of their deeds in former lives through some mechanism of inherent justice. Rather, it is God who brings about the just recompense, and as such, God *can* be called the author of evils that people experience and *can* in theory be charged for them. Nevertheless, even in this interpretation, the link between evils performed in former lives and the evils experienced in the present remain just recompenses, wherefore ultimately it is the person, and not God, who is to blame. In fact God or the

---

[42] See concise presentation in Morris, *Anthropological Studies of Religion,* 77–78.

[43] Weber, *Sociology of Religion,* 147.

[44] Weber, quoted in Morris, *Anthropological Studies of Religion,* 77. Weber elsewhere characterizes his interpretation of the idea of karma as follows: "It is held that within the ethical mechanism of the world not a single good or evil action can ever be lost. Each action, being ineradicable, must necessarily produce, by an almost automatic process, inevitable consequences in this life or in some future rebirth." Weber, *Economy and Society,* 533.

[45] O'Flaherty, *Origins of Evil in Hindu Mythology,* 4–5.

[46] Ibid., 7.

gods are themselves subject to the laws of karma, although they may help
inspire humans to improve their karma.

In the Buddhist version of karma more emphasis is placed on one of
the primary goals of religious observance, namely to escape the cycle of
karma that always—whether it improves to the better, or deteriorates to
the worse—is considered intrinsically an evil. Sakyamuni Buddha both indicates the diagnosis for the evils experienced as well as the therapy to
avoid its continuation. Fundamental to the teaching on evil are the "Four
Noble Truths" that constitute the enlightenment of the Buddha. The first
noble truth (Dukkha) refers to the sufferings that all living things experience in various ways, and that according to the second noble truth (Samu-
daya) are caused by attachment, craving, or selfish desire. The third noble
truth (Nirodha) describes the way out of the circle of reincarnations into
the state of nirvana, whereas the fourth (Magga) indicates *how* to follow
this way, that is, through the "Eightfold Path." Thus the entire reflection of
the Buddha regards human suffering and provides the cure for it as the
cure for a disease in four steps by identifying (1) the disease, (2) its cause,
(3) the question whether it is curable, and (4) the prescribed cure. As such,
although the depiction of life is pessimistic, Buddhists consider their answer to the problem of evil to be neither optimistic nor pessimistic, but realistic.[47] It is a matter of debate whether the indicated way of escaping the
evils that humans suffer here by aspiring for nirvana constitutes some sort
of free-will theodicy or not.[48]

2.2 Predestination

Weber saw in both Islam and Calvinism a clear-cut faith in a God who determines all that happens in the world and indeed in eternity: God's decrees are ominous to the point of "double predestination," where God not
only wills for some or even all to enter heaven as in "single predestination,"
but even decides for some to be damned. Humans cannot understand the
decrees of the *Deus absconditus*—Latin for "the hidden God"—but must ab-
stain from the project of identifying the justice in God's actions. Thus, in
this interpretation, both Islam and Calvin are faced with an epistemic
stalemate that jeopardizes theodic endeavors. The Qur'an itself seems
throughout to suggest a doctrine of absolute divine control, that is, in 70:34:
"He guides whomsoever he wills and leads astray whomsoever he wills," or
in Sura 61:5: "And when Musa said to his people: O my people! why do you
give me trouble? And you know indeed that I am Allah's apostle to you; but
when they turned aside, Allah made their hearts turn aside, and Allah does
not guide the transgressing people."

There are, nevertheless, different interpretations of Islam[49] as well as

---

[47] Bowker, *Problems of Suffering in Religions of the World,* 237; Rahula, *What
the Buddha Taught,* 17.

[48] Gerlitz, "Theodizee I – Religionsgeschichtlich," 213.

[49] Such an interpretation is found among the Qadarite and Mu'tazilite inter-
pretations of Islam. See Ibid., 211. See also Schuon, *Understanding Islam,* 2; Watt,
Free Will and Predestination in Early Islam; Montgomery and Marmura, *Politische
Entwicklungen und theologische Konzepte,* 87.

Calvinism.[50] According to these, both Islam and Calvinism believe that human freedom is compatible with or at least concurs with divine control and predetermination. This in turn allows for a free-will theodicy implying that, although God is in control, at least some of the evils occurring in this life and indeed the entry of men and women to heaven and hell results from their own free and responsible choices.

2.3 The Dualism of Zoroaster

Weber considered dualism to be an efficacious approach to the problem of evil, providing a valid systematization of the magical pluralism of spirits in good and bad. Religious dualism is usually identified with an explanation of the universe as the outcome of two eternally opposed but coexistent principles, one good, one evil. This theory can be traced back at least to the *Zend-Avesta* ascribed to Zoroaster, who probably founded or reformed Medo-Persian religion in the sixth century BCE. In this view, the world is the outcome of the struggle between Ormuzd, infinite light, and Ahriman, the principle of all evil. As we shall see, this notion has inspired various systems of thought, including Judaism, Christianity, and Gnosticism. But Manichaeism in particular continued this trend. Its founder, Manes, added a third principle that emanated from the good principle in the living spirit that in turn brought about the material world in which good and evil are mingled. To Manes, matter was essentially evil and therefore could not be in direct contact with God.

3 The Old Testament

The Old Testament constitutes a large and diverse collection of sacred writings featuring a variety of views on the origins and reasons of suffering. It continues to form an important source of inspiration for believers today, not just as the Christian backdrop of God's salvific work in Christ, but as a sacred text in itself that shapes the images of God both among Jewish, Christian, and Muslim believers. Although the New Testament portrays Christ's salvation as bringing about a fundamental improvement in the relationship between God and man, the Old Testament depiction of the often angry and vengeful God surfaces in many examples of negative religious coping, even among Christians. In order to address these forms of negative religious coping it is important to evaluate in more detail what the Old Testament and later the New Testament actually have to say about suffering. In doing so, we can obviously not evaluate all sources, nor can we account for the different theological schools and layers of redaction. Rather, methodologically we shall look at the Bible as a unified body of sacred writings for the historical impact they have had on the way believers think about human suffering. A detailed evaluation of what the Old and New Testaments have to say about suffering is obviously beyond the scope of the present paper. All we can do here is to indicate different approaches in the Biblical writings to human suffering.

---

[50] Bernhardt, *Was heisst "Handeln Gottes"?* 111–12.

3.1 Origins of Suffering

When evaluating Old Testament views on suffering, exegetes commonly differentiate between the theologies we find prior to and after the great exile, when the people of Israel were taken as captives to Babylon. In texts believed to have been written before the great exile, God is seen as the source for all that occurs in the world. However, texts written during and after the exile feature more dualistic trends in which God and his plans are countered with the disruptive influences of personified evil.[51] Such trends were further enhanced in early Judaism, especially in apocalyptic litera-ture. However, while there is a battle between good and evil, to the Old Testament writers God remains the source of all there is and merely allows Satan to exercise some power within the range of his overall purposes.[52]

The book of Job maintains a unique position in the Old Testament with regard to suffering. Job's sufferings transcend the limits of acceptable measures, given the admitted innocence of Job. The primary theodic com-fort found in Job seems to be that God is able to restore fully those he has allowed to be broken (42:10–17). From a more spiritual perspective, we could argue that Job comes to know God at a deeper level through suffering so that he at the end of his trials can say: "I had heard of you by the hearing of the ear, but now my eye sees you; therefore I despise myself, and repent in dust and ashes" (42:5–6).

The book's primary conclusion to the theological questions arising from sickness is that humans do not have the means to inquire about God's pur-poses; this constitutes the theme of chapters 38–41, which begin: "Who is this that darkens counsel by words without knowledge? Gird up your loins like a man, I will question you, and you shall declare to me. Where were you when I laid the foundation of the earth?" Nevertheless, the extent of the book's questioning the justice of God is unique to this period of Old Tes-tament writings. It offered "the ancient world's most significant theodicy and is still one of the best ever written."[53]

3.2 Reasons for Suffering

In summary, pre-exilic Old Testament theology emphasizes God as the cause for all events, whereas post-exilic theology begins to feature more dualistic perspectives, in which God's just and good actions are countered by the disruptive influence of personified evil. However, in both periods, God is seen to be in control and able to employ the events of history in the overall scheme of his just purposes. These general conclusions on the *whence* of disease and calamities naturally lead to reflections on the *why* of these events.

An important number of passages indicate human sins as the single most important answer to the question. The Old Testament mostly por-trays suffering as God's just retribution for infidelity to the Law and to the covenant between God and his people.[54] Disease in particular is seen as

---

[51] Kelly, *Problem of Evil in the Western Tradition,* 21–22.

[52] Köhlmoos, "Theodizee II – Altes Testament."

[53] Kelly, *Problem of Evil in the Western Tradition,* 19.

[54] Leviticus and Deuteronomy deal at length with the faithfulness of God's promises and his bountiful reward for virtue. And yet, as McDermott writes, "what

God's punishment for sin, and in various passages different diseases are listed as possible forms of retribution:

> The Lord will afflict you with the boils of Egypt, with ulcers, scurvy, and itch, of which you cannot be healed. The Lord will afflict you with madness, blindness, and confusion of mind; you shall grope about at noon as blind people grope in darkness, but you shall be unable to find your way. (Deut 28:27–29)[55]

The Old Testament presents at least four different views on God's punishment for sin: (1) suffering as a result of individual sins, (2) suffering as a result of the sins of ancestors, (3) suffering as a result of the sins of the people of Israel, and (4) suffering because of Adam's fall.[56]

However, although the retaliatory theology has been shown to influence the way many sufferers have thought about God, a number of factors constitute modifications to the general assessment of the relationship between sin and suffering in the Old Testament: (1) the problem of innocent suffering, (2) the conviction that suffering can be fruitfully transformed, (3) that suffering can be a means of testing, (4) that people can suffer vicariously for others, (5) that there is an inherent economy in sin and suffering, and (6) that God's love is always greater than his justice.

4 The New Testament

Interestingly, these themes continue in the New Testament. Thus, the problem of evil does not appear in an entirely different manner in the New Testament in comparison to the Old. Rather, themes found in the Old Testament are refined and enhanced in the New and put in the overall perspective of salvation in Christ. In fact, the issues of sin, redemption, and suffering are dealt with throughout the New Testament. Probably this is the reason why—although much has been written about the meaning of redemption, salvation, providence, etc.—"relatively little has been written specifically about the significance of suffering in the New Testament."[57]

And yet, as does the Old Testament, so too the New Testament provides examples of fruitful transformation of human sufferings. There is an important difference between godly and worldly grief, that is, grief sustained in God and without God, so that Paul can say to the Corinthians, whom he has rebuked and hence grieved: "Now I rejoice, not because you were grieved, but because your grief led to repentance; for you felt a godly grief, so that you were not harmed in any way by us. For godly grief pro-

---

must strike the reader is the vast disproportion in length between the blessings (28:1–14; cf. Lev 26:3–13) and the curses (28:15–68; cf. Lev 126:14–39) ... the horrifying catalogue for threats for disobedience would force most normal people to hesitate."

[55] See also Deut 28:22; 59–61. Some other passages depicting disease as the recompense of sin are Gen 12:17; 19:1; 20:18; 38:9–10; Exod 4:6–7; 9:8–10, 14ff.; 12:29; Lev 26:14ff.; Num 11:33; 12:9–14; 14:11–12, 36–38; 17:12–15; 25:3–9, 17–18; 31:16; 1 Sam 5:6, 11–12; 6:1–12; 6:19; 2 Sam 24:10–15; 1 Kgs 14:10–14; 2 Kgs 1:16; 5:26–27; 6:1–20; 15:3–5; 19:35; 1 Chr 21:7–14; 2 Chron 26:16–20; 21:14–15; 32:21.

[56] For more on the reasons for suffering in the Old Testament, see McDermott, *The Bible on Human Suffering.*

[57] Ibid., 15.

duces a repentance that leads to salvation and brings no regret, but worldly grief produces death" (2 Cor 7:9–10). Although the notion of God testing believers is less prominent in the New than in the Old Testament,[58] God still allows suffering in the lives of believers so they may have their branches pruned, come to their senses, and return to God.

Some passages in the New Testament seem to indicate a disruption of the retaliatory theology known from the Old. The most important of these passages is John 9:1–7. Jesus sees a man born blind from birth. The disciples ask him: "Rabbi, who sinned, this man or his parents that he was born blind?" Jesus denies any such link and says: "Neither this man nor his parents sinned; he was born blind so that God's works might be revealed in him. . . . As long as I am in the world, I am the light of the world."

Other passages are more reluctant to break with the sin-punishment theology even though they do point to the newness of the promises of salvation found in Christ. Although human beings are saved through union with Christ, sin still wrecks God's plans and causes the fair judgment of God. Thus, God remains the just judge who shall repay each human being according to her or his deeds.[59] Other passages confirm the theology that good will be rewarded and wickedness will be recompensed, but relegates such recompense to the future.[60] Such references are mostly made to Christ returning to judge the world, separating the sheep from the goats, the godly from the ungodly, the wheat from the darnel.[61]

It seems that the intensification of the weight of sin revealed both in the Old and the New Testaments provides a framework in which humans have to realize that they are not able to save themselves, and this may liberate them to abandon themselves to God and his mercy instead of seeking salvation in themselves. The way to salvation then, according to the New Testament and in particular Paul's letters, cannot be through the law, at least not the law alone, nor man alone without God's grace. Rather it will be a free gift of God because of the vicarious suffering of his Son: "For the wages of sin is death, but the free gift of God is eternal life in Christ Jesus our Lord" (Rom 6:23). The notion of redemption from the divisive effects of the Fall is present in too many passages to be mentioned here, for example, Romans 5:12–13:

> Therefore, just as sin came into the world through one man, and death came through sin, and so death spread to all because all have sinned. . . . But the free gift is not like the trespass. For if the many died through the one man's trespass, much more surely have the grace of God and the free gift in the grace of the one man, Jesus Christ, abounded for the many.

---

[58] See Ibid., 117–19.

[59] Even in Paul, who so extensively elaborated on the gospel of gratuitous salvation in Christ, we see a clear continuation of belief in the just recompense for both good and evil actions, see Yinger, *Paul, Judaism, and Judgment.* Acts 10:34–35; Rom 2:5–11; Gal 2:6; Eph 6:9; Col 3:25; 1 Pet 1:17. Ref. in McDermott, *The Bible on Human Suffering,* 86.

[60] Gal 6:7–10; 1 Pet 4:3–5, and others.

[61] Matt 13:37–43; 25:31–46; Jude 14–23, etc. Ref. in McDermott, *The Bible on Human Suffering,* 87.

A number of exegetes today argue that the vicarious redemption that the believer finds in Christ should be understood on the backdrop of a modified version of the notion of corporate personality, or rather corporate identity.[62] Salvation is found in life in Christ, in being united with him and hence in sharing all that is his. As Christ gave himself in an act of complete love, so the believer is called to respond to Christ in an act of love's complete surrender. "Love means personal union, for the lover joins himself to the beloved at that point where he is most himself, in his freedom."[63] In this sense, Paul is able to say, "it is no longer I who live, but it is Christ who lives in me. And the life I now live in the flesh I live by faith in the Son of God, who loved me and gave himself for me" (Gal 2:20). This concept of salvation, present throughout the New Testament and especially in the writings of Paul, has been summarized by E. P. Sanders, who writes that the "way to become a corecepient [of God's promises] is to *become one person in Christ.*"[64] The gift of being united with Christ, of sharing in his resurrection life, of being a branch on the vine (John 15:5) implies both being justified before God *and* Christ's life overflowing in the believer's life. Soteriology and sanctification are inherently related, and attempts at separating the two aspects have failed.[65] Both flow from God's gratuitous grace, so that only he can be said to be the true author of salvation. Thus, as the Christian reformed theologian Anthony Hoekema wrote in describing the New Testament view of salvation, "We should think, then, not of an *order of salvation* with successive steps or stages, but rather of a marvelous work of God's grace—a *way of salvation*—within which we may distinguish various aspects."[66]

The New Testament provides a number of significant reasons for suffering that we cannot investigate in this present text, but may only indicate: Suffering is ultimately an eschatological issue. We are not able to understand the origin and purpose of suffering while still on earth, but in faith we trust that all will ultimately work out for the good of those who believe. In this overall belief, believers can be comforted that their sufferings can be transformed semantically, so that their suffering may not be changed as such but be given another sense and therewith be alleviated.

At the root of this transformation dynamism lies the conviction that the believers share in the mystery of Christ's suffering. In this mystery, Christ shares in the suffering of believers, just as believers share in the sufferings of Christ. Some passages in the letters of Paul have furthermore contributed to a conviction that believers actually take part in the very paschal mystery, not by contributing to Christ's foundation of salvation, but to his work of "bringing it about" through the history of humankind. One passage of particular impact in this regard has been Colossians 1:24: "I am now rejoicing in my sufferings for your sake, and in my flesh I am completing what is lacking in Christ's afflictions for the sake of his body, that

---

[62] See Powers, *Salvation through Participation.*

[63] McDermott, *The Bible on Human Suffering,* 109.

[64] Ref. in Powers, *Salvation through Participation,* 123.

[65] See the discussions in the most influential update to the overly "Lutheran" interpretation of Paul in Sanders, *Paul and Palestinian Judaism.* Discussion hereof in Westerholm, *Perspectives Old and New on Paul.*

[66] Hoekema, *Saved by Grace,* 15.

is, the church."[67] As various commentators suggest, the "sufferings" described here are brought about by the "the apostle's bold teaching and divinely empowered labors as a minister of the gospel."[68] Thus, the most reasonable exegesis would be that Paul is talking about that which is missing in Christ's efforts of bringing about the reality of his sacrifice through his body, the church. "What is 'lacking' is not the atoning power of the cross but its manifestation in the Church as a present reality."[69] In suffering with and for Christ, Paul is able to avert some of the sufferings that other believers would have had to suffer. Walter Wilson thus summarizes the interpretation of the impact of Paul's teaching, and of Colossians 1:24 in particular: "By absorbing in his self and through his stewardship some of the predetermined measure of afflictions, the 'messianic woes,' the apostle performs a vicarious ministry, reducing the still deficient tally of sufferings that other Christians, like the readers, must withstand."[70]

5 Greek Thought

Reflections on the problem of evil are richly and diversely represented in Greek thought. Greek prose has influenced reflections on evil in literature ever since,[71] and such reflection is present in various strands of Greek philosophy.[72] Of particular interest are Greek debates on evil between the opposing monistic and dualistic schools in Greek thought.

As early as 500 BCE, the Eleatic School headed by Parmenides taught a fundamental unity of being. Pure being alone exists. It is *one, unchangeable,* and *eternal.* In this system of thought, which has reemerged in various forms through history and even resembles forms of monism in modern thought, the problem of evil can be considered mitigated since God is the source of all that is and even constitutes its deepest essence. God is thus not seen as a personal agent who chooses independently the fate of individuals. Rather, all that occurs in the world is the natural and necessary unfolding of the ground of being (as in some forms of pantheism). Obviously, this view of God helps to explain the origin of evil, but it raises serious questions about God's goodness, just as some have asked whether such

---

[67] It is a matter of debate among exegetes whether Colossians constitutes an "authentic" Pauline letter. Some believe it cannot be, others believe Timothy wrote it as Paul was not able to do so from prison, but that Paul may have added the last lines. Others again believe it is an authentic letter of Paul (Hay, "Colossians"). However, it is of less importance to our reading of the passage whether the letter is an authentic Pauline letter or not. That discussion changes nothing to its canonical status and the influence it as such has had on the history of Christian reflections on suffering. Finally, the teachings in Colossians on suffering are, as we shall see, consistent with those found in other Pauline letters, which makes it difficult to understand why Col 1:24 is brought forth as an argument against the authenticity of the letter.

[68] Wilson, *Hope of Glory,* 73.

[69] Fuller, *Preaching the Lectionary,* 488.

[70] Ref. to Kleinknecht, *Der leidende Gerechtfertigte,* 377–80.

[71] Greene, "Fate, Good, and Evil, in Early Greek Poetry"; Kahn, "The Problem of Evil in Literature."

[72] Gerlitz, "Theodizee I – Religionsgeschichtlich"; Karavites, *Evil—Freedom—and the Road to Perfection in Clement,* 18–28.

a monistic theory truly constitutes any theology of God at all, since a fundamental attribute of the idea of God is that of a personal agent.[73]

Opposed to this unity of being, Plato introduced the thought of an original duality, that is, between God and unproduced matter. The two co-existed since eternity. Unproduced matter was considered indeterminate and guided by blind necessity. Mind, on the other hand, was seen to act according to direction. According to this dualistic notion, evil is the result of the resistance of matter over which God does not have full control. As a result there is a clear dualism in humans so that mind and body (matter) are in opposition.

Plutarch followed Plato and criticized the Stoic ethics of obedience to what they believed to be universal reason governing the world. Like Plato, Plutarch was convinced that God could not be identified with nature in this way. This would entail that God was the source even of evil, which would contradict the goodness of God.[74] Some interpret the Platonic depiction of matter as being so passive and exempt of proper qualities that it can be viewed as a negation or privation of God, which would somewhat mitigate the dualistic opposition between God and matter.[75] Whichever interpretation is true, it is certain that Greek thought with its competing monist and dualist cosmologies constitutes fertile ground for theodic reflections.

Epicurus (342–270 BCE) founded a school at the same time as that of the Stoics, but diametrically opposed to theirs. His is considered the oldest philosophical formulation of the issue of theodicy,[76] at least in "analogous form":[77]

> God either wishes to take away evils and is unable; or he is able and is unwilling; or he is neither willing nor able; or he is both willing and able. If he is willing and is unable, he is feeble, which is not in accordance with the character of God; if he is able and unwilling, he is envious, which is equally at variance with God; if he is neither willing nor able, he is both envious and feeble, and therefore not God; if he is both willing and able, which is alone suitable to God, from what source then are evils? Or why does he not remove them?[78]

Many philosophers through the ages have quoted his famous formulation. Thus, Pierre Bayle, whose writings were the "direct occasion" for Leibniz's *Essays de Théodicée*[79] referred to the dilemma of Epicurus that the power and goodness of God in view of the realities of the world could not be founded simultaneously in a rational manner.

A later Greek thinker, Plotinus, developed the thesis of evil being the negation of Good. The core of the theory is present in Christian thinking

---

[73] See Gwynne, *Special Divine Action.*

[74] *De Stoicarum repugnantiis,* §§ 32–37. The criticism is directed against Chrysippus, *Treatise on the Gods.*

[75] Geyer, "Theodizee VI – Philosophisch," 233.

[76] Schumacher, *Theodizee,* 72.

[77] *theodizeeanalog.* Geyer, "Das Theodizeeproblem, ein historischer und systematischer Überblick," 9–14.

[78] *Letter to Lactanus,* quoted in Berthold, *God, Evil, and Human Learning,* 1–2.

[79] Janssen, *Gott, Freiheit, Leid,* 15; article "Epicurus," in Peter Boyle, *Dictionnaire historique et critique* [1695–1697] (Paris: Editions J.-M. Place, 1982).

from Augustine (who preferred the term *privation*) through Dionysius the Pseudo-Areopagite, Thomas Aquinas, and Leibniz to Karl Barth[80] and Charles Journet.[81] Plotinus was influenced by Plato, and some aspects of his thought resemble gnostic thinking. Thus, Plotinus's negative assessment of matter as the negation of good creates a distance between its formulation in Plotinus and in the majority of Christian thinkers, such that Charles Journet does not consider the Christian notion of evil as *privation* to be present in Plotinus.[82] Others have argued for more substantial Plotinic influence in Augustine, although the latter modified the concept to fit Christian cosmology.[83]

6 Augustine

St. Augustine lived and worked in the fascinating interface between Jewish and Greek thought, and much of his work can be seen as a creative and highly influential synthesis of the two. Augustine's thought cannot be easily classified; it is eclectic, as he often expounded his teaching when confronted with opposed views or heresies in the ancient church. But for the sake of acquiring a general overview, his contributions to the theodic project can be grouped into four areas of reflection: (1) the present world as the best possible one, (2) the nature of evil, (3) the quality and possible consequences of human freedom, and (4) suffering as just retribution for sin.

6.1 Whole-Part Order

Augustine holds an optimistic and positive view of the world. The world God created is glorious, reflecting the perfection of its maker. It must necessarily be less perfect than he, as there will always be an ontological difference between God and creation. As such, a part of the explanation for the evils in the world is that they are nothing but a natural part of the complete picture of creation, since creation must be deprived of some of the perfection of its creator. Nevertheless, it reflects an overall harmony as the "admirable beauty of the universe is made up of all things, in which even what is called evil, well ordered and in its place, is the eminent commendation of what is good."[84] Thus, what could appear as evil when seen in isolation can appear as a good or necessary component in the larger picture. Evil can serve a purpose of being the contrast to God's goodness as ugliness stands to beauty. This does not mean that the concrete evils that humans suffer are not truly experienced as authentic "evil" suffering, but from a metaphysical perspective they have a different status. Even the evils brought about by the devil (originally created as a good, free being) are employed by God's providential purposes in testing the good.[85] The ultimate

---

[80] Barth defines "das Sündige als das in Jesus Christus im Prinzip überwundene 'Nichtige'" (*KD* 3 §50, s.a. 3/1, §42)

[81] Geyer, "Theodizee VI – Philosophisch," 233; Journet, *The Meaning of Evil.*

[82] Journet, *The Meaning of Evil,* 27–28.

[83] See overview of theories of Plotinus's influence on Augustine in relevant articles of Fitzgerald and Cavadini, *Augustine through the Ages.*

[84] Augustine, *City of God* 10.11 (Loeb).

[85] Ibid. 11.17.

Augustinian depiction of what good might come out of God allowing the fall of angels and human beings is that it necessitated the far greater good of the salvation of Christ.[86]

6.2 Evil as Privation

Augustine modified Plotinus's theory of evil as negation by making it rather a matter of privation in those that are evil so that they are actually deprived of that purpose inherently natural to them. Evil is deficiency and no cause can be found for it.[87] It is therefore without independent status, and always parasitic on the Good.[88] In Augustine's thought, natural evils are not evils in themselves. Evil in the proper sense can occur only when created beings with the gift and power of free will use that power to turn away from what is their true purpose: "When the will abandons what is above itself and turns to what is lower, it becomes evil, not because that to which it turns is evil, but because the turning itself is wicked."[89] This perspective is at the heart of the matter in Augustine's theodic reflections, and his other theodic conclusions follow from it. The main reference point is always God—the ultimate good. Hence, the evil of a created free being is judged in terms of the falling away from her or his purpose—being in the truth, being in God—and this turning away from God is the essence of sin. Hence, only two kinds of evil exist as a result of humankind's turning away from God and his goodness: *sin and the consequences of sin.*

6.3 Free Will

Closely linked to these considerations on the nature and quality of creation are the theodic consequences Augustine draws from libertarian freedom. According to Augustine, the world God created would have been less perfect had he not created beings in it with the gift of free will:

> Neither the sins nor the misery are necessary to the perfection of the universe, but souls as such are necessary, which have the power to sin if they so will, and become miserable if they sin. If misery persisted after their sins had been abolished, or if there were misery before there were sins, then it might be right to say that the order and government of the universe were at fault. Again, if there were sins, but no consequent misery, that order is equally dishonored by lack of equity.[90]

One cannot accuse God for the freedom he allowed his creatures to enjoy, although he knew they could and indeed would use it to sin:

> Such is the generosity of God's goodness that He has not refrained from creating even that creature which He foreknew would not only

---

[86] Ibid. 14.27.

[87] Ibid. 12.7.

[88] Augustine, *Enchiridion,* chaps. 13–14.

[89] Augustine, *City of God* 12.6 (Loeb).

[90] Augustine, *On Free Choice of the Will,* bk. 3 (p. 9). Quoted in Plantinga, "The Free Will Defense," in Rowe and Wainwright (1998), 270.

sin, but remain in the will to sin. As a runaway horse is better than a stone which does not run away because it lacks self-movement and sense perception, so the creature is more excellent which sins by free will than that which does not sin only because it has no free will.[91]

Thus, the free-will argument follows from the argument of the best possible world: God could not have created a better world than one in which there are free created beings and such beings are capable of doing evil. But the world with the evils they commit is still better than the world in which there had been no free creatures and hence no evil.

6.4 Suffering as Necessary and Just Outcome of Sin

Much has been written on Augustine's complex notion of God's justice, and the concept is more subtle than its presentation has often been in the history of Christian ideas. According to Augustine, sin always is followed by its disruptive effects, namely suffering and death, as in a necessary web of cause and effect. This is the retributive notion of evil evident in the Old Testament and throughout the history of Christian thought in various forms and many authors, including Thomas Aquinas, Luther, and Calvin, and even mystics such as Hildegard of Bingen. John Hick has called this the "principle of balance"[92] which is believed to be active in the universe.

Augustine does indeed consider everything that occurs in the world as intricately involved with the activity of God, as he cannot see God to be excluded from anything that happens. Writing against the Manicheans and their teaching of a dual cause to the phenomena of the world (mentioned above), Augustine saw in God the unique cause of all things.[93] Obviously, this easily imparts to God the dual role of judge and executioner—the one who evaluates the actions of humans *and* who strikes justly those he has deemed worthy of punishment.[94] Nevertheless, a number of factors render Augustine's account more complex.

By portraying evil as that which is not, Augustine is able to tackle evil's charge against God without reverting to dualism: God is the creator of all there is; evil does not exist in its own right and does not have an origin of its own; it merely exists in parasitic form as falling away or being deprived from creation's intended purpose. And since such privation is possible only in those free willed creatures that decided to turn away from God, only created beings are to blame.

Augustine's writings allow for at least two interpretations of natural evil: one that denies and one that confirms a notion of natural evil. According to the first interpretation, there is no real "natural evil," for that "would be either a God-made evil or a Manichee, alien power coequal with the good," and obviously, Augustine cannot allow such an interpretation.[95]

---

[91] Quoted in Ibid.

[92] Hick, *Evil and the God of Love* (rev. ed.), 87–89.

[93] Augustine, *On Free Choice of the Will* 3.126.

[94] Augustine, *City of God* 22.24.

[95] Evans, *Augustine on Evil*, 98. "Augustine and other, present-day adherents of traditional theism, such as Jolivet and Journet, clearly teach that physical evil does not exist." Vermeer, *Learning Theodicy*, 24.

As we saw above, Augustine believes there are only two kinds of evils: sin and the consequences of sin. Therefore, natural evils cannot exist.[96] Physical pain is not an evil in itself. Rather, it is protective as it warns against bodily threats. Only sin causes humans to experience pain as suffering. Augustine writes that God loves order and hence accommodates evil by making something intrinsically good out of it. What seems disorderly to us God can incorporate in his overall order of goods.[97]

According to the second interpretation, natural evil does exist in Augustine's thought. Although it is sin that causes pain to be experienced as suffering, disorder has entered the created order. There is a necessary relation between the fall of humankind and the disorder and misery it created in the natural world, first and foremost by corrupting the will of humans. But, Augustine writes, even prior to the creation of humans, disorder and evil had entered the world through the rebellion and fall of Satan and the angels that followed him. Although Satan and the fallen angels are real and have the power to create real effects in the created world, they remain under his dominion but in their own proper "natural subsistence" in the world.[98] Just as God could not have created a better world than the one in which there were human beings who could potentially sin, so God could not have created it better than with the angels who were free to turn away from God. Demons differ from angels not by nature but by fault.[99] Thus, to Augustine, natural evil is due to the free actions of nonhuman spirits, except when it is the result of God's punishment.[100] For Augustine, the freewill argument operates even as the link between moral and natural evil.[101]

Both interpretations of natural evil are co-present in Augustine and are in principle reconcilable:[102] Both perspectives build on a conviction that retribution is rational as "the world moves away from punishment and suffering as irrational fate or demonic possession towards a standard of equitable justice and order."[103] Evil begins in the free will that falls away from its purpose, and hence the root of evil is found in the will that God created good and free with the risk that it might turn away from him. However, the outcome of such turning away from God is necessarily felt as real effects in all realms of the world, both moral and physical.

The notion of original sin, so clear in Augustine, explains why the distribution of suffering is not evenly tied up with the actual sins committed by people and why innocent children need to suffer. Through the sin of Adam, destruction came into the world of humans as collective punishment. No one is innocent and everyone deserves to be punished. One of the worst outcomes of this disorder is that further sin is committed. The one sin of Adam is continued in the multitude of sins that humans commit through time, for

---

[96] Augustine, *City of God* 11.22; *Confessions* 7.13.19.

[97] Augustine, *De ordine* 1.7.18. Quoted in Evans, *Augustine on Evil*, 96.

[98] Augustine, *City of God* 22.24 (Loeb).

[99] Ibid. 12.1.

[100] Hick, *Dialogues in the Philosophy of Religion*, 8; Plantinga, *The Nature of Necessity*, 192.

[101] See the discussion in the section on "The Devil and Natural Evil," in Vardy, *The Puzzle of Evil*, 56–57.

[102] Evans, *Augustine on Evil*, 95–96.

[103] Beker, *Suffering and Hope*, 45, 48.

through the sin of Adam the wills of all future humans became corrupted so that they are not able not to sin *(non posse no peccare)*.[104]

The question is obviously how to establish the causal nexus between moral and natural evil, that is, how the will of humans and angels can have such disastrous effects in nature as earthquakes. According to Gillian Evans, no matter how distant this theory may be from the modern mind, it was a natural theory for Augustine due to his Platonic inspiration. According to Plato's *Timaeus,* bodies are first moved by souls, whether good or evil. Any body that is thus moved will move other bodies in a chain of events, the source of which may no longer be identifiable by mere contemplation of the final outcome "as if Achilles pushed Hector and in falling Hector dislodged some of the leaves of a tree he chanced to strike."[105]

In this perspective, the link between moral and natural evil, between sin and suffering, is necessary and logical, almost inherent to the world itself.[106] The necessary outcome of falling away from God, who is the source of all happiness, can be nothing but unhappiness and suffering. This perspective is contained in the very notion of privation. Hence, for Augustine, privation is sin, but at the same time it constitutes suffering. As mentioned, his writings do provide an interpretation of suffering as God's direct punishment for sin. However, the depiction of sin and evil as the privation of good represents what we could call an ontological perspective of *immanent justice* where sin can but reap misery and suffering. Such a conviction resembles what we saw in the Old Testament, especially in the wisdom tradition, as well as in the New Testament. In Augustine, this Biblical synthetic view of life is enriched by Greek notions of evil as privation.

This notion is developed further in Christian mysticism, especially by Hildegard von Bingen, who considers suffering to be the groaning of the cosmos over the moral evils of humankind. Moral and natural evils are linked in an inherent way, grounded in the ontological union between God and creation with the human being as the micro-cosmos in the macro-cosmos of creation. Given this fundamental union between creation and the God who constantly upholds it, creation cannot be conceived of without the presence of its creator. Thus, human falling away from God in sin can but lead to suffering, almost in the form of an inherent necessity, whether this be effected by God—who must punish God's creatures, lest justice be impaired—or by the immanent justice of the divinely created and upheld order by which creation destroys itself when falling away from its own inherent purpose. Common for both of these options is that sin is evaluated against the standard of God's goodness.

7 From Augustine to Leibniz

Having traced some of the most significant Augustinian perspectives on evil, we shall briefly sketch a few highlights from Augustine's to Leibniz's handling of the problem.

---

[104] Augustine, *On Free Choice of the Will* 3.161–81.

[105] Evans, *Augustine on Evil,* 98.

[106] ". . . konsequente Strafe im Sinne eines immanenten Tun-Ergehen-Zusammenhangs verstandenen Leidsin den Mittelpunkt des Theodizeeproblems." Rosenau, "Theodizee IV – Dogmatisch," 224.

The notion of evil as privation of good is followed faithfully. Boethius advanced the argument very clearly in *The Consolation of Philosophy*, written during his imprisonment prior to his martyrdom: "And so evil is nothing. . . . [God] cannot commit evil."[107]

Aquinas also dealt with the problem of evil in various passages.[108] As did Augustine, he proposes four main perspectives on the issue: evil as privation, whole-part order, free will and the necessary link between moral and natural evil.[109] Thomas's teaching on natural evil requires particular attention. He treats moral and natural evil as distinct realities. God does not perform any morally evil acts, but he permits his creatures to perform them because he respects their freedom. However, natural evil is permitted by God to such an extent that one gets the impression that he believes God sometimes is its direct cause due to the principle of plenitude, the term employed by Thomas for the best of possible worlds.[110] Evil is permitted by God because only God sees the overall picture and knows which is the best of possible worlds. Further, quoting Augustine, Aquinas writes that God is so powerful that he can even make good out of something evil. Hence the presence of evil constitutes a challenge that allows for God's greatness to be glorified as he is able to turn it into something good.[111]

Martin Luther and John Calvin equally followed Augustine, in particular in his assessment of the sinfulness of humankind and the justice of God. Like Augustine, they believed that all evils come from the fall that damaged humans so much that they cannot even work on their own redemption, but are in everything dependent on the grace of Christ. Luther in particular repudiated the project of providing a philosophical theodicy. It is not God but humankind that is to be justified. Raising the problem of evil as an accusation against the goodness of God's decrees only further reveals the sinfulness of humanity. Only faith can provide the horizon in which the human being is justified and saved from the ultimate evil of not being able to satisfy the justice of God.

The *Deus absconditus* rules *supra nos* in ways not intelligible to women and men, but in the salvation of the *Deus revelatus* in Christ, God has become man *pro nobis*. Hence, although no analysis may ever reveal God's reasons for allowing so much suffering, Christian faith says that he will make everything good for those who put their trust in him. This perspective that emphasizes faith (and revelation) over reason (and philosophy) has continued to the present, especially in the neo-orthodox theology of Karl Barth and his school.[112]

One can trace the Augustinian influence all the way to and even beyond Leibniz. However, important changes start to occur in the fabric of thought that eventually led to the Enlightenment. One of these changes is the basic approach to theology and faith: Pre-Enlightenment theology always aimed toward a higher good than any goods in this world, mainly

---

[107] Boethius, *The Consolation of Philosophy*, 66–68.

[108] See Aquinas, *On Evil*.

[109] See Aquinas, *Summa Theologiae* 1.a, qq. 48–49, in *Collected Works*.

[110] Kane, "Principles of Reason."

[111] Aquinas, *Summa Theologiae* 1.48.A2, in *Collected Works*.

[112] Barth, *Church Dogmatics*, 3/1–3. See Haga, *Theodizee und Geschichtstheologie*; Rodin, *Evil and Theodicy in the Theology of Karl Barth*.

that of union with God. Theology was thereby ultimately theocentric, with human beings finding freedom through union with God.[113] With Leibniz and the Enlightenment the perspective became anthropocentric, with freedom found in the liberation of reason from any determining and potentially inhibiting environments.

As Tilley, Surin, and others argue, theodicy in the pre-Enlightenment period consisted in addressing specific evils *from within* a Christian perspective. However, with the Enlightenment, theodicy became an attempt to provide reasons *for* such a perspective. This anthropocentric shift emphasized temporal over eternal goods or evils. Human faith in our own ability to judge between goods and evils, and ultimately judge the rationality of maintaining faith in God grew to new heights of faith in human rational powers. Already Blaise Pascal (1623–1662) lamented philosophers' growing rationalization of the concept of God. He contrasted the God of Abraham, Isaac and Jacob with the God of philosophers that "subjects faith to reason in a no-longer rationally grounded act."[114] This God of philosophers was not the God of worship and love in the same way as the God of Abraham, Isaac and Jacob, but the God who could contribute to the human quest to conquer the world of reason. Thus, what seemed irrational could more easily create distance between humanity and God than was the case before. An attempt at a general characterization of the change from pre- to post-Enlightenment theodicies would amount to the following:

In pre-Enlightenment reflections on the problem of evil, God is the one offended by the apostasy of angels and humans, as well as the damage this apostasy has done to the beautiful world God provided for his creatures. The offence is so great that only God's Son can amend it by dying on the cross. Conversely, in post-Enlightenment theodicies, it is not God but humans who are offended by the evils found in the world. Before this shift, the belief was that God's creatures had sullied creation and brought such evils to the world that only a Savior God could redeem it. After the shift, God was charged with the evils in the world in such a way that it could lead to atheism. Before, the problem of evil could cause God to lose faith in humankind. After, it could cause humans to loose faith in God.

In Christian Europe prior to Leibniz, God's just anger for what humans had made of creation by sinning was considered so great that it required nothing less than the sacrifice of Christ to appease it. This was the only *reasonable* framework for theodic reflections. They were anchored in a confessional context closely related to Christian notions of atonement. After Leibniz, doctrine was exchanged for reason as the ultimate ground for theodic equations. Furthermore, because of the Enlightenment's quest for emancipation from religious confessions, theodic reflections were detached from any system of conviction based on divine revelation.[115] Divinely

---

[113] Janssen, *Gott, Freiheit, Leid*, 9.

[114] "die Vernunft in einem nicht mehr rational begründbaren Akt dem Glauben unterwirft." Ibid., 15.

[115] "Leibniz unterscheidet sich von den theodizeeanalogen Deutungen der Antike und des Mittelalters durch das bemühen, wissenschaftliche Aufklärung mit einer nicht mehr konfessionsspezifischen Apologetik zu verbinden, in der die Vernunft das Dogma als Bedingung von Theologie ablöst." Geyer, "Theodizee VI – Philosophisch," 234.

revealed premises were exchanged for logical argumentation. They were "true" only to the extent that they reflected logical deductions. As we shall see, in Leibniz's theodicy, even God's power and creativity were subject to a set of possibilities the boundaries of which were determined by logical and rational necessity. Thus, against Descartes and Spinoza, Leibniz taught that there was a "standard of goodness and perfection that exists independently of God and that God creates the world because it is good, a world which is good not just because it is the creation of God."[116]

8 Leibniz

The shift from pre- to post-Enlightenment theodic reflections occurred in and around the work of Leibniz more than in any other thinker. As we shall see, the theodic strategies for encountering the problem of evil are surprisingly faithful to those found in the tradition he draws from. It is the setting that is so different. Leibniz was trained in the discipline of law and this is evident in his theodic universe. God is in the seat of the accused, charged for being the cause of evils in the world. The judge is human reason, and the defender and the prosecutor both advance arguments in favor of their respective case: God is innocent vs. God is guilty.

8.1 The Best of All Possible Worlds

As did Augustine, Leibniz held a highly optimistic view of the world. This optimism was anchored in his rationalistic and equally optimistic views of God, to whom nothing but the best imaginable qualities could be attributed: divine wisdom, omnipotence, foreknowledge. In his infinite knowledge, he would logically have in front of him a host of possible worlds to create. In his goodness he would logically choose the best among these options: "God has chosen the most perfect world, that is, the one which is at the same time the simplest in hypothesis and the richest in phenomena."[117] Hence, no matter how the world might look from a human perspective, by necessity it would have to be the best of possible worlds. Unlike both Aquinas and Malebranche, who taught that God could improve the world by adding good to it, Leibniz taught that this was impossible, and that God could not have created the world better than it is.[118]

In Leibniz's thought, even God was constrained by logical necessity when creating; only certain combinations of things can exist together or be, in Leibniz's word, "compossible."[119] The idea that God could simply eliminate elements experienced as evil failed to ignore that "the universe is all of a piece, so that if the smallest evil that comes to pass were missing, it would no longer be this world."[120] Leibniz's belief in the perfection of God's initial creation was so great that he concluded God would never tinker with what he had created and especially not perform miracles in the sense of breaking the established laws. This would contradict the conviction that

[116] Garber, "Leibniz." Ref. to Leibniz, *Discourse on Metaphysics,* §2.

[117] Leibniz, *Discourse on Metaphysics,* §6.

[118] Ibid., §§3–4.

[119] Ref. in Cottingham, *Western Philosophy,* 260.

[120] Ref. in Ibid.

God had created the most perfect of possible worlds, and Leibniz therefore considered the notion of miracles repugnant. The best possible world was one where God would never have to intervene in the unfolding of its pre-established trajectory. Even if it appeared that God would improve the world by miraculously preventing individual accidents, such interventions would suggest that the world was not as good as it could have been, wherefore God would allow the accident to happen as it was part of his overall purpose—the wisdom of which only he could see.

Leibniz believed that the world was so harmonious from the outset of its creation that it could only follow the order God had laid down in its inmost nature. Leibniz actually expressed caution with regard to a too mechanistic worldview and warned against the World Machine (*machina mundi*) of mechanical philosophy because he did not consider it to disclose "the innermost nature of things."[121] Nevertheless, he exemplified his conviction of the world harmony by analogy with a recent invention of his time—the clock: The clock of the world unfolds as if by necessity toward the greater good. The unfolding of the "clock of the world" occurs as "the first mind . . . in its wisdom, establishes things from the beginning so that there is rarely need of extraordinary concurrence . . . for the conservation of things."[122] Even natural evil is part of the best of possible worlds. It is determined by laws which also define the best possible consequences, and as Augustine taught, evil must always be judged teleologically in terms of the best possible whole.

8.2 Evil as Privation

As St. Thomas and other theologians and philosophers had done, Leibniz distinguishes between three kinds of evil: moral, physical (natural), and metaphysical. He explains each as follows:

> Evil may be taken metaphysically, physically and morally. *Metaphysical evil* consists in mere imperfection, *physical evil* in suffering, and *moral evil* in sin. Now although physical evil and moral evil be not necessary, it is enough that by virtue of the eternal verities they be possible. And as this vast Region of Verities contains all possibilities it is necessary that there be an infinitude of possible worlds, that evil enter into divers of them, and that even the best of all contain a measure thereof. Thus has God been induced to permit evil.[123]

Although Augustine identified creaturely privative weakness in Adam before the fall, he had focused mainly on moral evil (sin), that is, the falling away from God. This evil he considered the sole cause of all others, for the original weakness did not by necessity lead Adam to sin, and from sin to suffering and evil. Leibniz, on the other hand, placed more emphasis on metaphysical evil. By this he meant the limitation of creation vis-à-vis its creator. Although Leibniz considered the present world to be the best of possible worlds, he nevertheless also entertained the conviction that the created order had to be imperfect so that there would remain a qualitative

---

[121] Quoted in Mercer, *Leibniz's Metaphysics,* 308.

[122] Quoted in Ibid., 239.

[123] Leibniz, *Theodicy,* 1.21, p. 136.

difference between God and creation. For instance, God is the cause of God and owes his existence to no one, but creation owes its existence to a cause beyond itself, namely God. God could not have given humankind all degrees of perfection without making the creature God:

> And when it is said that the creature depends upon God in so far as it exists and in so far as it acts, and even that conservation is a continual creation, this is true in that God gives ever to the creature and produces continually all that in it is positive, good and perfect, every perfect gift coming from the Father of lights. The imperfections, on the other hand, and the defects in operations spring from the original limitation that the creature could not but receive with the first beginning of its being, through the ideal reasons which restrict it. For God could not give the creature all without making of it a God; therefore there must . . . be different degrees in the perfection of things, and limitations also of every kind.[124]

According to Leibniz, this original limitation antedates original sin and any moral evils and sins angels or humans may have committed. It constitutes the privation of its origin—namely its creator, its origin, its God: "there is an original imperfection in the creature before sin, because the creature is limited in its essence; whence ensues that it cannot know all, and that it can deceive itself and commit other errors."[125] These limitations that led to original sin as well as the subsequent multitudes of individual sins that followed—are necessary to maintain the right balance between Creator and creature. Thus, for Leibniz, even original sin and the subsequent corrupted will of humankind are natural and necessary parts of the best of possible worlds.

This obviously endangers God's innocence vis-à-vis evil in the world. Yet God is not to blame for the evils in the world, as the world must be deprived of some of the goodness of its creator by a law of necessary logic that not even God's will can overrule. It could not be different if God were to create, and it is a greater good that God create than not create, even with the limitations and evils inherent in creation.

8.3 Free Will

As in the main trends of theodic reflections prior to him, Leibniz taught the absolute advantage of having a world with created beings with free will. Some people had suggested to Leibniz that God could have created a world with less suffering if he had not created them with free will.[126] Leibniz rejected such a utilitarian idea that God would only act according to what would provide the highest degree of happiness. Rather, he argued, God would act according to what he necessarily knew to be the best, seen from his divine perspective:

> the author is still presupposing that false maxim advanced as the third, stating that the happiness of rational creatures is the sole aim

---

[124] Ibid., 1.31, p. 141–42.

[125] Ibid., 1.20, p. 135.

[126] Feinberg, *The Many Faces of Evil*, 54 n. 26.

of God. If that were so, perhaps neither sin nor unhappiness would ever occur, even by concomitance. God would have chosen a sequel of possibles where all these evils would be excluded. But God would fail in what is due to the universe, that is, in what he owes to himself. [127]

Although creatures could choose evil, they are nevertheless necessary in the best of possible worlds. Just as creation had to be limited at large vis-à-vis the creator, so the will of humankind had to be limited, which in turn led to moral evil or sin. According to Leibniz, such sin is real, but it cannot be avoided due to the necessary limitations of created beings.

8.4 The Link Between Moral and Natural Evil

As we saw, Leibniz taught that metaphysical evil antedated both moral and natural evil and was the real cause of both. One could have expected that this would have decreased human guilt, but Leibniz follows Augustine faithfully and confirms that natural evil is just punishment for moral evil and sin. He did not employ much space to describe why this is so and how the link between sin and suffering comes about. As Susan Neiman suggests, "Leibniz held the connection between moral and natural evils to be too self-evident to warrant serious question."[128] Moral evil consists in the wrong choice, which he equates with sin.[129] Such wrong choice comes about because of our confused perceptions. Leibniz was convinced that the link between moral and natural evil would become clearer with time and that at present it was a matter of trust that the present arrangement of things ultimately would prove to be the best possible.

9 Theodicy after Leibniz

The impact of Leibniz's thought and in particular his theodicy is evident from its influence on intellectual optimism in the Enlightenment. But it equally prompted forceful criticism.

The Lisbon earthquake of 1755 shook the ground under Leibnizian optimism. The disaster itself was of an impressive magnitude. A third of the city's 275,000 inhabitants were killed. But, as various authors observed, a series of coincidences made it appear particularly evil and irreconcilable with the notion of a good (Christian) God who takes good care of his creatures. The quake hit *Lisbon,* a Roman Catholic city known for the religiousness of its inhabitants and its Christian art. It occurred on a Christian holiday—All Saints' Day—during service. Thus the majority of victims were worshippers and many of them died under collapsing church spires. Fearing fires, falling debris, and the possibility of further quakes that could cause other buildings to fall over them, tens of thousands of citizens took refuge on the waterfront just to be killed by the wave that hit

---

[127] Ibid., 55.

[128] Neiman, *Evil in Modern Thought,* 22. See also Cottingham, *Western Philosophy,* 260.

[129] Leibniz, *Theodicy,* 411, and 16.

the city thirty minutes after the first earthquake. Conversely, those who chose to sin on this religious feast day in the brothels in the city's eastern outskirts survived. As Martin Schönfeld observes: "Overnight, Leibniz's theodicy was turned into a bad joke."[130]

Partly because of this disaster, and partly because he had a different cosmology, Voltaire tore apart Leibniz's optimism. In his *Poème sur le Désastre de Lisbonne* (1756), he attacked it directly. Three years later, in his *Candide* (1759), he polemically noted: "If this is the best possible world, then I do not want to know what the worst looks like."[131] And in the *Philosophical Dictionary* (1764), Voltaire quoted Lactantius who in turn quoted the theodic passage in Epicurus. Voltaire did not find Lactantius's solution convincing. Much less did he find Leibniz's depiction of the world reconcilable with the facts of history. According to Voltaire, only someone living in luxury as Leibniz had done could propose such an idealized image:

> Leibniz . . . did mankind the service of explaining that we ought to be entirely satisfied, and that god could do no more for us, that he had necessarily chosen, among all the possibilities, what was undeniably the best one. . . . What! To be chased from a place of delights, where we would have lived for ever if an apple had not been eaten! What! produce in wretchedness wretched children who will suffer everything, who will make others suffer everything! What! To undergo every illness, feel every sorrow, die in pain, and for refreshment be burned in the eternity of centuries! Is this really the best lot that was available? This is not too good for us; and how can it be good for god?

> Leibniz realized that these questions were unanswerable: so he wrote thick books in which he did not agree with himself.

> A Lucullus in good health, dining well with his friends and his mistress in the house of Apollo, can say laughingly that there is no devil; but let him put his head out of the window and he will see unhappy people; let him suffer a fever and he will be unhappy himself.[132]

David Hume—who promoted the Epicurean dilemma in simplified form, placed in the words of Philo in the *Dialogues Concerning Natural Religion* (1779)—further enhanced Voltaire's criticism of the Leibnizian optimism with relentless analysis. David Hume's version of the problem of evil, seen as its "most succinct formulation" has had tremendous influence: "Is he willing to prevent evil, but not able? Then he is malevolent. Is he both able and willing? Whence then is evil?"[133]

---

[130] Schönfeld, *The Philosophy of the Young Kant,* 75.

[131] Quoted in Ibid.

[132] Voltaire, *Philosophical Dictionary,* 68–69.

[133] Hume, *Dialogues Concerning Natural Religion* (J. C. A. Gaskin trans.). Quoted in Surin, "The Problem of Evil," 192.

Immanuel Kant criticized all former attempts at a theodicy in his *On the Failure of All Philosophical Attempts at a Theodicy* (1791).[134] It was written after the three Critiques, but before *Religion within the Boundaries of mere Reason* (1793). Kant was less optimistic than Leibniz with regard to the range of reason. He denied that human reasoning could judge God and his works in any responsible project of combining the notions of God, evil, and reason. Kant dismissed as hubris the conviction that finite reason and theistic concepts of God could converge.[135] He published a number of essays on the Lisbon earthquake in which it is clear that he was more interested in the scientific side of the event, *how* it happened, than the metaphysical reflections on *why* it happened. As terrible as such disasters were, they did not have divine causes—they were accidents.[136] Further, in the second of the essays with the provocative title "On the Use of Earthquakes," Kant enumerated the good side effects he believed could come from such devastating events as earthquakes.

Kant did not share Leibniz's view that the universe was rationally arranged. He wrote from a teleological perspective. Like Leibniz, Kant ordered the divine attributes in three areas: (1) goodness, (2) omniscience and omnipotence, and (3) holiness. But unlike Leibniz, Kant found each of these to be challenged by the empirical fact of disteleology *(Zweckwidrigkeit):* "A. the holiness of the creator contrasts with moral evil in the world. B. The goodness of the Creator contrasts with woes and suffering. C. The justice of the Creator contrasts with the impunity of the guilty."[137] Hence, according to Kant, all theodic attempts had failed.

In the place of theodic solutions, Kant emphasized human responsibility for moral conversion. Books One and Two of his *Religion within the Boundaries of Mere Reason* have been of great importance to the development of theodicy.[138] Book One represents a philosophical counterpart to the Christian doctrine of original sin, whereas Book Two parallels the Christian doctrine of redemption, reinterpreting them both in ways that have taken theodic reflections further away from their pre-Leibnizian habitat. This is evident in the reception of Kant's ideas in the works of J. G. Herder, J. C. F. Schiller, F. X. von Baader, F. W. J. Schelling, and F. Schleiermacher.[139] Some scholars in recent Kantian research suggest (in contrast with his early, mainly Lutheran, biographers) that Kant disliked Christianity and was really more oriented towards pantheism. They argue that he and his earliest biographers sought to cover it up, both because it was dangerous to his career and because his ideas would be received with greater difficulty in the aftermath. However, they see such antipathy in his reinterpretation of a number of fundamental Christian ideas.[140] Christian-

---

[134] Kant, *Kant on History and Religion, with a Translation of Kant's "On the Failure of All Attempted Philosophical Theodicies."*

[135] Rosenau, "Theodizee IV – Dogmatisch," 235.

[136] Schönfeld, *The Philosophy of the Young Kant,* 75.

[137] Caygill, *A Kant Dictionary,* 389. Ref. to Kant, *Kant on History and Religion,* 285.

[138] Kant, *Religion within the Boundaries of Mere Reason.*

[139] Rosenau, "Theodizee IV – Dogmatisch," 235.

[140] In the context of censorship, writers tend to become circumspect and avoid trouble by publishing anonymously, or making oblique remarks instead of direct

ity "survived" in Kant due to his assessment (or reinterpretation) of it as "the only moral religion" vis-à-vis all other religions, which he considered to be servile attempts to manipulate supernatural powers.[141]

Whereas original sin is seen as inherited, Kant views radical evil as self-incurred by each human. Instead of being guided by the categorical imperative (Act only according to that maxim by which you can at the same time will that it would become a universal law), humans make the satisfaction of their own ends the first priority in their actions. Overcoming radical evil requires a "change of heart" which can only occur in the reordering of one's principles of choice.

The question then becomes, what will be the result of our moral culpability? In response to this question, Kant reinterpreted the Christian doctrine of vicarious atonement through the death of Christ (Book Two of *Religion*). Because of his analysis of morality as autonomy, he rejected the doctrine of vicarious atonement.[142] Instead he saw in Christ a role model, and considered Christ's work an "exemplary" one in which we recognize just moral behavior. Christ had adhered to this principle of moral rightness and Kant considered it fundamental to the "religion of reason."[143]

Kant's reinterpretation of the Christian notions of original sin and vicarious atonement have had important consequences for the history of theodicy; it contributed to the theodic dilemma in the modern form, first established by Leibniz. Instead of God being offended by the sins of his creatures, it is humans who are offended by God's "sin"—evil in the world. Of course, Leibniz believed he was able to find enough evidence in favor of God's innocence, so that he could let God go free. He did so for two reasons: (1) because he was utterly convinced of the logic that the world had to be the best of possible worlds, and (2) because he held on to a notion of original sin with natural evil as just retribution for sin. Voltaire challenged Leibniz's first reason. Kant challenged the second.

The changes from Leibniz to Voltaire, Hume, and Kant altered the entire situation. We may or may not *believe* in the notion of paradise, original perfection, a fall away from it through sin, and the cosmology in which it became the constituent theodic component. But whether we consider the changes positive or negative, a historical overview of the development of theodicy must take into consideration which consequences the departure from the traditional beliefs have had for theodicy.

10 Development as Theodicy

Much like Augustine and Leibniz, G. W. F. Hegel considered the evils suffered by humankind in its historical development to be partial and necessary in the overall picture of good. However, Hegel criticized Leibniz for having undertaken a theodicy with "indeterminate abstract categories." In-

---

statements, or later retracting earlier "dares" if they have second thoughts. Kant did all three, but it was easy for later readers of more liberal ages to miss such subtleties. Schönfeld, "Kant's Philosophical Development." Ref. to Kuehn, *Kant,* esp. 2–16, 328, 69–79, 82, 92.

[141] Chidester, *Christianity,* 491.

[142] Sullivan, *Immanuel Kant's Moral Theory,* 270.

[143] Rossi, *Kant's Philosophical Development.*

stead, he saw a theodicy in history. At the conclusion of his *Philosophy of History* he wrote: "The history of this world is this process of development and the actual coming-into-being of spirit, underneath the variable dramas of its histories—this is the true theodicy, the justification of God in history."[144] This developmental theodicy has had enormous impact, partly because of Hegel's authority, partly because it picks up on developmental theodic aspects that have been co-present with the Augustinian-Leibnizian attempt to free God from any responsibility for the evils in the world. In developmental theodicies, God allows destruction and suffering as means towards the refinement of both biotic and non-biotic nature, and especially human character, both individual and collective. This notion was applied for instance by Karl Marx, who secularized Leibniz's metaphysical optimism even further. Following Hegel's conviction that humankind develops through the hardships of history, Marx saw the suffering of the striving classes as the necessary sacrifices for the development toward a society in which equality would eventually prevail.

Such approaches take a theodic strategy that comes across as the most significant modern approach to the problem of evil. It is a view that sees destruction and suffering as fruitful catalysts in the development towards greater goods, that is, those of the development of individual human beings, of collective human history, of the emergence of the world, or of biological evolution. Such strategies already became evident with the rise of scientific theories of a natural order that involves destruction and subsequent suffering as part of its inherent fabric. They were further advanced by evolutionary biology and its conclusion that destruction is a necessary requirement for the refinement of races and species.

The developmental approach to theodicy has been received and developed further in John Hick's *Evil and the God of Love,*[145] a book that has had considerable influence. According to Hick, Western theodic reflections have suffered from the dominance of the Augustinian paradigm: God allows free will for its intrinsic value and the subsequent fall has thwarted nature and caused the evils it contains. For Hick, the primary problem with the Augustinian reading is the plausibility of its premises:

> the logical possibility that it would establish is one which, for very many people today, is fatally lacking in plausibility. For most educated inhabitants of the modern world regard the biblical story of Adam and Eve, and their temptation by the devil, as myth rather than as history; and they believe that so far from having been created finitely perfect and then falling, humanity evolved out of lower forms of life, emerging in a morally, spiritually, and culturally primitive stage. Further, they reject as incredible the idea that earthquake and flood, disease, decay, and death are consequence either of a human fall, or of a prior fall of angelic beings who are now exerting an evil influence upon the earth. They see all this as part of a pre-scientific world view. . . . those of us for whom the resulting

---

[144] Hegel, *The Philosophy of History,* 11.569 and 11.42.

[145] Hick, *Evil and the God of Love.* Hick's thoughts are also expressed in various articles and book chapters, i.e., Hick, "An Irenaean Theodicy."

theodicy, even if logically possible, is radically implausible, must look elsewhere for light on the problem of evil.[146]

As is evident, Hick is heir to the Enlightenment revisions of theodicy (and theology in general) that we have seen in the foregoing paragraphs, with its rejection of many of the basic maxims in which Christian reflections on the problem of evil used to be embedded, and he does nothing to hide this. However, it is important for him to argue that his developmental approach has roots in Christian tradition as well. As an alternative to the Augustinian paradigm, Hick points to another tradition that he considers to have been present throughout Christian history, but that has received less attention, that is, that of the early Hellenistic Fathers of the church, in particular St. Irenaeus. Hick acknowledges that Irenaeus was firm in his belief in original sin and the fall, but he maintains that Irenaeus sees it "as a relatively minor lapse, a youthful error, rather than as the infinite crime and cosmic disaster which has ruined the whole creation."[147] Thus, in Hick's reinterpretation of Irenaeus the theodic approach is one that does not hinge "upon the idea of the fall, and which is consonant with modern knowledge concerning the origins of the human race." For it hinges "upon the creation of humankind through the evolutionary process as an immature creature living in a challenging and therefore person-making world."[148]

According to Hick, man could only develop a truly free search for God in love if he would *not* have full knowledge of him. If he did, the disproportion between God and creature would be so great there would be no freedom, for "what freedom could finite beings have in immediate consciousness of the presence of the one who has created them ... and who claims their total obedience?"[149] Only at an epistemic distance can God's creatures come to seek him in freedom. They would further have to be in an imperfect state themselves to freely choose the greater good, namely God. Against Alvin Plantinga, and agreeing with Antony Flew and J. L. Mackie,[150] Hick argues that creatures could *not* have been created morally perfect and yet free. Had they had moral perfection, they could have but constantly chosen good, and hence would not have been free. Further, there would have to be resistance and all sorts of frightful challenges in the soulmaking world, for only in such a challenging environment would humans beings be able to grow in character. Therefore, God could only create the world as we know it, with all its imperfections, including the human condition and the human morally imperfect character. Only in this way, Hick argues, could God gradually create children for himself out of human animals. Thus, he even speaks of two stages of creation: the initial establishment of the imperfect world we know, and the development of human animals into children of God who freely seek union with him. With these requirements, the world

---

[146] Hick, "An Irenaean Theodicy" (1981 ed.), 40–41.

[147] Ibid., 42.

[148] Ibid., 41.

[149] Ibid., 42–43.

[150] Hick's classification of these authors is not convincing.

can by necessity only be one that must "provoke the theological problem of evil."[151]

As we shall see later in this book, it is legitimate to ask whether *so* much evil (for instance, the Holocaust) is necessary to these developmental goals and whether the good of development justifies the evils humans suffer. Hick believes it does. He argues that God would have to adopt a general strategy of truly leaving his creation free and never intervening: "if we take with full seriousness the value of human freedom and responsibility, as essential to the eventual creation of perfected children of God, then we cannot consistently want God to revoke that freedom when its wrong exercise becomes intolerable to us."[152] He is aware that this conviction may be of little avail to humans who are suffering, but "if God were to remove such suffering, it would convert the world from a person-making into a static environment, which could not elicit moral growth."[153]

Following the Christian conviction that Christ will return to perfect all things, Hick furthermore argues that such a development of God's creation will continue and be perfected only in an eschatological future. There are many things we still do not understand about God's allowing evil in the world, but we may rest assured by Christ's promise that he will perfect all things.

A number of theological issues in Hick's theodicy complicate its reception into a classical Christian setting. It seems difficult to find any purpose for the atoning work of Christ when Hick believes that humankind must develop "all the way" to God. Therewith he believes in a continued development through what he calls "vertical reincarnation (i.e., reincarnation in an ascending series of environments beyond this earth)."[154] Further, Hick adopts the essence of *apocatastasis,* that is, that God would ultimately be able to have the whole of creation restored and reconciled to himself.

Despite these issues, Hick provides theological expression to a conviction that God creates a world that from the outset is chaotic but through various forms of creativity renders it fruitful without distorting the freedom and autonomy of either creatures or natural causality. He is not the first or only philosopher to propose such a view, and his soul-making theodicy is not perfect, but he "is widely regarded as the most influential and persuasive exponent of such an approach."[155]

11 Conclusion

The problem of evil has challenged faith in a good, just, and powerful deity throughout the ages of religious belief. The problem has been considered great when focusing on the evils that stem from what humans do to other humans, although evils caused by humans, so-called moral evils, take away some of the burden from God. The predicament is greater with those evils that cannot be directly traced back to human activity, so-called natural

---

[151] Hick, "An Irenaean Theodicy" (1981 ed.), 48.

[152] Ibid., 49.

[153] Ibid., 50.

[154] Hick, *The Center of Christianity,* 116. In this perspective even the title of one of his books does not surprise: Hick, *The Myth of God Incarnate.*

[155] McGrath, *Christian Theology,* 292.

evil. The chapters in the present volume address this dilemma from the perspectives of contemporary philosophy and theology, informed by the natural sciences.

The problem has been raised in various forms so that we should really speak of the *problems* of evil. Likewise, history has seen numerous attempts at solving the dilemma or at least casting sufficient light upon it to prevent believers from losing their faith. In this short history of the development of the problem of evil we have therefore merely scratched the surface. Nevertheless, we have seen a number of recurring traits in the Bible and variously throughout the Christian tradition. This holds in particular for the way in which theodicists have reflected upon (1) the causes of suffering, (2) the reasons for suffering, and (3) the possible purpose of suffering.

First, the evils of the world raise the question of their causal origin. Either the emphasis has been on God's omnipotence, with the frequent conclusion that he must be the direct cause of all that happens in the world, including those events we consider evil. Or the emphasis has been on God's goodness, with the frequent dualist conclusion that there must be two sources of all that occurs in the world: one good (God) and one evil (the Devil). These two perspectives have sometimes been combined so that evil would come from the evil one, but never outside of God's overall control. Rather, God would merely permit the Devil to afflict God's creatures, but such affliction would nevertheless be within the overall vision of his ultimate eschatological victory. This same eschatological perspective has often been proposed as the ultimate horizon by those thinkers that situated evil's origin in the haze of mystery.

Second, sometimes replacing the question of the causal origin of evil, sometimes adding to it, have been the considerations of the *reasons* for suffering. The primary Old Testament explication has been the rebellion of Satan and/or human beings. Suffering has thus been presented as a just recompense for sin. Sometimes God was seen as the judge *and* the one carrying out the just retribution. Other times the equation of sin and suffering was linked up with a law of cosmic justice that was inherently logical *and* efficacious, as in one of the Psalms: "Their mischief returns upon their own heads, and on their own heads their violence descends" (Ps 7:16).

Third, humans have thought often about the possible *purposes* of suffering. Suffering has been seen as enabling existential transformation of human existence. Repeatedly, suffering has been presented as the key to newfound faith and knowledge, in particular when suffering was presented as divine testing: God's purpose in testing his faithful was that they grow in faith and come out of the trial strengthened and wiser than before it.

In the Christian tradition the dilemma of human sin and subsequent judgment has been deemed so great that only a God—that is, God's son— could save humans from their self-caused predicament. This belief is at the core of the mystery of the cross that not only redeems humans from eternal damnation, but also in some sense redeems human suffering itself: Humans do not suffer on their own. Christ entered the realm of human existence and experience to the fullest extent of human death in order to both suffer *with* God's children and *for* them. In doing so he called them to unite their sufferings with his, therewith including human sufferings in his own, giving it new meaning and purpose. Not surprisingly, to many Christian thinkers Christology has therefore taken the place of theodicy.

Many of these thoughts have been followed by postbiblical thinkers. They have often been enriched by Greek material, as in the famous example of Augustine, who more than any other single thinker has influenced Western reflections on the problem of evil.

Things changed dramatically with Leibniz. Ironically he did not in any remarkable way shift the *content* from the traditional questions and answers related to the problem of evil. However, in an unseen way the *focus* changed with Leibniz. Before Leibniz, humankind was in the seat of the accused with God on the throne of judgment. With and after Leibniz, God now was in the position of the accused, with humans—especially human reason—in the position of judge. The perspective is entirely different. Whereas the content and the conclusions are largely the same, God is now the accused. Evidence for and against God's culpability is brought before the judge. Leibniz's system weighed the evidence in favor of God's innocence. This changed with later thinkers of the Enlightenment. They kept Leibniz's basic perspective with God accused for the existence of suffering, but they came to different conclusions, so that God was able to escape human accusations with greater difficulty. Thus, Enlightenment philosophers challenged theodic reflections tremendously. As we shall see, these difficulties largely constitute the overall point of departure of contributions in the present volume.[156]

---

[156] I wish to thank H. Lundbeck for the grant that secured the research for this text and the participation at the conference in Castel Gandolfo, the papers of which are gathered in this volume.

THE PROBLEMS OF THEODICY: A BACKGROUND ESSAY

Terrence W. Tilley

1 Introduction

Too often those who work in the area of analytical philosophy of religion
and those who work in the area of theology and natural science display
remarkable lack of awareness of the work in each others' fields. Both of my
chapters in this volume are attempts to remedy that a bit. Here I want to
share a summary of my reflections on the modern problems of evil and
theologians' and philosophers' responses to those problems, especially those
in the analytical tradition, over the last twenty-five years or so. The concern with logical and rhetorical differences between various texts helps
clarify the problems of theodicy and suggests other responses to the
problems of evil.

   I also want to utilize a more "analytic" mode of discussion of the problem of evil than is typical of the style for CTNS/VO participants' work in
discussions of theology and the natural sciences. This will be that case in
my chapter titled "Towards a Creativity Defense of Belief in God in the
Face of Evil."

2 Theodicies and Defenses

I have shown elsewhere that there are crucial logical and rhetorical differences between constructing defenses of religious faith in the face of challenges, especially the modern problem of evil, and constructing theodicies
to respond to that problem.[157] Each practice has a different purpose,
burden of proof, and structure of argument. Being clear about the
differences is crucial for properly discussing the problem of evil in creation
today.

   In constructing a *defense,* a defender of religious belief is not trying to
warrant a claim that God exists. The defender believes in God on other
grounds, e.g., religious experience, authority, etc. Nor is the defender trying
to warrant a claim that any given event or class of events is not genuinely
evil; the defender may, and probably must, admit that there are genuine
evils in the world and may even recognize as evil all the states of affairs
that opponents challenging their religious convictions recognize as evil. Nor
is the defender trying to give a theory that accounts for the evil in the
world. Rather, the defender seeks to show that an attack against the religious believer's claims attempted by the opponent does not affect, and

---

[157] These paragraphs develop the discussion in Tilley, "Use and Abuse of
Theodicy," particularly 307–9. There I argued that neglecting the distinction
between theodicies and defenses has led to confusions in critiques of the work of
Alvin Plantinga; such mistakes still can be found in the literature, e.g., in the
discussion of Plantinga's work in Kenneth Surin, *Theology and the Problem of Evil.*
In 1984, I accepted John Hick's critique of the "Augustinian theodicy" (cf. my "Use
and Abuse of Theodicy," 310 n. 23) in his magisterial *Evil and the God of Love,* and
still do. But there I did not clearly distinguish—as I did later in *Evils of Theodicy*—
between an "Augustinian theodicy" and Augustine's authoritative teaching, which
takes the form of a defense, not a theodicy.

should not affect, the quality of assent religious believers give to their beliefs. The defender does this by showing that the set of beliefs, specifically the belief that "God is all-good, all-knowing, and all-powerful" and "there are genuine evils in the actual world" are compatible. To show this compatibility, a defender need only find a possibly true proposition which, when conjoined with the proposition expressing belief in God, entails the proposition that evil exists; that shows belief in God compatible with belief that evil exists.

In confronting a religious believer, the opponent bears the burden of proof in attempting to overthrow an established belief. To be successful, this opponent must show that the believer cannot rationally hold both beliefs: that God is omnipotent, omniscient, and omnibenevolent, and that evil exits. To do so, the opponent must show that there is no proposition which, when conjoined with belief in God, entails that evil exists. Hence, a defense succeeds if it rebuffs the attack. Because a defense is not used to show those beliefs true, but to defend them from an attack of incompatibility, a defender does not have to demonstrate that the set of challenged beliefs is true, or that the conjoined belief is true, but only that all are possibly jointly true, that is, that they are not contradictory and do not entail propositions that would contradict whatever else the believer accepts. A successful defense is the rebuttal of a challenge.

In constructing a *theodicy,* on the other hand, the theodicist tries to show "what God's reason is for permitting evil. At bottom, he [sic] says, it's that God can create a more perfect universe by permitting evil" in it.[158] The theodicist in fact attempts to construct a theory that gets God "off the hook" for the evil in God's creation. Reinhard Hütter put the problem in sharp and accurate terms: "In humanity's utmost presumption lies the seed of the fall from the modern daydream of freedom into the postmodern nightmare of freedom. It is important to remember that part of the modern daydream was a fundamental exchange of juridical positions between God and humanity, that is, between bench and dock. In C. S. Lewis's apt description: "The ancient man approaches God (or even the gods) as the accused person approaches his judge. For the modern man the roles are reversed. He is the judge: God is in the dock. He is quite a kindly judge: if God should have a reasonable defense for being the god who permits war, poverty and disease, he is ready to listen to it. The trial may even end in God's acquittal. But the important thing is that Man is on the Bench and God in the Dock."

"During reason's trial of God, the god of the deists died in the dock. And with this god dead, only humankind is left to blame for the miseries that we inflict upon each other. Theodicy turns into anthropodicy. That is, in the face of evil and suffering, it is now humanity, instead of God, that needs to be acquitted."[159]

So the purpose of a theodicy is not to show that an opponent's attempt to show the believer's claims about God and evil incompatible. Rather, the theodicist must make the argument for a theory that gets God off the hook.

---

[158] Plantinga, *God, Freedom and Evil,* 27.

[159] Hütter, "Bound to Be Free," 25. Theodicy is the task of acquitting God for responsibility for evil; finally it really acquits humanity. Defenses, on the other hand, do not place God in the dock, but show that God is not to be placed on trial.

Here the burden of proof is not on the attacker, but on the theodicist who attempts to show that "the facts which give rise to the problem of evil" do not truly count against belief in God. "Accordingly, theodicies proceed by bringing other facts and theories into account so as to build up a wider picture which includes the fact of evil but which is such that it is no longer more natural to infer from it that there is no God than that there is."[160] A successful theodicy would demonstrate either that a person can justly move from the data of the actual world, including its evils, to the claim that an omniscient, omnibenevolent, omnipotent deity created it or that the actual world is as good as or better than any other possible world or that its picture of the world is at least as plausible as other available alternatives. In contrast to a defense, which only attempts to show the compatibility of a set of beliefs, a theodicy has the more difficult task of warranting the beliefs which comprise it.

It is important to distinguish theodicies from defenses. Unfortunately, many who have not followed the philosophical debates among the analysts have failed to do so. The consequences are serious.

3 Some Historical Notes on Design and Theodicy

Theodicy originated as an explanation of an anomaly in the argument from design.[161] Each form of the design argument is rooted in the natural science of its time. David Hume effectively demolished the modern argument from design in his *Dialogues Concerning Natural Religion* (1779). Perhaps the most up-to-date form of a design argument uses the anthropic principle derived from scientific cosmology. Because the proposition that there is genuine evil in the actual world is obviously true, the contemporary version of the design argument also has an anomaly, a contemporary problem of evil, analogous to the problem imbedded in the early modern form of the argument.

These historical notes are relevant to the present discussion because, like the earlier "designists," contemporary "designists" also commit theodicy, that is they somehow transubstantiate evil into good. Typically, they argue that evil is necessary in creation—God had no choice in the matter. Because it is necessary in the universe and God therefore had to create a universe with such evil, that makes it somehow necessary for good, and thus in itself becomes even something of an instrumental good—hence the transubstantiation.[162] This move, of course, makes "necessity" greater than

---

[160] Hick, *Evil and the God of Love* (rev. ed.), 371.

[161] The following paragraphs depend on the fuller argument in my *The Evils of Theodicy*.

[162] It also seems to be the case that modern and contemporary theodicists create what is now an oxymoron, "natural evil." In scientific discourse, what is natural is what is by nature: it is neither good nor evil; it just is. Once philosophers and theologians, however, view some event, action, person, or state of affairs as "good for something" they value or view something as "bad for something" they disvalue, they come to talk of natural good and natural evil. But this fundamentally changes the use of the word *natural* from a factual use to a normative one. Of course, this can be done, but it seems that the move is made so facilely that the previous question of why something natural is perceived as good or evil, rather than just natural, is ignored. I will say more about this in my later chapter.

that which no greater can be conceived. It is just the virtue of defenses, rather than theodicies, that they show this "necessity" to be one of many possibilities, do not make it necessarily greater than God, and thus do not put God in the dock. Let me explain.

First, many theologians who are interested in the dialogue between theology and natural science find the argument from design a central and necessary strategy in theological debate. In the absence of the authority of the religious authorities, the authority of evidence and inference becomes central, and the burden of proof shifts away from those who would argue *against* the existence of God.[163] The burden of proof is now on the theologians who would argue *for* God's existence. God's design of the world, no longer evident to all, must be demonstrated. It must be the foundation on which religious thought can be erected. However, when God is seen as the "creator and benevolent governor of the world," and when the world God created seems hardly benevolent or well-governed, the supporter of the design argument must address challenges about the adequacy of this foundation for religious thought. Specific theoretical issues become pressing, especially those concerning the meaning of divine providence and the "Enlightenment" problem of evil, epitomized in one of Philo's challenges in Hume's *Dialogues:* "Epicurus's old questions are yet unanswered: Is God willing to prevent evil, but not able? Then, he is impotent. Is he able, but not willing? Then, he is malevolent. Is he both able and willing? Whence, then is evil?"[164]

Second, in the context of the design argument, "evil" became a theoretical term abstracted from specific instances of sin, suffering, and violence. In short, evil became a bloodless abstraction. While this seemed a good thing, I argue this is quite deceptive. William King, Anglican archbishop of Dublin, in *An Essay on the Origin of Evil* (written in 1697; published in Latin in 1702; ET, 1731) is the author of the classic fully developed Enlightenment theodicy.[165] Although philosophers, theologians, and poets had been attempting to deal with concrete evils in God's world, King now drives home the use of "evil" as an abstraction from any particular evils, as a part of the design. King typically divides evil into moral evils brought about by wrong elections, natural evil, and the imperfection of being. King states, in evident dependence on Augustine, what has become a standard categorization. Yet in a book written by the Anglican Archbishop of Dublin, less than a decade after the Glorious Revolution, there is no reference to the previous century of religious wars or to the evils in and/or of those wars. Even martial metaphors are remarkably rare in King's text. There is no mention of the fact that King himself was imprisoned during the struggle less than a decade earlier. Nor is the redistribution of land from the Irish peasant to the English landowner, which began with the settling of English into Ireland on the plantations created in the sixteenth

---

[163] For a sketch of the collapse of traditional religious-intellectual authority, see my *History, Theology, and Faith,* 70–76.

[164] *Dialogues,* 198.

[165] I see no way to show that one text is the "originating event" for the discourse of theodicy, the "ur-theodicy" or the "first theodicy." One could dub one of a number of specific texts as the "first theodicy," but I am unable to see how such a determination would not be arbitrary.

century and was decisively ratified by the victory of King William at the battle of Boyne in 1690, an issue. Specific events, e.g., the Lisbon earthquake of 1755, would later vivify the rhetorical force of the theoretical "problem of evil," and give its challenge a bite that went beyond the sphere of theoretical debates. Nonetheless, "evil" would remain an abstract theoretical term in the discourse of philosophical theism.

Moreover, it is only in the Enlightenment context of the need for a design argument that the abstract problem of evil becomes a central issue. Surin put the issue elegantly: "Pre-seventeenth century Christian thinkers were certainly not unaware of the conceptual difficulties that these antinomies [between divine omnipotence and worldly evils] generated; but, unlike their post-seventeenth century counterparts, they did not regard these problems as constituting *any* sort of ground for jettisoning their faith."[166] For Augustine, Boethius, Aquinas, and the great reformers, the conceptual problems of evil were small anomalies in intellectual visions of great breadth and depth, if not theological systems of great strength. "Evil" was not a single problem to be solved, but an aspect of various issues.

For Aquinas, for instance, there is no single problem of evil. In the *Summa,* evil is an objection to the existence of God (*ST* 1.2.3, obj. 1 and reply) and is a problem for the theology of the creation (*ST* 1.47–49). Other problems are sprinkled throughout the *Summa.* Aquinas does divide evil affecting creatures with volition into two types, *poena* and *culpa* (roughly, injury/pain and fault/sin; *ST* 1.48.5), but this has the purpose of showing that temptation is not evil. He explicitly does not reduce all evil to these, but recognizes that corruption and defect are also evils. Certainly one can construct a Thomistic resolution to the problem of evil,[167] but "the problem of evil" in the modern sense was not Thomas's problem.

Theodicy proper, as a resolution to the problem of evil, is a discourse practice, then, which emerges in the Enlightenment. Within this realm of discourse, two different shapes of "the problem of evil," and of responses to it, can be discerned in the eighteenth-century debates. In the first part of the century, discussions of worldly evil presumed the reality of an all-good, all-benevolent God. G. W. F. Leibniz's *Theodicy* (1710) presumed that God has these attributes. The issue is to understand how it is possible that there is evil in the world God created. King set up the problematic explicitly:

> [T]o point out a Method of reconciling these Things [evils] with the Government of an absolutely perfect Agent, and make them not only consistent with Infinite Wisdom, Goodness and Power, but necessarily resulting from them . . . then we may be supposed to have . . . answered all the Difficulties that are brought on this Head, against the Goodness, Wisdom, Power and Unity of God.[168]

---

[166] Surin, *Theology and the Problem of Evil,* 9 (italics in original).

[167] For example, Journet, *The Meaning of Evil.*

[168] King, *An Essay on the Origin of Evil,* reprinted four more times in English in the eighteenth century, 80.

Even Immanuel Kant's negative evaluation of theoretical theodicy[169] published in 1791 showed that the task was not demonstrating the existence of God, but of showing how worldly evil is consistent with the reality of the god of providential theism—presumed to be an appropriate representation of the Christian God. In short, theodicy was a discourse generated as much by the design argument used as an independent argument for the existence of God—independent of all religious experience, practice, or tradition.

Third, as the century progressed, however, the problem acquired an important new shape and significance. The design argument was becoming, at best, a dubious foundation even for deism. Authorities no longer had the authority to render basic religious beliefs plausible, even as "mysteries." Evidence was required to give them any probability. In this context, a new problematic emerges. As Jeffrey Stout put it,

> The problem of evil is no longer a problem of figuring out what God is up to, given all the theology we already believe. In a context shaped by the new probability, it is a problem of figuring out what kind of God—*if any*—is plausible as an explanation of the origins of the universe as we find it. Given the existence of earthquakes, plagues, and the suffering of innocent children, the existence of a supremely perfect personal God seems unlikely.[170]

The theodicist must then bear the burden of proof to show that it is *not* improbable that God exists.

Hence, the discourse practice of theodicy, responding to the "problem of evil," generates two different forms of argument in the context of Enlightenment theism. One is the problem of explaining how the evil in the world is consistent with an accepted belief in God, and the other is one of the plausibility of belief in God, given the evidence of the world including its sufferings *(poenae)* and individuals' sins *(culpae)*.

Interestingly, the work of a preeminent contemporary theodicist shows itself as shaped by *both* these problems. In material composed for the first edition of *Evil and the God of Love,* John Hick wrote:

> For us today the live question is whether this [evil] renders impossible a rational belief in God: meaning by this, not a belief in God that has been arrived at by rational argument (for it is doubtful whether a religious faith is ever attained in this way), but one that has arisen in a rational individual in response to some compelling element in his experience, and decisively illuminates and is illuminated by his experience as a whole. The aim of a Christian theodicy must thus be the relatively modest and defensive one of showing that the mystery of evil, largely incomprehensible though it remains, does not render irrational a faith that has arisen, not from the inferences of natural theology, but from participation in a stream of religious experience which is continuous with that recorded in the Bible. (244–45)

This conception of the project of theodicy is in the tradition of the projects of Leibniz and King, demonstrating how evil can be shown to be *coherent*

---

[169] Immanuel Kant, "On the Failure of All Attempted Philosophical Theodicies," 284.

[170] Stout, *Flight from Authority,* 123.

with and incorporated into an accepted theistic faith. Yet in the same book, in material added for the revised edition, the same author also wrote:

> In other words, the facts which give rise to the 'problem of evil', if taken by themselves, point away from rather than towards the existence of God. Accordingly, theodicies proceed by bringing other facts and theories into account so as to build up a wider picture which includes the fact of evil but which is such that it is no longer more natural to infer from it that there is no God than that there is. But this very procedure acknowledges that evil, but itself, *does* count against there being an infinitely good and powerful creator. (371)

This second conception of the project of theodicy responds to the problem of the plausibility of theism. It seeks not to answer difficulties believers have with their faith, but to show to anyone that belief in God is *plausible* despite the "natural inference" from the reality of evil to the nonexistence of God.

I have argued (in "The Use and Abuse of Theodicy") that Hick's project is incoherent. The two conceptions of theodicy have different presuppositions, argument structures, and goals. The former presupposes religious faith in the god of theism, must respond to challenges to its consistency, and is directed to believers; it is a defense. The latter presupposes epistemic neutrality about God, must bear a heavier burden of proof against challengers, and is directed to any reader, whether they presume the reality of God or not; it is a theoretical theodicy. But even if Hick's project were not incoherent, the key point is that it, like the theodicies of Leibniz, King, Journet (as well as those of David Griffin and Austin Farrer) is structured by the discourse practice of theodicy which presumes the task is either to understand the attributes of, or to argue for the reality of, a "unipersonal" god of Enlightenment theism despite the abstractly conceived moral and natural evils of the world. The task is not to enable folk to believe in the trinitarian God of Christianity, nor to confront the concrete reality of evils.

If the theodicy project were successful, it would only render plausible belief in a creator and designer of the universe. It is not at all clear that this is sufficient to overcome the implausibility of faith in the triune God worshiped in the Christian churches. These theodicy projects make sense only on the presumption that Enlightenment theism is the expression of belief in God proper to Christianity. But this presumption has increasingly been called into doubt by those contemporary theologians who do not construe doctrines of the Trinity, Christology, and soteriology as auxiliary to basic theism, but as constitutive of Christian belief in God. Indeed, the retrieval of the centrality of the Trinity, rather than the unipersonal God of the Enlightenment, is a characteristic of much contemporary Christian theology.

A corollary of this view is that the "classic theodicists" of the Christian tradition cannot have been doing theodicy—in the sense of responding to the "Enlightenment" problematics. In *The Evils of Theodicy,* Chapter 5, I argued that understanding Augustine as a theodicist is mistaken. One of his discourse practices is a defense *(Enchiridion).* Another is confession. And even in the *City of God,* the problematic is not the plausibility of *theism,* but a hermeneutic of history. Surin puts it bluntly: "Augustine's treatment of this 'problem' has as its proper locus a *theology of history* which views history as both a work and a sign of God's providence."[171] Augustine did not write a theodicy, although theodicies can be constructed from his writings.

Similar claims can be made about other major Christian authors. I also argued, in Chapter 6 of *The Evils of Theodicy,* that Boethius was also not doing theodicy in writing *The Consolation of Philosophy,* but therapy for those ill-treated by fortune. Aquinas did not write a treatise on theodicy (although he did address a *quaestionem disputatem, De Malo,* to evil) but treated evil in the contexts of other problems, as noted above. Theodicists typically extract Luther's and Calvin's "discussions" of evils from texts written for purposes other than to resolve the problem of evil and force them to deal with an Enlightenment problematic.[172] Moreover, the inspiration for the "Irenaean theodicy," supposedly opposed to the Augustinian themes of the mainstream traditions, comes from an author who was not doing theodicy at all, but also proposing a theology of history and arguing against gnostic dualism. The problematic for Irenaeus was not a "kind of 'soul-making' or anything resembling a theodicy. . . . The real problem of evil, for Irenaeus, arises in connection with the struggle to love God truly in Christ."[173] Irenaeus's project was to inoculate Christians against what he saw as the comforting, but delusive, intellectual illness of gnosticism. In short, constructing theodicies is not a Christian discourse practice before the Enlightenment.

The root problem that generates theodicy is the transformation of the intuition of design to an argument from design. Even Hume has Philo and Cleanthes acknowledge the intuition or vision of design in the final part of the *Dialogues.* Even Philo agrees:

> [Cleanthes]: The most agreeable reflection, which it is possible for human imagination to suggest, is that of genuine theism, which represents us as the workmanship of a Being perfectly good, wise, and powerful; who created us for happiness, and who, having implanted in us immeasurable desires of good, will prolong our existence to all eternity. . . .
>
> [Philo]: These appearances . . . are most engaging and alluring; and with regard to the true philosopher they are more than appearance. . . .
>
> If the whole of natural theology, as some people seem to maintain, resolves itself into one simple, though somewhat ambiguous, at least undefined, proposition, *that the cause or causes of order in the universe probably bear some remote analogy to human intelligence:* If this proposition be not capable of extension, variation, or more particular explication: If it afford no inference that affects human life, or can be the source of any actions or forbearance: And if the analogy, imperfect as it is, can be carried no farther than to the human intelligence: and cannot be transferred, with any appearance of probability, to the other qualities of the mind: If this really be the case, what can the most inquisitive, contemplative, and religious

---

[171] Surin, *Theology and the Problem of Evil,* 12.

[172] For an example of this see Griffin, *God, Power and Evil* (1st ed.), 101–30.

[173] Surin, *Theology and the Problem of Evil,* 13.

man do more than give a plain, philosophical assent to the proposition as often as it occurs; and believe that the arguments, on which it is established, exceed the objections which lie against it?[174]

The intuition is obvious even to the skeptic. The argument proves nothing much beyond the intuition.

4 Theodicy as a Discourse Practice: The Evils of Theodicy

Theodicy is a discourse practice that is "impractical." That is, it is a purely theoretical practice responding to theoretical problems, not a practical theory responding to actual problems in religious practice. Theodicies do not respond to complaints or laments. They are not addressed to people who sin and suffer. They are addressed to abstract individual intellects which have purely theoretical problems of understanding evil. Given the intellectual contexts as sketched above, the purpose of constructing a theodicy seems *purely* theoretical.

But in their interminable pursuit of theory, theodicists devalue the practical issues. Numerous examples of the neglect of the practical litter theodicies, but consider how even a religiously sensitive writer of our era, Farrer, sets up the problem of theodicy by marginalizing suffering:

> The practical problem is pastoral, medical, or psychological, and differs from case to case too widely to allow of much useful generalization. We are concerned with the theoretical problem only. If what we say is neither comforting nor tactful, we need not mind. Our business is to say, if we can, what is true. So far from beginning with the sufferer and his personal distresses, we will attempt to get the issue into perspective, and sketch the widest possible view.[175]

Farrer addresses an appendix on "imperfect lives," to the "death of speechless infants" and "the survival of imbeciles,"[176] but offers no answers to people baffled by any actual and seemingly unlimited ills.

Charles Journet also refuses to approach "the concrete, existential and specifically religious aspect of the problem of evil, or to listen to the cry of man afflicted by pain, the supplication of a Job or Jeremiah, overwhelmed by unbearable trials who call upon God to come out of his silence" in favor of the "metaphysical difficulty which no one can elude."[177] He finishes his neo-Thomist approach with a chapter on the right attitude to evil in which he cites numerous spiritual writers. Journet concludes his book with the truly comforting final words that "if ever evil, at any time in history, should threaten to surpass the good, God would annihilate the world and all its workings."[178] At this high level of abstraction, not only are practical concerns that generate problems for religious believers ignored, but they are so marginalized and distorted that the possibility of God's destruction of the whole world God created is rendered *a good thing!*

---

[174] Hume, *Dialogues,* 224, 227.

[175] Farrer, *Love Almighty and Ills Unlimited,* 7.

[176] Ibid., 181–91.

[177] Journet, *The Meaning of Evil,* 59, 60.

[178] Ibid., 289.

Griffin presents his work as different from this tradition of "impractical" theodicy. As part of his opposition to those who eschew theodicy in favor of practical and existential approaches to practical evils, he claims that the theoretical issues at least partially constitute the practical problems of evil because "the questions people raise about evil always contain theoretical dimensions."[179] Indeed, the expectation of comfort, safety, and divine care that belief in the doctrine of divine providence creates in people is a result of theological renderings of providence. The theoretical problem is part of the existential problem, and "a theoretical problem can only be met with a theoretical solution."[180] So Griffin introduces his theodicy project.

Yet the main "practical" application of his theodicy seems to be the removal of "the basis for that sense of moral outrage which would be directed toward an impassive spectator deity who took great risks with the creation."[181] The God Griffin portrays took great risks in creating, but is not merely a spectator; Griffin's God does not only risk what is created, but God also risks God's self. Because God shares the risks inherent in the development of higher intensity and deeper feeling, that somehow removes the cause of moral outrage. But is this so? Why does *sharing* a risk with others make it moral to induce the risk? Does this alleviate our moral outrage at some drunken driver who "shares the risk" of getting home safely with others in the car? God has also risked those I love. If they turn out to be some of the "losers" in the great divine wager, whatever the cause of their destruction, if they are the victims of the process, then that's just the way the dice of creation rolled. The real practical problem for me, then, would be how to forgive God for playing dice with the lives not only of those dearest to me, but also of all the life in the universe (not to mention my own life).

The practical issue relevant to Griffin's theodicy is how we can come to forgive God for taking the risks that have destroyed so many so miserably. Yet Griffin offers no suggestions on how we might be able to do so. Griffin's conclusion simply puts God on the risky hook along with creation. If we're all on the hook, and God put us there, then God is not a spectator. But God, then, cannot do what is necessary to resolve the destruction that can emerge from this risk. If God could save those marginalized, betrayed, destroyed, or crushed by the practically perpetual process, God might be forgiven for initiating it. But Griffin's God cannot. Griffin's impracticality finally undermines his initial linkage of practical and theoretical problems.

In writing theodicies, individuals detach themselves from the realities of sin and suffering. The purpose of most theodicies is to show why the sufferings people endure and the sins they commit do not count against belief in God. Yet can any writer get into a position to be so detached? And who is their audience? No one is without some sin. No one lives without some suffering. Yet theodicies are not only produced by, but also directed to, people detached from sin and suffering. Theodicists even encourage the reader to "try to be the most dispassionate," as Griffin put it.[182] Can and should anyone be so dispassionate, so detached from the realities of evil? If George

---

[179] Griffin, *God, Power and Evil* (1st ed.), 16.

[180] Ibid.

[181] Ibid., 309.

[182] Ibid., 16.

Eliot's novel *Adam Bede,* for example, portrays what is required to over-come the real evil of human "denaturing," accepting the recommendation of detachment when considering evils may render one oblivious to the com-mitment and constancy needed to overcome some evils.[183]

Thus, theodicists' very attempts to be detached are not the ways to solve the problem, but are part of the problem:

> A theodicist who, intentionally or inadvertently, formulates doc-trines which occlude the radical and ruthless particularity of human evil is, by implication, mediating a social and political practice which averts its gaze from the cruelties that exist in the world. The theodi-cist . . . cannot propound views that promote serenity in a heartless world.[184]

The theodicist encourages readers to "distance" themselves from the evils of the world in order to "understand" them. Yet should readers distance themselves from their own sins and refuse to own them? Is it not just those who will not or cannot own their lives—including their sins—who are fated to continue the process of dehumanizing victimization without reconcilia-tion? Should readers distance themselves and be serene in the face of all their suffering, including their victimizing victimization? Perhaps so—but such attitudes must be differentiated from the advocacy of quietistic fatal-ism, the inculcation of masochism, or the promotion of escapism so often the upshot of prescribing serene distantiation from real evils.

Theodicies are part of the Enlightenment obsession with reducing the muddy and mixed to the clear and distinct. But in doing so, the theodicist idealizes the reality of evils. No theodicist claims to have explained fully how God allows evil in the world or eliminated the "mystery" of evil by such distancing. For all theodicists, the practical problems remain. Indeed, they are exacerbated by the practice of distancing oneself from the reality of evils to "understand" how evil does not count against belief in God. The practice of theodicy valorizes the spotless hands that write about evils without being sullied by them.

Some theodicists do not even claim to solve the theoretical problem. In the face of evils they give only a theoretical possibility. This theory is sup-posed to give readers grounds for hope. Even if theodicies did give war-ranted answers to the theoretical problems, the *real* problem is not theo-retical. Theoretical hopes ought not to be raised lest they be shattered when those encouraged to hope confront the realities of massive destruction alone while the theodicists keep cerebrally beyond the fray.

If theodicy is "impractical," what sort of discourse practice is it?

At first glance, theodicy seems to have the assertive illocutionary point. Although theodicists carefully cultivate a style that requires the reader to pay homage to the fear and trembling with which theodicists approach their work, they see their task as representing with sufficient accuracy the way things are. Some are more modest than others, but none denies that his or her purpose is to get matters straight, to show things how they really are (despite appearances), to tell *the truth.* Richard Swinburne provides a

---

[183] See Tilley, *Evils of Theodicy,* chap. 8.

[184] Surin, *God, Power and Evil,* 51.

clear example by first formulating a question and then supplying a (presumably true) answer:

> What then is wrong with the world? First, there are painful sensations, felt both by men, and, to a lesser extent, by animals. Second, there are painful emotions, which do not involve pain in the literal sense of the word—for example, feelings of loss and failure and frustration. Such suffering exists mainly among men, but also, I suppose, to some small extent among animals too. Third, there are evil and undesirable states of affairs, mainly states of men's minds, which do not involve suffering. For example, there are states of mind of hatred and envy; and such states of the world as rubbish tipped over a beauty spot. And fourth, there are the evil actions of men, mainly actions having as foreseeable consequences evils of the first three types, but perhaps other actions as well—such as lying and promise breaking with no such foreseeable consequences. As before, I include among actions, omissions to perform some actions. If there are rational agents other than men and God (if he exists), such as angels or devils or strange beings on distant planets, who suffer and perform evil actions, then their evil feelings, states, and actions must be added to the list of evils. [185]

Swinburne's discourse—and the discourse of all theodicists—is flatly assertive, for all theodicists' purpose is to portray what is true, "the facts, and nothing but the facts," in good positivist, "scientific" fashion.

Nonetheless, there are peculiarities about theodicy as a discourse practice. One has been noted: it is an impractical practice. A second is that theodicy has become a discourse with a home more in the academy than in the churches. It is not institutionally bound; theodicy, like prayer, does not require that the speaker have a *specific* role within any religious tradition or institution. Nonetheless, those participating in that practice now usually have *some* role—critic, student, researcher, teacher—in (or towards) one of the institutions of academia—college, university, seminary. People in pastoral work are notoriously absent from the discourse practice of theodicy, unless they write what is clearly derivative.

Contemporary academic theodicy has also come to construe certain approaches and certain texts as its tradition. Farrer mentions the tradition, and even feels comfortable within the constraints it imposes: "There is nothing new to say on the subject. Only the fashions of speech alter, and ancient argument is freshly phrased." [186] Surin describes part of this tradition in more lucid terms:

> There is a sense in which scholarly reflection on the "problem of evil" is a highly ritualistic activity, which requires the theodicist to focus here attention on a number of canonical or "sacred" texts (Leibniz's *Theodicy,* Tennant's *Philosophical Theology,* Journet's *Le Mal,* and so forth), and to engage in certain professional rituals (writing doc-

---

[185] Swinburne, "The Problem of Evil," 83; compare Swinburne, *The Existence of God,* 180–224, and King, *Essay on the Origin of Evil,* iii.

[186] Farrer, *Love Almighty and Ills Unlimited,* 16–17.

toral theses, submitting articles to learned journals, and producing books!).[187]

Yet rarely, if ever, are "practical" or "spiritual" texts on God and suffering admitted to this tradition (Job, Boethius, Julian of Norwich, Simone Weil). Nor do literary texts, texts of protest, or even biblical texts contribute substantially to the tradition of theodicy. Theodicy is a practice of the detached from the lives of people. It is purely academic and fully at home in the academy, where detached, "value-free" research is still valorized. And the practice constrains it practitioners to retrace the paths marked out in the past.

Beyond this academic focus, a third peculiarity of contemporary theodicy is that theodicists also rewrite the past as if all writings on suffering and sin were not only commensurable, but also contributions to the conversation constituted in the Enlightenment. Hume and Augustine—to note two of the "canonical" authors always cited—are often taken, respectively, to be *asserting* the Enlightenment problematic and *answering* it. Alternatively, Augustine is taken to have embedded answers in his texts which would have been answers to Enlightenment problems, if only he had been able to read Leibniz, Hume, Kant, et al. Theodicists mine an earlier author to "find" answers—and then often displace the "theodicies" they "find." Contemporary theodicists work hard to *create* a commensurability, to *make* a tradition by construing pre-Enlightenment authors like Job, Augustine, and Boethius as participants in the discourse practice of theodicy, along with King, Leibniz, Hume, and the modern authors. Yet the force of their writings, the purposes of their writings, and the rhetoric of their writings is not that of theodicy.

A fourth and most telling peculiarity is the way theodicists talk of evil. More than other critics, Kenneth Surin has brought out the fact that theodicies render "evil" an abstraction.[188] But the language he uses to reveal this is strained to the breaking point. He has claimed that theodicies obscure "the radical and ruthless particularity of human evil" and are a "practice which averts its gaze from the cruelties that exist in the world. " But how is this possible? How can theodicies' *talk* and *writing* avert "its" gaze from cruelties in the world? How can theodicy "provide—albeit unwittingly—a tacit sanction of the myriad evils that exist on this planet"? How can a form of writing be *unwitting*? How can theodicy silence "the screams of our society"? Does it "outshout" them? The strain of Surin's syntax, his personification of theodicy as an agent, and his invocation of a moral judgment that these sorts of theodicies are "not worth heeding" suggest that theodicies may have a power that is practically demonic and which should be fought. The strains in Surin's portrayal of the power of theodicies make sense if one recognizes that to do theodicy is not to participate in merely an assertive discourse practice, but to declare that evils are mere abstractions to be dealt with only by theory.[189]

---

[187] Surin, *Theology and the Problem of Evil,* 49.

[188] Subsequent quotations are from Surin, *Theology and the Problem of Evil,* 50–52.

[189] This is the central claim argued for in *The Evils of Theodicy*. There, the tool used is speech act theory, and theodicies are construed not as merely assertives, but

However, if they fail to declare "evil" what is truly evil, their declarations create a reality in which what is truly evil is not evil. If theodicists misdeclare evil, they create an inconsistent and finally destructive discourse. I cannot make the argument here, having already done so at length in *The Evils of Theodicy,* but can and do raise the question whether the practice of cosmic theodicy that is characteristic of the chapters of this volume does not do the same evil work as classic theodicies.

5 Defense: Showing that One Can Rationally Believe in God and that Evil Exists

The philosophical debates over the problem of evil have shifted substantially in the last thirty-five years. What has emerged is a different discourse about God and evil, the defense. Various versions of the free will defense (FWD), especially that of Alvin Plantinga, have been so successful that the "logical problem of evil" can legitimately be claimed to be solved.[190] At the very least, the burden of proof has profoundly shifted. The burden of proof is now on a philosopher who wants to show the inconsistency of the propositions that "God as traditionally conceived exists" and that "there is genuine evil in the actual world." Unfortunately, theologians have not caught up with this development for the most part as they still participate in the practice of constructing theodicies.

The "evidential problem of evil" remains a topic of debate among philosophers. But the debate now seems almost stalemated. Numerous philosophers remain unpersuaded by Plantinga, among others, about the ways to "weigh" the evidence of gratuitous evils and the relevance of the evidence to accepting theism. This debate seems more to involve the background beliefs a person holds rather than either the reality of evils or their evidential value.[191]

---

declarations, a significantly different speech act with significantly different power than mere assertions.

[190] Solved, that is, for theists who accept a libertarian notion of free will and who find that one can talk of God's creative power and agency as in some way comparable to or in competition with, rather than being analogous to, human creativity and power. Whether there is a version of the free will defense that can be compatible with a "compatibilist" notion of freedom is not clear, at least not to me; the problems with the FWD lead to my own subsequent chapter.

[191] One might even say that we have returned to the point at which Hume left us some two hundred thirty years ago. Hume's Philo put it this way: "Is the world considered in general, and as it appears to us in this life, different from what a man or such a limited being would, *beforehand,* expect from a very powerful, wise, and benevolent Deity? It may be strange prejudice to assert the contrary. And from thence I conclude, that, however consistent the world may be, allowing certain suppositions and conjectures, with the idea of such a Deity, it can never afford us an inference concerning his existence. The consistence is not absolutely denied, only the inference" (Hume, *Dialogues* 11.4, p. 205). The evidence does not clearly support the belief in the existence of God without "certain suppositions and conjectures," which affect how a person takes the evidence. What we have learned—and what Hume may have obscured—is that not only do believers have such conjectures that affect which propositions they accept, but also nonbelievers operate with their own set of conjectures.

Some have argued that this whole debate is miscast, that it misrepresents the shape of religion as practiced. Foremost among these "anti-theodicists" are D. Z. Phillips and others who take a Wittgensteinian approach to the philosophy of religion. I have argued elsewhere, however, that to take an "anti-theodicy" position does not mean that one must reject defenses of the consistency of religious belief such as the FWD; my differences from Phillips and others are differences about how philosophically to construe "religion."[192] While there are serious conceptual, practical, and religious problems with constructing theodicies, as shown above, I have not in the past argued against the defense of religious belief from charges of inconsistency.

Nor, in the end, will I do so here. While I am often in agreement with Phillips regarding philosophers misconstruing religion in their discourse, my own construal of religion and philosophy of religion differs both from Phillips and from those who do philosophy of religion as if it were limited to the disciplines of natural or philosophical theology.[193] While, *pace* Phillips, I find that some exercises in natural or philosophical theology ought to be done by philosophers of religion, philosophical *theology* ought not be construed as the whole of the philosophy of religion. Other approaches to religious belief and practice should be utilized as well. For those of us concerned with logical and rhetorical issues as well as theological issues, the sort of thinking done about "cosmic evil" may be of use to constructing a more robust defense.

Why should we bother? If the logical problem of evil is indeed "solved," by the FWD, what need have we to do more? From my perspective there seem to me three "religious" problems with the FWD. Some philosophers may find that these are not properly philosophical problems. If one thinks of philosophical problems as logical problems or pure conceptual problems, one may well not think these to be philosophical problems. However, if one thinks that philosophers of religion—including authors of FWDs—ought to take account of concepts as they are actually deployed in religious traditions, then these may well be problems of interest to some philosophers.

These problems are can be construed as tensions between the philosophical concepts and moves in the FWD (especially as it has been developed by Plantinga) and religious forms of the beliefs the defense defends (which can be shown with reference to his *Warranted Christian Belief*). Here I want to explore one common problem addressed to the FWD: the allegation that theists logically must believe that the devil exists for their belief set to be consistent (the defense's diabolical problem). In my subsequent chapter, I address the problem of the libertarian notion of freedom for religious believers (the defense's liberty issue) and the problem of the erasure of social evil or social sin (the defense's social disease). Considering these problems leads me to suggest a possible useful extension of the FWD.

---

[192] For the basic views, see Phillips, "Problem of Evil," 103–21, 134–39; Tilley, *Evils of Theodicy*; "The Philosophy of Religion and the Concept of Religion: D. Z. Phillips on Religion and Superstition"; "'Superstition' as a Philosopher's Gloss on Practice: A Rejoinder to D. Z. Phillips." A response from Professor Phillips appears in the intervening pages of the same issue. Phillips reiterates his view in his recent *The Problem of God and the Problem of Evil*.

[193] My own view of the matter is in *The Wisdom of Religious Commitment*, especially chaps. 2 and 3.

6 Resolving the Diabolical Problem Alleged to Afflict the FWD

In an article in *Religious Studies* in June 1985, Wallace A. Murphree constructed an argument which sought to prove that theism cannot survive without a belief in the devil.[194] He argued that if one accepted a free will defense, such as Plantinga's, one is committed to affirming the existence of the devil. He argued that other forms of theism which do not affirm the reality of the devil are incoherent, but that "fundamental theism" which ascribes diabolical responsibility for natural evil is coherent. But it is also implausible. If arguments like Murphree's were correct, theists who do not believe that there is a devil as classically construed could not utilize the FWD, or at least not the codicil that Plantinga develops to account for natural evil.

Murphree's argument utilizes two fairly standard propositions to state the problem:

> *P*: God is omniscient, omnipotent, and wholly good.

> *Q*: The world contains absolute contingent evil (of some sort or other).

He then argues that three and only three candidates are available to show *P* and *Q* compatible:

> R_1: Evil came about by chance.

> R_2: Evil obtained by natural law.

> R_3: Evil came about by free choice.

If any of these propositions are possibly true, they could be marshaled to show the compatibility of *P* and *Q*. However, R_1 and R_2 are incompatible with what Murphree calls "common sense theism." Hence, a theist is committed not only to the possibility, but to the truth, of R_3, "as there are no more *R* claims available."[195] But only a fundamentalist belief in theism with a devil has enough free creatures to explain all the evil in the world. Thus, only "fundamental theism" is coherent.

Murphree's argument fails either if R_1 or R_2 can be shown to be compatible with theism or if there are other *R*-type propositions available to show the compatibility of *P* and *Q*. If either of these arguments is successful, a theist would not be stuck with R_3 as true and as needed for accounting for all the kinds of evil in the world.

Some might wish to argue that Murphree's "common sense theism" is hardly up to the task of rejecting R_1 and R_2. Forms of theism compatible with these propositions, such as a theism that strongly distinguishes pri-

---

[194] Murphree, "Can Theism Survive Without the Devil?" Murphree is not alone in thinking that theism requires the devil. John Wesley evidently thought so, and Swinburne has also noted that a diabolical origin for natural evil is an ancient part of the Christian tradition and suggested that it "may indeed be indispensable if the theist is to reconcile with the existence of certain animal pain" ("The Problem of Evil," 93).

[195] Murphree, "Can Theism Survive Without the Devil?" 239. An evident typographical error at this point has R_2 rather than R_3.

mary and secondary causality or a theism that argues that the "natural laws" of the world are good enough for God to have created them, could be developed. However, the more straightforward path is to find some R_n propositions that would show that R_3 is not the only option. That is the path I will follow here.

For example, consider the following:

> R_4: It was not within God's power to create actual entities which could not resist God's power.

While few traditional common sense theists would be persuaded that R_4 is true, many theists could accept R_4 as possible, since they can see that forms of process theism or other accounts of the relationship of God and the world could also be possibly true. Another possibility is

> R_5: It was not within God's power either to create perfect creatures or to create imperfect creatures which did not suffer absolute, contingent evil.

The advantage of R_5 is that it does not presume or assert the existence of a devil. The first half of the disjunct is a staple of much Christian theology. The second half can also be understood both as suggesting the reality of a metaphysical evil of imperfection and as suggesting that imperfect creatures cannot avoid being affected by other entities. Rather, R_5 can be construed as claiming that it is possibly true that a "half-finished world" populated by imperfect creatures is a good world, perhaps even better than no world at all. Similarly, one might find another possible Irenaean principle would solve the compatibility issue, such as

> R_6: It was not within God's power to create a world in which people could develop noble characters and which did not contain absolute, contingent evils.

> R_7: It was not within God's power to create a world in which entities could exercise creativity and which did not contain absolute, contingent evils.

R_6 and R_7 respectively assert the possibility that there are worlds in which people develop character through surmounting adversity or participate in the life of God by exercising divinely given creativity in the face of evils.

7 Conclusion

The "diabolical" problem is fairly easily dismantled then. While some free will theists may indeed believe that there is a Devil responsible for many of the evils in the world, that belief is not necessary for the free will defense to succeed. However, R_7 points to a new form of a defense that fits with the work being done in relating theology to the natural sciences, and the subject of my subsequent chapter. The "creativity defense" also finds a way to resolve the issues associated with the libertarian orientation of the FWD and with its ignoring social evil.

THE USE AND MEANING OF THE WORD "SUFFERING" IN RELATION TO NATURE

Wesley J. Wildman

1 Introduction

All complex beings suffer. In fact, the capacity to suffer is as good a measure of biological complexity as any, especially because visceral reactions to physical injury, conscious pain, and emotional distress are visible and measurable. The problem of other minds may prevent us from knowing what pain feels like to another being, but it does not seriously interfere with the conclusion that suffering is widespread in nature. But precisely how widespread is suffering? And in what forms? There is considerable confusion and conflict surrounding these questions. A satisfying account of the characteristics of suffering in nature should gear itself to the various levels of emergent complexity. Such an account should also be multidimensional, covering issues important for understanding suffering in nature, such as types of suffering, responses to suffering, and causes of suffering. [196]

The conclusion of this discussion can be stated here. "Suffering" is a more useful category than "evil" to frame the initial phase of the problem because "suffering" is more neutrally descriptive and does not prejudge the moral character of natural disasters, predation, and the like. To render the various problems of evil as particular interpretations of suffering would help keep their guiding moral assumptions clearly in view.

2 Types of Suffering in Nature

A good place to begin is in the middle of the complexity scale, where we encounter a controversial question about suffering in plants. There is some evidence that plants possess biochemical injury mechanisms despite the lack of a central nervous system. [197] For example, some plants release a hormone called jasmonic acid when under attack or sick, and this hormone triggers reactions that minimize damage, defend the plant from further injury, and initiate repairs. There is also evidence that some injury mechanisms extend across the boundaries between individual plants. For example, an acacia tree responds to having its leaves chewed by releasing airborne chemicals (such as a volatilized form of jasmonic acid) that trigger the release of a foul-tasting chemical in neighboring acacias' leaves, irritating the digestion of foraging insects and animals and driving them off. Does the phrase "pain response" fairly describe such chemical mechanisms in plants? Probably not: pain suggests a conscious sensation and that requires a central nervous system, which plants do not have. Yet the fact that

---

[196] Throughout this paper I will be presenting basic science that would be covered in any introductory textbook in biology or biochemistry or cosmology. I only cite special sources when I make an unusual statement or in order to direct readers to a valuable internet resource.

[197] See "Take Two Aspirin for Re-Leaf," from Reuters Online, August 5, 1999, http://flatrock.org.nz/topics/science/dont_wilt_have_a_pill.htm (accessed August 5, 2005).

aspirin inhibits the production of jasmonic acid, just as it blocks the production of the pain-response substance prostaglandin in injured animals,
indicates a biochemical and probably an evolutionary relationship between
the two situations. This suggests that pain is a fair *analogy* for plant injury, even if plants lack other dimensions of pain familiar to animals, including elements that we should demand of any literal usage of "pain" to
describe plant injury. This is the basis for a distinction between conscious
pain and physical injury as forms of suffering in nature.

At the low end of the complexity scale, there is a question about the extent of the idea of suffering as physical injury. Can we use injury to describe the effects of tectonic plate movement on Earth's crust, the effect of
colliding asteroids on planets, the collisions and gravitational disruptions
inevitable within merging galaxies, or the wrenching expansion of spacetime in the big bang itself? Plant responses to injury are integrated and efficacious, in a way that requires complex biochemistry developed in a long
evolutionary history, and these features are absent in such physical systems. The argument for using "physical injury" outside the biological realm
is difficult to make, accordingly. Yet there are two possible paths along
which such an argument might possibly move, one concerning an analogy
with death and the other an analogy with disruption of biological integrity,
and we shall consider each in turn.

One line of argument involves noting that some nonbiological processes
seem loosely similar to dying in the biological realm. A star has a "life
cycle" that begins with a cloud of gas and "culminates" with one of several
spectacular fates. We sometimes speak of the explosion of a star, or the
gravitational collapse of a star into a black hole, as "death." But is death a
form of suffering as physical injury or merely a natural stage of anything
with a life cycle? I shall argue that death is not a direct cause of suffering
even in the realm of living beings so it certainly is not a form of suffering in
the nonbiological realm, despite the effectiveness of analogies such as "life
cycle" to describe the origin and dissolution of stars. What about cases
when the normal life cycle of a star is disrupted, as when galaxies merge
and two stars collide? Is this "premature death" a form of suffering through
physical injury? This case can be considered in what follows.

The other line of argument involves the idea of integrity. This updates
Aristotle's final causes, whereby the form of a thing expresses its purpose
and function. Integrity registers the fact that a complex physical process
may have zones of equilibrium within which it functions relatively stably
and supports complex and interesting behavior. The idea is more metaphysically neutral than Aristotle's final causes because it does not require a
first cause to determine and knit together the purposes of physical objects
and processes. It is also less susceptible to idiosyncratic value judgments
because integrity can often be quantified through analyzing information
flow: systemic equilibrium maximizes richness of information, and this defines integrity for a nonbiological process. Yet using a virtue word to describe such zones of systemic functioning can also express a profound aesthetic judgment: the integrity of physical systems is valuable and beautiful
because it supports emergent complexity. The human interestedness of
such aesthetic judgments is obvious but may also draw our attention to objectively achieved value in nature; a metaphysics of value is necessary for
interpreting such possibilities. Of course, integrity is not applicable to all

natural processes; some are not sufficiently systemic but are mere agglomerations of matter or flows of activity lacking deep patterns or significant information. Such processes are subject to change but they do not suffer integrity disruption or physical injury.

The best case for injury in nonbiological systems of nature turns on showing that there is a kind of physical integrity to some nonbiological systems that can be disturbed in a way that is loosely similar to the way injury disrupts the integrity of a biological organism. Returning to the case of an exploding star, we might say that a star has a kind of physical integrity deriving from its physical constitution as a sphere of gas dense enough to support nuclear fusion and to perform its light-emitting function, which has the potential to sustain ecosystems within a surrounding planetary system. We might further say that this integrity is destroyed if the star explodes and thus that the star suffers a fatal physical injury in the explosion. If we were persuaded that the life cycle of stars is relevant to assessing their integrity, we might want to limit our solar injury claim to cases of untimely demise, as in the merging galaxy scenario introduced above.

This best case for linking integrity disruption and physical injury is strained. We can distinguish what we can't identify, however, and we certainly have here the basis for a distinction between physical injury and integrity disruption, with the former being a species of the latter in the biological realm. We should determine whether a physical system can suffer injury based on the richness of its systemic connectedness rather than merely the existence of systemic integrity that can be disrupted. The biological realm has the potential to realize connectedness richly and densely enough to speak meaningfully of injury. The nonbiological realm lacks this potential. This distinction also furnishes a rationale for limiting suffering to physical injury and denying that suffering occurs in the simpler, nonbiological forms of integrity disruption.

An important test case for this distinction and the associated limitation in the idea of suffering is ecological damage. Physical injury requires a high degree of systemic integrity as well as injury responses that serve the end of maintaining structural integrity and restoring it where possible. It presumes the presence of life because we cannot injure a dead animal or a dead plant. It presupposes nutritive and environmental needs because these are conditions of structural integrity. The idea of physical injury probably does not in itself assume the capacity for reproduction, but evolutionary constraints are such that reproduction is almost always present with life, nutrition, and environmental dependence. Thus, we reach the outer edge of the idea of physical injury when we ponder the question of environmental damage. There are ecological cycles and regimes of equilibrium, which means that an ecosystem sometimes tends to return to a state of equilibrium when disturbed from it, and that suggests a certain degree of structural integrity. Moreover, ecosystems are dependent on sunlight for energy, which suggests nutritive needs and environmental dependence. But ecosystems are not alive in the right sense to speak of integrity disruption as injury because their operating parameters are enormously wide; they lack the dense structural integrity and biochemically regulated form of being that is life. They change and adapt but do not suffer. Mars once had flowing water and now does not. This is change through integrity disruption, to be sure, but not suffering through physical injury. From a descriptive point of view we can acknowledge that the new Mars environment cannot support many or perhaps any forms of life, which is the basis for saying that past systemic integrity has been disrupted. From a valuational point of view, we might think of such a change as bad or undesirable or ugly. But such changes are not suffering in the form of physical injury; suffering can only exist in the context of intensely structured, biochemically regulated forms of being.

This is a controversial conclusion among some conservationists, for whom effective political action seems to depend in part on stimulating compassion in typically self-centered human beings through the rhetoric of a "suffering" environment, thereby springing them out of their obtuse and ingenuous assumption that everything is just fine. Elaborate worldviews sustain the idea of a suffering environment and make it conceptually robust. Disruption of environmental integrity is variously framed as harming the spirit of the Earth, as injuring a vast and diverse organism, or as exploiting a generous and fertile mother. Most people have deep sympathy for a suffering planet described in such terms because they impute recognizable human characteristics to nature. Yet it is also possible even for a fervent nature mystic to appreciate nature without taking such imagery literally. Nature is not disvalued or the disruption of its harmonious integrities trivialized when we insist on a denser form of systemic connectedness to justify attributions of suffering through physical injury.

At this point, we have discussed four categories corresponding to emergent levels of reality. From the simplest to the moderately complex, they are sheer change, integrity disruption, physical injury, and conscious pain. I have argued that the latter, conscious pain, is clearly suffering. Conscious pain is possible for many animals. There are ample biochemical and behavioral signs of this. But sensation of any kind requires a central nervous system and some degree of awareness, so conscious pain would not be possible for plants or for simple animals lacking sufficient neural complexity for sensory awareness. Yet simpler creatures may be capable of physical injury, providing that they have sufficient density of biochemical connectedness. Physical injury also falls under the heading of suffering, though it is suffering's least intense form. Integrity disruption without physical injury is not suffering because the requisite systemic connectedness is absent, and this is even more the case with sheer change. It is important to note, however, that the disruption of certain systems, particularly ecologies, can be the cause of enormous suffering to plants and animals. The ordinary operation of ecological systems can also cause great suffering, as in the case of tsunamis and earthquakes due to tectonic plate movement, and this is equally true of sheer change when it works out badly for living creatures. We will return to causes of suffering below.

At the high end of complexity lies another form of suffering, emotional distress. This can be an exquisitely agonizing form of suffering for creatures capable of it, but there are not many creatures with the requisite neural complexity. Emotional distress requires cognitive, psychological, and social capacities that need not be present in creatures capable of conscious pain, but conscious pain in response to physical injury routinely involves emotional distress in creatures able to suffer in this way. Emotional distress does not require physical injury, but physiologically it involves many of the same neural processes as pain from physical injury, so it presupposes and builds on the evolutionarily more basic capacities for injury and pain. The main question about emotional distress is which animals are, in fact, capable of it. Almost all human beings suffer emotional distress, and we would think any person incapable of it was profoundly defective. But other mammals seem to grieve the death of a parent or child or mate, and seem to be concerned and solicitous about their own or others' physical injuries or illnesses. The social interactions of some primates, such as chimpanzees, are extremely intricate, and clearly bear the marks of a simple and often distressing emotional life. Thus, it seems fair to assume that suffering in the form of emotional distress appears among higher mammals and reaches a special intensity in higher primates.

The most extreme forms of emotional distress concern peculiarly human existential anxieties. The pertinent measure of the degree of such emotional distress is the phenomenon of suicide. A precise definition matters here: suicide is when a healthy animal capable of reproducing deliberately takes actions foreseeably guaranteed to result in immediate death.[198] Individuals of no other species, *when healthy* and *capable of reproducing,* act so as to end their lives, *foreseeing* the result of their actions. A whale might follow a beached leader into danger, but this is because of trusting habits crucial in a social species, not an attempt to end life out of grief, as is sometimes alleged. An animal fiercely defending its young against a deadly predator may die, but often enough this sort of aggression succeeds by driving off the predator, so death when it occurs cannot be foreseen and is not suicide. Sick or injured animals will often act so as to achieve a faster death, but this is not suicide because the animal in question is not healthy. Some social species, such as ants, have sterile members who sacrifice themselves for the sake of the colony, but this does not meet the definition of suicide because such creatures cannot reproduce, nor presumably are they capable of foreseeing. As defined, suicide is confined to human beings; as far as we know, no other species is capable of regarding existence as a problem that can be resolved through deliberate death.[199]

Because existential anxiety and its dark companion suicide are definitive for being human, it has been common to suppose that there is a sharp distinction between it and emotional distress, corresponding to a qualitative distinction between human beings and other animals. The standard Roman Catholic view, for example, is that human beings possess an ontologically immaterial soul in addition to their physical bodies and brains, and that God infuses this soul at conception. There are compelling reasons to adopt such a supernaturalist hypothesis, including especially the resulting crispness of moral discourse about the value of human life. But nei-

---

[198] See "Does Any Animal Besides Humans Commit Suicide?" on *The Straight Dope* website, Feb. 1, 2001, http://www.straightdope.com/mailbag/mbugsuicide.html (accessed Aug. 5, 2005).

[199] See "Can and Do Animals Commit Suicide?" at http://www.bizarremag.com/ask_bizarre.php?id=208 for a review of controversial cases, including debunking reports of alleged animal suicide. It is widely known now that lemming suicide by jumping off cliffs is a myth. This report suggests that a certain Disney film was a factor in spreading this story and claims that the makers of the film intentionally herded lemmings onto a ledge, forcing them off the edge, presumably to make them behave as it was believed they were supposed to behave in the wild. If true, this is an example of monumental human stupidity leading to stunning cruelty.

ther the interpretation of exotic experiences nor scientific study of human brains, minds, and groups seems to require this hypothesis. I think there is every reason to conjecture that the neural and social requirements for the exquisite suffering of existential anxiety are quantitatively but not qualitatively different from those for emotional distress. With existential anxiety, we have the most evolutionarily advanced form of suffering in nature that we know, its peculiar agony emerging from and depending on more basic forms of achieved complexity, and flowering only in what I shall call, in the sense of nonsupernaturalist theologians such as Paul Tillich, the dimension of spirit.

I have distinguished four basic forms of suffering. In increasing order of complexity requirements these are physical injury, conscious pain, emotional distress, and existential anxiety, corresponding to the realms of biology, sensation, cognition, and spirit. In less complex situations, outside the biological realm, suffering is not possible, but integrity disruption and sheer change remain important as potential causes of suffering. In more complex situations than we experience, we can only imagine what might obtain with regard to both emergent complexity and suffering. In all things, intensity of suffering co-emerges with the complexity of nature.

3 Responses to Suffering in Nature

A multifaceted appreciation of suffering also calls for distinguishing a number of types of response to injury, pain, distress, and anxiety. These responses are not limited to individual beings reacting to their own suffering but include the possibility of corporate and species and interspecies responses. They do not necessarily presuppose conscious awareness but allow for biochemical responses without attendant sensations. The resulting array of suffering responses is not a strict hierarchy of levels, therefore, though there are strong dependence relations among some aspects due to their relationships with the stricter hierarchy of the four basic forms of suffering.

First, the biochemistry of injury response supports detection of integrity breaches, self-protection, damage control, wound repair, and automatic signaling in many living beings including most animals (internally especially by prostaglandin and externally especially by communication), many plants (for example, by means of jasmonic acid, both within and in some cases between plant organisms), and possibly some other much simpler creatures as well.

Second, a sexually reproducing species whose individuals possess diverse biochemical injury responses may be subject to selection pressures if some injury responses confer differential reproduction advantages. This gives biochemical injury responses a genetic and historical dimension, along with the possibility of different forms of response both among species and even within a single species.[200]

---

[200] For an example of the latter, consider the fact that male and female mice (also rats and probably humans) have different neural pathways for pain response, which probably will inspire the creation of painkillers customized to sex at some point in the near future. See "One Man's Pain May Be Another Woman's Agony," from *The Dominion,* March 8, 2000, http://flatrock.org.nz/topics/science/dont_wilt_have_a_pill.htm (accessed August 5, 2005).

Third, moderately complex central nervous systems permit the capacity for reflexive habit formation and automatic behavior modification through memory and aversive learning, as when a fish learns to avoid a particular plant through being stung, or a curious puppy learns to steer clear of porcupines through being pricked.

Fourth, some social species are capable of a more complex type of behavior modification through socially-supported mimicry, as when some dolphins teach their offspring to explore the seafloor with sponges over their noses to avoid potentially fatal attacks on the snout from dangerous creatures such as stingrays hiding in the sand.[201]

Fifth, some creatures can also form cognitive attitudes to suffering. A variety of animals with cognitive powers seem to do this. Chimpanzees have different warning calls for different types of predators, which variously trigger flight into trees (when the warning is about a lion), flight down from the top of trees (when the warning is about an eagle), and flight out of one tree altogether (when the threat is about a snake in their midst). Human beings form extremely complex attitudes to suffering, as when we write poems about it, report on it in newspapers, and speak about it as bad or tragic. Many human activities presume cognitive attitudes to suffering.

Sixth, the capacity for cognitive and emotional attitudes to suffering promotes in some creatures the capacity for compassion, even when the type of suffering involved has never been felt directly by the compassionate one. Compassion for suffering creatures depends on a host of conditions, including possessing one's own biological capacities for suffering response, mirror neurons that partially recreate in a witness the experience of the sufferer, memories of one's own past suffering, the cognitive ability to imagine injurious or painful or distressing or anxious circumstances, and even evolutionary pressures toward altruism. It certainly reflects a high degree of sociality: a group can cultivate compassionate responses to suffering within its members, including especially corporate identification with suffering. Such responses are present to different degrees in different individuals.

Seventh, in sufficiently adept species, the compassionate response produces a desire to intervene. Intervention can involve alleviating suffering through healing injury, easing pain, soothing distress, and calming anxiety. It can involve avoiding injury and pain through protection or rescue, and minimizing distress and anxiety through physical contact or friendly advice. Sometimes intervention places the one intervening at great risk, particularly when the protection of offspring is the goal, yet this behavior is quite common in nature. Cognitive attitudes are crucial in determining which kinds of suffering we focus on and those we marginalize or do not notice. Social animals with warning calls cannot recognize or meaningfully alert fellow animals to every possible danger. Human beings are constrained in their interventions both by variability in intensity of compassion within members of the species and by the way their cognitive attitudes to suffering direct their attention toward some forms of suffering and away from others. For example, we develop technologies to heal or avoid injury

---

[201] See "Mama Dolphins Teach Their Babies," on *Animals in Translation* website, dated June 8, 2005, http://animalsintranslation.blogspot.com/2005/06/mama-dolphins-teach-their-babies.html (accessed August 5, 2005).

for some house pets but not masses of chickens on factory farms; we create medicines to alleviate the pain of disease in some situations but not others; we strategize about how to avoid distress and prevent suicide for some people but not others. These constraints are particularly important in the way human beings come to take responsibility for alleviating suffering: they require both an accurate grasp of where suffering actually occurs and a lively sense of compassion that motivates intervention; both are difficult to obtain.

Eighth, social species with highly developed cognitive powers engage in what sociologists of knowledge have termed the social construction of reality. The traditions, practices, and beliefs that result serve to disseminate prevailing attitudes to suffering, stabilize strategic responses and produce opportunities for refinement of interventions, legitimate action and inaction in face of suffering, and produce rich tradition-based narratives and theories to give an ultimate explanation of the reality and prevalence of suffering. The religious aspects of this response go to the heart of the form of suffering I have called existential anxiety. Less complex forms of suffering are not eased much by stories and theories, yet existential anxiety positively requires creative, tradition-borne thoughtfulness.

Ninth, at the higher levels of emergent complexity, there emerges the possibility of cultural evolution, chiefly in human societies.[202] This produces socially organized methods of avoiding and alleviating suffering, such as economic arrangements or technologies of healing, and these take shape as cultural traditions that pass to future generations as a set of stable social practices inviting refinement and further innovation. Cultural evolution requires socially constructed realities to confront both challenges and competition. The challenges may take the form of threats to survival, scarcity of food and water, changing environmental conditions, and natural disasters. At a corporate cognitive level, challenges may appear in the form of plausibility difficulties with basic narratives and theories in face of apparently contradictory experience. The competition needed for selection may take the form of cognitive dissonance upon encountering other worldviews. This seems to have been relatively less important in the history of culture, however, than the effects of innovation, whereby new insights and technologies force the abandonment of older practices as less efficient or the rejection of older ideas as somehow deeply mistaken.

4 Causes of Suffering in Nature

A multifaceted approach to suffering requires some understanding of the causes of suffering, particularly because moral judgments about suffering in nature crucially depend on our understanding of how it occurs and whether anything can be done about it. In this case, I think the analysis is best served with a series of distinctions, some nested, some overlapping.

---

[202] There is a lot of debate over whether cultural learning applies to nonhuman animals. For an interesting discussion of this issue in the case of chimpanzees, see Christophe Boesch and Michael Tomasello, "Chimpanzee and Human Cultures," *Current Anthropology* 39, no. 5 (December 1998): 591–; available on CogWeb's Evolutionary Psychology webpage, http://cogweb.ucla.edu/Abstracts/Boesch_Tomasello_98.html, dated December 1998 (accessed August 5, 2005).

First, suffering in nature may be caused by nature or it may be caused by supernatural beings, discarnate entities, and gods. The Hebrew Scriptures recount several spectacular examples of God personally visiting devastating suffering upon entire cities, as in the destruction of Sodom and Gomorrah; upon all people, animals, plants, as in the Great Flood; and upon human beings, as in the plagues of Egypt. Sometimes the Bible represents God as rightfully causing suffering by the hand of others, as in Jael's execution of Sisera with mallet and tent peg, the Israelite subjection of the native inhabitants of what came to be Israelite territory, and Satan's tormenting of the faithful but unfortunate Job. God visiting destruction by the hands of others is a common biblical motif when prophetic literature interprets the suffering of God's chosen people: the destruction of the northern kingdom of Israel is by God through the Assyrians, and the exile of the kingdom of Judea is by God through the Babylonians. Many world cultures carry forward living beliefs in suffering through demon possession, through haunting by the anguished souls of the dead, or through torment from supernatural monsters. By contrast with all of this, some suffering in nature is caused by nature, not by anything supernatural.

One sharp view on the question of natural versus supernatural origins for suffering in nature is simply that there is no supernaturally caused suffering because there is no supernatural realm that affects the realm of nature. This view holds that all causes of suffering are within nature, that all events (even strange ones) are natural, and that there are no supernatural beings or discarnate entities. The theistic versions of this kind of naturalism affirm God in one way or another as the ground of being, after which God is implicated in every natural event, and thus in every moment of natural suffering as much as in every moment of creative response to suffering. The distinction between natural versus supernatural origins for suffering in nature is not required, therefore, but it has been and still is extremely important in many religions and cultures, partly because it is a key conceptual element in the narrative construction and theory-building activities that we use to furnish a satisfying explanation for suffering.

Second, nested within the first distinction is a second: natural causes for suffering in nature are lower level, same level, or higher level. The levels refer to the various levels of emergent complexity in nature and correspond to the four realms of suffering as existential anxiety (in the realm of spirit), suffering as emotional distress (in the realm of cognition), suffering as conscious pain (in the realm of sensation), suffering as physical injury (in the biological realm)—and also the realm of no suffering despite integrity disruption and sheer change (in the rest of nature). Lower-level causes of suffering are devastating earthquakes or storms or cosmic collisions that affect plants and animals. This includes cosmic rays that cause point mutations in DNA with injurious consequences to an organism or its offspring, plants that sting or cut or feed on hungry or unlucky animals, and animals that bite or frighten human beings. Same-level causes include plants competing for sunlight and soil nutrients, animals injuring each other over breeding rights or preying on each other for food, and human beings behaving in cruel or hurtful or neglectful ways toward one another. Higher-level causes are animals feeding on plants, and human beings killing almost any living being for food and dominating habitable environments so as to endanger animals and plants through malnourishment or toxins.

Suffering has different tones depending on whether its cause is lower level, higher level, or same level. In particular, higher-level causes sometimes suggest intentionality or deliberation that is lacking in lower-level causes, as when human beings hunt a lion, or the lion hunts a gazelle. Lower-level causes suggest misfortune through the intersection of independent causal chains, as when a lion steps on a thorn or dies of thirst and hunger in a drought, or when vast hosts of plants and animals are destroyed by an asteroid colliding with the Earth.

Third, overlapping the second distinction is one that pertains only to the realm of cognition, and opens up questions of responsibility for and power over the causes of suffering in nature. Some causes of suffering are beyond deliberate, effective, compassionate intervention, while others are not. As applied to human beings, this distinction is a dynamic one. New medical technologies and new understandings of the workings of nature and society place more causes of suffering under the control of human beings and so subject to deliberate, effective, compassionate intervention where the determination to intervene exists. For example, complex societies and economies have many effects, some of which are unintended and disastrous. With greater understanding of economic and social life, human beings have become better at predicting such side effects and better at managing some of them—this is part of the meaning of modern managed economies. Many other suffering-causing effects remain outside the realm of deliberate, effective, compassionate intervention, such as the impact of modern capitalistic economies on traditional cultures. Many more suffering-causing effects that could be targets of compassionate intervention are not targeted for lack of will to alleviate suffering, as when cruel forms of factory farming persist even in nations that could afford to cause less pain to the animals they rear for food, or when wealthy nations do not intervene decisively to alleviate needless disease and starvation caused in part by the predatory economic and agricultural practices of those same nations.

Fourth, some causes of suffering are instances of evil and others are not. This idea is difficult to stabilize because of the fluid semantic scope of the word "evil" but the distinctions already introduced help to some degree. In terms of the third distinction, evil requires the presence of a high degree of control and responsibility, and so can only apply when causes of suffering are within the reach of deliberate, effective, compassionate intervention or avoidance. This may arise, on the one hand, with the deliberate infliction of avoidable suffering, as with predation. But most human beings would not be willing to say that deliberately hunting and injuring an animal for food is evil by itself, so two other elements are required. The most important necessary condition is that the suffering must be understood and yet still enjoyed or discounted; in other words, cognition and cruelty and selfishness must be present. The bird that "cruelly torments" a dying mouse does not meet this condition because the bird probably does not understand the suffering it causes, even if the bird is capable in some sense of savoring the struggles of the vainly writhing mouse. We would be less inclined to make an exception of the human being that deliberately tortures an animal for the sake of breaking taboo by being in the presence of the mysterious mana of pain, and thereby exercising godlike control over another being's experience and life. And we would certainly call evil the deliberate and cruel infliction of pain by human beings when, in terms of the second distinction, the cause is same-level rather than higher-level, which is to say when the victim is capable of more dimensions of pain and the interpreter understands this, as when a human being deliberately tortures another human being for the pleasure only of the torturer. On the other hand, causes of suffering may be evil when suffering could be alleviated but is not. The necessary condition in this case is that the degree of control must be high, so that intervention is minimally difficult, as when a powerful person witnessing an act of cruelty does nothing to intervene even though intervention would surely succeed.

The most frustrating and sinister forms of evil arise in two situations. On the one hand, in particularly rigid forms of social organization, intervention can be easy to accomplish but personally costly. For example, in every instance of group killing, it is easy to alleviate suffering in a given circumstance by simply not inflicting it or by preventing someone else from inflicting it, but the price of intervention is often such severe punishment that few are willing to try to stop the group killing as it unfurls like a vast black banner. This might apply to human or chimpanzee societies, though it is rare among chimps and sadly common, even in the extreme form of genocide, among human beings. On the other hand, in particularly loose forms of social organization, and now perhaps only in relation to human beings, intervention on a small scale is relatively easy but utterly ineffective in changing things on a large scale, as when a single hungry child can be fed but poverty and avoidable disease persists on a massive scale. The sinister character of the evil in this case consists in human beings deceiving themselves about how much control they really have: in fact we now have enough control to intervene on a scale as massive as the problem itself, but we rationalize neglect by saying that hunger and avoidable disease is an uncontrollable side effect of complex societies and economies. This rationalization was partially correct at one time, but it is true no longer, and the difficulty of the task does not mask the evil of self-deceiving neglect.

Fifth, how do causes of suffering relate to death? On the current account, decay and dissolution is the natural and inevitable fate of all forms of complex self-organization, from planetary systems and ecologies to the individual beings within them. We reserve the word "death" in its literal sense for the realm of living creatures: it is the process of decay and dissolution that ends life. As the natural and inevitable fate of all living beings, death is not a form of suffering in itself. This would not change even if aging and death were to become subject to delay through creative medical or artificial intelligence technologies; in that case, aging and decrepitude would be avoidable forms of suffering, at least temporarily, but death would still be simply the end of life. Indeed, this is precisely the situation of modern social organization and medical technologies, which together have vastly increased human life span, and may continue to do so. Similarly, even so-called "untimely" or "unnatural" death is not suffering in itself; these qualifications merely redescribe the cause of death in value-laden ways. The key distinction here, therefore, is between death itself, which is not a form of suffering or a direct cause of suffering, and the physical injury, conscious pain, emotional distress, and existential anxiety associated with death, whether untimely or unnatural or otherwise; death in its psychological and social and natural connections is a profound cause of suffering. Among human beings, the anticipation and denial of death are

psychologically potent, culture-conditioning phenomena that produce
stupid acquisitiveness and heartless deflection of responsibility. Relentless
grief and the compassionate identification with those who grieve can be
agonizing. Watching the dying can be terrifying, especially when it is from
violence or disease. By contrast with this endless turmoil of creatures
capable of existential obsession with death, death in itself is not a terrible
thing, despite its definitiveness. It is merely the final confirmation of a
creature's finitude, of its naturalness—indeed, of its life.

Sixth, we appear to require a distinction among intra- and extra-
organismic causes of physical injury and conscious pain. Predation, natural
disasters, and even viral illness are extra-organismic causes because the
source of suffering derives from outside the suffering plant or animal. Cancer and disintegration are intra-organismic causes because the source of
suffering is inside the suffering plant or animal. In practice, intra-organismic causes can be triggered by external conditions, as when carcinogens in
the environment or food render an organism more vulnerable to cancer, so
the distinction is not clear-cut. This reflects the fact that every creature is
deeply entangled with its context, biochemically, environmentally, and
evolutionarily.

Seventh, and finally, can knowledge of causes of suffering illumine
questions of the frequency and necessity of suffering? Indeed they can. Decay and dissolution are inevitable features of complex self-organization and
also forms of injury, relative to the ideal of health or integrity. So suffering
due to injury is universal in the biological realm. The necessity of predation
means that all animals and cellular organisms must inflict suffering to live,
even if not all creatures are the victims of predatory feeding. In the realm
of sensation, conscious pain is startlingly common. Some creatures may
escape extreme pain for the whole of their lives, but this must be exception-
ally rare given the conditions of life. In the cognitive realm, distress is utterly universal, though to greatly varying degrees and in many different
modes. This is due to the social embedding, psychological complexity, and
emotional capacities of creatures in the cognitive realm: this is a potent
combination of factors. In the spiritual realm, the very idea that there may
be creatures that somehow transcend existential anxiety stimulates religious hope to an extraordinary degree, yet existential anxiety may not be
universal in another sense: not all human beings may achieve this kind of
sensitivity, perhaps because of neurological deficits or trauma. There certainly is evidence of sociopathic human beings who appear to lack the
capacity for guilt, so the universality of existential anxiety must be a similarly complex issue.

Knowledge of the causes of suffering suggests, in sum, that all creatures suffer in some respects and cause suffering in other respects. All
creatures capable of a given level of suffering do experience it, though to
varying degrees. The safest generalization is this: suffering is universal,
inevitable, and frequently intense.

5 The Meaning and Use of the Term "Suffering"

We might elaborate forms of suffering, responses to suffering, and causes of
suffering in intricate detail. Indeed, this has been done movingly and often
in literature—particularly the variations and causes of emotional distress,

the elaborate human responses to existential anxiety, and the intensification of the causes of suffering into the realm of evil. But the four types of suffering, the nine facets of response to suffering, and seven ways of thinking about the causes of suffering in natural creatures serve as a basic framework for determining the meaningful use of "suffering" and related terms. Indeed, I have been using key words in accordance with a specific policy, which I now state compactly.

Where the biochemical basis for "injury response" is lacking, application of any suffering words is straightforwardly tropic. Human beings frequently impute human-like suffering responses to aspects of natural reality that do not have the capacity for them. Thus, the apostle Paul's reference to creation being subjected to futility and in bondage to decay, groaning in travail and longing for freedom (Rom 8:18–23), when interpreted within a modern naturalistic cosmology rather than in Paul's own ancient worldview of a cosmic battle between supernatural spiritual powers, is an inspiring poetic reference to the fact that suffering is inevitable and often tragic. But the cosmos does not literally suffer hurt or pain or distress; entropic systems do not groan in travail or long for liberation, no matter how much sympathetic and imaginative human beings do this on their behalf. I have also argued that ecosystems do not suffer, even in the sense of physical injury, because they are not sufficiently densely biochemically connected. We reserve the phrase "integrity disruption" for ecological damage, which leaves it available for moral debate as a cause of suffering.

In cases where some of the facets of suffering response are present but not others, we should understand the meaning of "injury," "pain," "distress," and "anxiety" in terms of the facets actually evident, taking care not to ascribe other dimensions of meaning, except self-consciously for the sake of poetic expressiveness. In cases of nonnatural creatures experiencing emotional distress, as when angels weep, presumably this refers to the response of compassionate co-suffering, but in isolation from any biochemical roots. I will not consider this possibility beyond noting its importance in religious piety. My concern is with suffering in nature. More specifically, in terms of facets of suffering response, I have used "injury response" to refer to situations in which only the first two facets are present—biochemical response and evolutionary conditioning of biochemical responses to injury. We should reserve "pain response" for the various higher kinds of reactions to pain: aversive learning, social learning, emotional and cognitive attitude formation, compassion, intervention, social construction, and cultural evolution. We should use "distress response" and "anxiety response" only for the more complex pain responses, when specifically emotional suffering is in view, whether caused by physical injury or not. We should use "suffering response" to describe injury responses, pain responses, distress responses, and anxiety responses collectively, just as we use "suffering" to refer to all of injury, pain, distress, and anxiety. This is slightly prejudicial, admittedly, because "suffering" is a loaded word for human beings. But all of the wealth of suffering words—such as "affliction," "agony," "anguish," "misery," "torment"—suggest pain sensation or emotional distress more directly than "suffering" does, so "suffering" may be optimally neutral and thus the best general word available in English. Moreover, the foregoing should clarify the breadth of situations to which "suffering" applies and prevent inappropriate attribution of aspects of suffering that are not actually present.

We should limit "evil" to the realm of cognition, where suffering can be controlled or avoided and yet is caused cruelly or ignored selfishly, with several conditions discriminating less from more evil. When people use "natural evil" in reference to anything else—such as earthquakes, tsunamis, tornadoes, and a host of other situations of lower-level causes of suffering—it is not literal usage but analogical. The analogy is between human beings and imagined supernatural beings that control and wield natural disasters as punishment or as perverse forms of play. Death is not a form of suffering, in itself, but, among cognitive creatures, its social and psychological embedding causes a formidable array of intense forms of suffering.

These are stipulations about how to use suffering words as much as they are descriptions of the extent of suffering in nature. Nevertheless, I intend this presentation to create a burden of demonstration for those making alternative proposals for the extent of suffering in nature, and substantively different recommendations for how to use suffering language.

The types of suffering distinguished above and the associated word-use recommendations are summarized in the following diagram. The social dimensions of suffering are not reflected in the diagram. This is because I assume they are causes rather than types of suffering, much as I treat death as a cause rather than a type of suffering. Yet death and the social dimensions of suffering are no less important for that reason. Other diagrams could register the various collective manifestations of suffering. This diagram focuses on distinctions supportable from biochemistry and neurophysiology. A partial exception is this: the biochemical basis for suffering in plants is related to the biochemical basis for suffering in animals but the lack of a central nervous system makes the two types of suffering quite different. The recommendation to include plant injury as a form of suffering is partly practical; this usage engages literature in which this kind of usage is common.

Emergent Levels	Types of Suffering		Word Use Recommendation
[Supra-human]	[??]		
Human (self-conscious)		Existential anxiety	
		Emotional distress	SUFFERING
Pre-human (conscious)	Conscious pain		
Life (pre-conscious)	Physical injury		
Inanimate (matter)	Integrity disruption		
	Sheer change		

VARIETIES OF THEODICY:
AN EXPLORATION OF RESPONSES TO THE PROBLEM OF EVIL BASED ON A TYPOLOGY OF GOOD-HARM ANALYSES

Christopher Southgate and Andrew Robinson

1 Introduction

Theodicy, the task of affirming the righteousness of God in the face of the existence of evil, is notoriously one of the most difficult problems in religious thought. Many responses to the problem rest on arguments to the effect that there are certain goods which are related to, and thereby provide a justification for, the existence of the harm in question. As a way of understanding the basic structure of responses to the problem of evil we offer here an analysis of the possible ways in which these goods and harms may be understood to be related. We suggest a classification of different types of good-harm analysis, which in turn provides a typology of strategies available to the theodicist. We hope that this typology of theodicies will both illuminate the character of various possible responses to the problem of evil and also show where multiple strategies are at work within the same argument.

A somewhat neglected element in the problem of evil is the issue of the suffering of *nonhuman* creatures.[203] The difficult task of the theodicist becomes even more formidable when the biological world is viewed from a Darwinian perspective according to which, both across all currently living species and far into the evolutionary past of the Earth, it is seen that the very process that has given rise to such a diversity of ways of being alive is accompanied by pain, suffering, and extinction.[204] In the latter part of the article, therefore, we use examples from the nonhuman world to illustrate different forms of good-harm analysis at work. We show how, in attempting to wrestle with this wider evolutionary perspective, theodicists sometimes construct responses which dynamically combine different types of strategy in subtle and potentially fruitful configurations.

1.1 Good-Harm Analyses and the Logical Problem of Evil

Thomas F. Tracy in this volume distinguishes between "thin defenses," which merely set out to show the logical compossibility (Tracy's term) of the existence of suffering and the existence of a good God, and "thick defenses" which endeavor to elaborate an account of the God-world relation which is not merely logically possible but sufficiently coherent and plausible to at-

---

[203] Theologians have been aware of the issue, though, at least since Aquinas, who writes: "Since God, then, provides universally for all being, it belongs to His providence to permit certain defects in particular effects, that the perfect good of the universe may not be hindered, for if all evil were prevented, much good would be absent from the universe. A lion would cease to live, if there were no slaying of animals; and there would be no patience of martyrs if there were no tyrannical persecutions." (Aquinas, *Summa theologica* 1.22.2.)

[204] For an analysis of the burden on the theodicist of considering "evolutionary evil," which is termed in this volume "bio-physical" or sometimes more specifically "biological" evil, see Southgate, "God and Evolutionary Evil."

tract the adherence of the faith community as having a claim to the truth. Several elements are required for a fully articulated thick defense or theodicy. For example, within a theistic tradition an account must be given of whether (and if so how and to what extent) God acts in particular events in the world. An excellent example of such a "module" is the argument of Clayton and Knapp in this volume. It does not seek to be a full defense or theodicy; it seeks to give an account of special divine action which would make a theistic defense against the problem of evil sustainable. Again, a Christian thick defense will necessarily contain an account of how eschatology affects our understanding of the problem of evil.[205] But a fundamental "module" in many theodicies, and one which is logically prior to either of those exercises, is an analysis of the relation between the goods that are considered important and the harms or "evils" that give rise to concern.[206]

We call this type of analysis a "good-harm analysis" (GHA). A thin defense may rest solely on a good-harm analysis. A thick defense must incorporate into a coherent framework a range of other elements (such as an account of providence, eschatology, and the relation of harms to major doctrines such as creation, atonement, and the status of humanity). This framework can then be used to dissolve or resolve the problem posed by the apparent harms, seen in relation to the apparent goods, by reference to an understanding of the relationship between God and the creature(s) concerned.

The role of good-harm analyses in the task of formulating a defense or theodicy can be expressed more formally by considering the apparently mutually incompatible premises which give rise to the "logical" problem of evil[207]:

> P1: Evil exists;
> P2: God is good;
> P3: God is omnipotent;
> P4: A good being will always eliminate evil as far as it is able.

Theodicies and defenses may be developed by rejecting or modifying any of these propositions. Rejection of P1 would involve a denial of the reality of evil; rejection of P2 would be a denial of the essential goodness of God. Neither of these positions is tenable within mainstream Christian theology. The third premise offers more scope for the development of Christian responses to the problem of evil. For example, some versions of process theology modify P3 by suggesting that, by virtue of the divine nature rather

---

[205] The recent work of R. J. Russell provides a fine example. See his "Eschatology and Physical Cosmology" and his "Natural Theodicy in an Evolutionary Context."

[206] By "harms" we mean (1) the actions of moral agents that lead to suffering or other loss of value in living organisms, human or nonhuman, and (2) actions, events, processes, or states that cause either suffering in organisms sufficiently sentient to be aware of their injury, or loss of value to the biosphere as a whole. Both of these types of harm are often referred to in the literature as "evils." Where possible we avoid this latter term as tending to confuse suffering with maleficent moral action. Wesley Wildman in this volume gives a helpful list of what harms to living systems should count as concerns for the purposes of theodicy.

[207] See for example Mackie, "Evil and Omnipotence," 25–26.

than by voluntary self-limitation, God's power over nature is limited. Likewise, dualist understandings of creation posit an eternal opposition between good and evil which the sovereignty of God cannot guarantee to overcome. In general, from the point of view of Christian thought, the problem with responses that rest on modifications of P3 is that they tend to run the risk of surrendering belief in a God who is sufficiently powerful to be a legitimate object of worship and ground of hope. If approaches based on modifications of P3 are deemed inadequate, then the burden falls on attempts to modify P4; in other words, to show that there are circumstances in which a perfectly good, all-powerful being might choose to allow the existence of evil. The theodicist's task then centers on showing why God may have "morally sufficient reasons"[208] for permitting evil or suffering. The role of a good-harm analysis within such a defense or theodicy is to provide the basis for an argument that, for a particular instance of evil or suffering, there is a corresponding good which amounts to a morally sufficient reason for God to allow the harm in question.

1.2 Terminology and Presuppositions

The purpose of this article is to offer a classification of the different possible forms of good-harm analysis, and to show how this classification helps clarify the strategies available to the theodicist—especially the theodicist who is concerned with nonhuman as well as human suffering. We do not primarily intend here to offer an evaluation of these good-harm analyses, or the strategies of defense to which they give rise. We shall, however, illustrate and test the classification with reference to recent work on the problem of "bio-physical evil." We thus offer a framework which is intended to facilitate exploration of what the disvalues and negativities associated with the suffering of creatures mean for our understanding of God and God's relationship to the creation.

We acknowledge the distinction that is sometimes made between theodicies, as being accounts that endeavor to explain the relationship between a good God and evils in the world without relying on a prior commitment to faith, and defenses, as being explorations within communities of faith of God-language in relation to suffering in the world.[209] However, in practice we do not find this an easy distinction to maintain. We write from within the Christian tradition, and our discussion below of the varieties of response to bio-physical evil draws on the resources of that tradition. We

---

[208] Pike, "Hume on Evil," 41.

[209] So Tilley, *Evils of Theodicy,* 130–31. The terminology of theodicies and defenses as it has grown up in the field must be agreed to be unfortunate. *Defense* seems—far from being an intracommunity activity—to imply engagement with attackers from outside the "defending" community. The term *theodicy* connotes engagement with God and God's justice, something that might seem more characteristic of reflection within a faith community than of apologia beyond its borders. However, it would be idle to try and reverse these terms at this juncture. What we are terming "thin" and "thick" defenses are really logical ("thin") and systematic ("thick") recourses within the theological community in the face of the fact of moral evil, and human and nonhuman suffering, in the context of a world—at least in Judaism and Christianity—that is taken to be both created by God and "very good" (Gen 1:31).

hope that the classification clarifies the nature of possible Christian responses to suffering, and will thus help such responses to be articulated. *However, the structure of our classification does not, in itself, presuppose a specifically Christian response to the problem of evil.* Accordingly, our discussion of good-harm analyses and their application to the problem of evil may be considered relevant to both theodicies and defenses, whether "thick" or "thin." In order to avoid unnecessary repetition we will, in general, use the term *defense* where we are talking about specific types of argument, and *theodicy* where we are referring more generically to the task of responding to the problem of evil.

We are mindful of the fact that, for good reasons, many people regard the task of attempting to theorize about evil as, at best, misguided and, at worst, a betrayal of the victims of suffering. According to this view the proper response to evil would be practical rather than intellectual, religious rather than philosophical.[210] At first sight it might be expected that any attempt to classify the possible varieties of theodicy will inevitably fall squarely within the theoretical/philosophical tradition, and our proposal may initially appear to do so. It is therefore worth remarking at this stage that one advantage of our classification may be to show why the so-called theoretical and practical approaches to the problem of evil are not necessarily as incompatible or mutually exclusive as is often supposed.

2 Three Categories of Good-Harm Analysis

We propose three types of logical relation between goods and harms, each giving rise to a distinct *category* of good-harm analysis (GHA). The three categories of GHA that we propose are:

1. Property-consequence GHAs: a consequence of the existence of a good, as a property of a particular being or system, is the possibility that possession of this good leads to it causing harms.
2. Developmental GHAs: the good is a goal which can only develop through a process which includes the possibility (or necessity) of harm.
3. Constitutive GHAs: The existence of a good is inherently, constitutively, inseparable from the experience of harm or suffering.[211]

We now explore these categories in more detail.

---

[210] See especially Tilley, *Evils of Theodicy,* and Surin, *Theology and the Problem of Evil.*

[211] Alternative names for these categories, which perhaps lack the immediate clarity of our chosen terms but which emphasize the relationships between the categories, are (1) aetiological GHAs: the good is a property whose existence has the potential to *cause* harms; (2) teleological GHAs: the good is a *goal* which can only be produced by a process that may or must give rise to harms; (3) axiological GHAs: the good and the harm are related in a way that hinges on questions concerning the *intrinsic value* of particular experiences or states of affairs.

2.1 Property-Consequence GHAs: A Consequence of the Existence of a Good as a Property of a Particular Being or System Is the Possibility that Possession of This Good Leads to It Causing Harms

The paradigmatic example of a defense based on a property-consequence GHA is the free-will defense: the *existence* of the property of free will in humans (a good) gives rise to the *possibility* of its deliberate or accidental use in such ways as may cause harm. Another example would be a defense constructed around the notion of sentience. The possession of the property of sentience may be regarded as a good, but (so the argument would go) sentience inevitably gives rise to the *possibility* of the experience of suffering.

These initial examples immediately suggest two points that will be true of our classification in general. First, to place two different analyses together within the classification does not imply that they necessarily give rise to similar (or equally plausible) defenses of God in respect of suffering; what they share is a similarity of logical structure. Free will is different from sentience, and a defense based on one may be quite different in many respects from a defense based on the other. Nevertheless, and this is the second point, the similarity of the logical structure of the underlying good-harm analysis may give rise to some parallels regarding the strengths and weaknesses of the argument eventually deployed. For example, there has been a highly complex debate about whether it would be logically possible for God to create beings with free will who would never choose to use their free will for evil ends.[212] Similarly, a defense based on the phenomenon of sentience might be asked to show why it could not be held to be logically possible for God to create sentient beings who could be guaranteed never to have unpleasant experiences.

2.2 Developmental GHAs: The Good Is a Goal which Can Only Develop through a Process which Includes the Possibility (or Necessity) of Harm

The distinguishing characteristic of developmental good-harm analyses is that a good can only come about through a process which also gives rise to at least the possibility of suffering. So if part at least of the underlying purpose of God the creator in respect of this element of creation was to give rise to the *development* of this good, then the harm, or at least the possibility of the harm, would follow.

It may be that (i) the process of development itself involves the harm or (ii) the process gives rise to the harm as a "by-product." These are two significantly different subcategories of developmental GHA. The corresponding defenses may be designated developmental-instrumental and developmental-by-product defenses respectively (see 4.2 for further discussion). What links them together is that the good arises (can only arise) via a *process*. Put another way, the key distinction from the property-consequence GHAs is that, whereas in property-consequence GHAs the possibility of suffering only arises once the good in question exists, in developmental GHAs the possibility (or even inevitability) of suffering is associated with the *process* by which the good is produced.

---

[212] Cf. Plantinga, "God, Evil and the Metaphysics of Freedom."

2.3 Constitutive GHAs: The Existence of a Good Is Inherently, Constitutively Inseparable from the Experience of Harm or Suffering

The basis of the constitutive analyses may be illustrated with an example from aesthetics. Some argue that the existence of beauty is only possible where there is also the possibility of ugliness. The argument is not that beauty arises as a consequence of ugliness, or that ugliness happens to arise in whatever processes lead to the production of beauty, but that the existence of beauty is *inherently inseparable* from the existence of ugliness.[213] Similarly, philosophers in the existentialist tradition may see the value of life (a good) only in relation to the inevitability of death (a source of suffering). In both these examples the disvalue or suffering has in some (perhaps somewhat mysterious sense) a constitutive (essential, defining) relationship to the good in question.

Another example of an articulation of a constitutive approach is offered by Wesley Wildman in this volume. Wildman considers that the ultimate is not either goodness or evil but contains and gives rise to both. To address ultimate questions of meaning is therefore not to be able to disentangle what leads to harms from what leads to goods. This highlights a general point about the constitutive category of good-harm analysis. In property-consequence and developmental GHAs the values assigned to the goods and harms in question are not fundamentally in dispute. The arguments tend to hinge on how the relevant goods and harms are connected and whether the balance of good to harm is justifiable. In contrast, constitutive GHAs are less concerned with comparing two sides of a "balance sheet," and more with calling into question our presuppositions about the nature of the values themselves. A further point illustrated by Wildman's proposal is that, as noted above, the actual content of different defenses underpinned by the same category of good-harm analysis may be widely divergent. Wildman's conclusion is that the problem of evil must lead us to abandon the notion of the loving personal creator of the Judeo-Christian tradition. On the other hand, as we shall explore in more detail below, the constitutive category of good-harm analysis may also be used to develop some specifically Christian responses to suffering.

3 Nine Types of Defense

Our classification of *types* of defense arises out of consideration of how the different *categories* of GHA (property-consequence, developmental, constitutive) may be combined with different *references* of analysis. By the reference of a GHA we mean the answer to the question: who benefits from the good, and who experiences the suffering which that good allegedly justifies? The beneficiary and the sufferer are not necessarily the same.

All three of the above categories of GHA may be applied to purely human affairs, where both goods and harms are experienced by human individuals.[214] When the reference of a GHA is restricted to humans we call this a human GHA. When the problematic is extended to include nonhuman creatures an additional dimension arises. One possible way of

---

[213] Whether or not this is a valid argument in aesthetics is not the issue here.

[214] This is not to say that it is necessarily the same individuals who benefit as those who suffer.

including the nonhuman world is to consider cases in which a good experienced by human beings is being related to, and may provide the justification for, suffering experienced by nonhuman creatures. We call this an anthropocentric GHA. The widest extension of any good-harm analysis is where both the good and the suffering may be experienced by any creature, human or nonhuman. We call this a biotic GHA.[215]

For each of the three *categories* of GHA defined above (property-consequence, developmental, constitutive) there are thus three possible *references* for the GHA: human, anthropocentric, and biotic. This gives rise to nine different *types* of defense (table 1). For example (top left box) the category of property-consequence GHA applied with a purely human reference gives rise to the family of human property-consequence defenses.[216]

	Human Reference	*Anthropocentric Reference*	*Biotic Reference*
Property-consequence GHAs	Human property-consequence defenses	Anthropocentric property-consequence defenses	Biotic property-consequence defenses
Developmental GHAs	Human developmental defenses	Anthropocentric developmental defenses	Biotic developmental defenses
Constitutive GHAs	Human constitutive defenses	Anthropocentric constitutive defenses	Biotic constitutive defenses

Table 1: Types of Defense

4 Varieties of Defense

We are now in a position to illustrate the types of analyses identified in table 1 by mentioning one or two representative varieties of defense corresponding to each type. We emphasize again that each type of GHA may lead to a potentially large number of different varieties of defense, all of which share an underlying logical structure of analysis but which may differ greatly in philosophical, theological, and rhetorical strategy. Our examples are representative of the nine *types* of defense, but they do not consti-

---

[215] We note that process theologians often extend the range of experience to include entities beyond what would be regarded by science as living organisms. We ourselves consider that interests—goods and disvalues—can only be experienced by living organisms (for an analysis see Rolston, *Environmental Ethics*). Hence our use of the term "biotic." For the purposes of this classification this term should be taken to include all experiencing entities to which goods and disvalues may be assigned.

[216] Note that our terminology shifts here from that of "GHAs" to "defenses" because the focus of our attention will now turn to the application of the categories of GHA to specific types and varieties of defense. It should be remembered that, for example, "human property-consequence defense" is shorthand for "defense based on a property-consequence good-harm analysis with a human reference."

tute a comprehensive survey of the numerous possible *varieties* of defense corresponding to each of these types.

4.1 Property-Consequence Defenses

As we mentioned above, the best known variety of defense based on a property-consequence good-harm analysis is the free-will defense. The classic free-will defense is a justification of human suffering on the basis that it is the result of the misuse of human free will (a human property-consequence defense). This type of strategy, often traced back to the work of Augustine,[217] has been comprehensively explored, most notably by Alvin Plantinga.[218] Plantinga's project would be an example of what was referred to above as a "thin defense" in the sense of concentrating on the logical possibility of harms being compatible with the existence of a perfectly good God (see section 1.1).

A variety of anthropocentric property-consequence defense would involve the extension of the free-will model to cover the suffering of nonhuman creatures; it would regard the misuse of human free will as the explanation (and justification, in terms of theodicy) for that suffering. Descriptions of the nonhuman creation as altered by the effects of the Fall of human beings would be in this category. To be human is to have freedom, which is a great good. The way humans chose, and continue to choose, to use that freedom has caused, and continues to cause, the existence of suffering throughout the creation. It is a move which, in a Darwinian perspective which recognizes pain and suffering as phenomena within deep evolutionary time, is capable of offering an explanation of only a tiny fraction of the total suffering ever experienced by nonhuman creatures.[219] (Of course, this is not to minimize the seriousness of humankind's contribution to ecological evils, as noted by Don Howard in this volume.)

A further extension of the classical free-will defense would seek to justify the suffering of non-human creatures on the basis that it arises from the misuse of a certain freedom of action possessed either by themselves or by other non-human creatures. This is a biotic property-consequence defense. The case of a predator playing with its prey, such as when a doomed sealion is tossed into the air by orcas,[220] might be one such example.[221] We explore this case more fully in section 6.1. Another biotic property-consequence defense might be advanced to the effect that sentience, even of a

---

[217] Though Augustine was not addressing the problem of evil in its modern form (see Surin, *Theology and the Problem of Evil,* 12; Tilley, *Evils of Theodicy,* 133).

[218] See e.g., Plantinga, "God, Evil"; Tilley, *Evils of Theodicy.*

[219] Cf. Peacocke, *Theology for a Scientific Age,* 222–23, Polkinghorne, *Reason and Reality,* 99–101.

[220] Rolston, "Naturalizing and Systematizing Evil," 67.

[221] It is worth noting again (see section 2.1) that to recognize this as a property-consequence defense is not to claim that it is *identical* to a free-will defense. The predator is not guilty of a conscious decision to undertake an act that may cause suffering. Rather, the analysis underlying the defense is a property-consequence one insofar as the created order has had the freedom to evolve in such a way that not only do predator species exist, but that the manner in which they manifest their evolved freedom may cause apparently unnecessary suffering to other creatures.

limited degree, is a good the possession of which leads to the possibility of suffering.[222] Process metaphysics arguably combines both of these approaches, in that it regards *all* entities as possessing some degree of both freedom and sentience.[223]

4.2 Developmental Defenses

As we indicated above, there is a distinction to be made between developmental analyses which regard the sufferings of creatures as *instrumental* to the learning process that is regarded as giving rise to the good, and those which regard those sufferings as a *by-product* of the process. The paradigmatic variety of a human developmental defense is what we might call the pedagogical defense. According to this variety of theodicy, personal suffering is necessary to our individual moral or spiritual development. John Hick refers to a world containing such educative suffering as a "vale of soul-making," and traces the roots of this argument back to Irenaeus of Lyons.[224] From the modern era it is much informed by the thought of Schleiermacher. In this model the harms are instrumental in the formation of the good of moral virtue.

A variety of anthropocentric developmental defense would be the argument that the suffering and waste of the evolutionary process is necessary to the development of an eventual good such as the appearance of freely-willing agents. This is the approach that Keith Ward seems to take in his book *God, Chance and Necessity* (1996). He writes:

> evidence of purpose is to be found almost everywhere in the universe. We can make a good guess at what that purpose is—namely, the generation of communities of free, self-aware, self-directing sentient beings. . . . The question to ask of the theistic hypothesis must be: is this purpose a worthwhile goal, and could it be a purpose proposed by a God who is supremely perfect? . . . For if the existence of suffering is a *necessary* condition of the realisation of a worthwhile goal, then even a creator God could not eliminate it, and choose that goal.[225]

A biotic developmental defense would identify the relevant good in outcomes, manifest in *nonhuman as well as human creatures,* which have developed through the evolutionary process. For example, we might identify adaptive fit to an environment as a good that justifies the suffering inherent in the process (natural selection) which produces such fitness. Such an emphasis on adaptiveness would suggest an *instrumental* understanding of harms within evolutionary process. Much of the writing of Holmes Rolston

---

[222] See e.g. Tracy, "Evolution, Divine Action."

[223] See e.g. Griffin, "Creation Out of Nothing," 120.

[224] For a summary see Hick, "An Irenaean Theodicy." As Hick conceded from the first, an eschatological extension of this theodicy is essential to its integrity. Given the situation and nature of some human persons during their life on Earth, the person-making process cannot be complete in every case during this present life. We discuss the eschatological aspect of theodicies in section 5.4.

[225] Ward, *God, Chance and Necessity,* 191.

III on evolutionary suffering is of this kind.[226] However, shifting the focus to the goods of complexity and intelligence as emergent within the evolutionary process, as in Murphy's paper in this volume, may shift the understanding of biophysical evil to that of a *by-product* of the production of the good that is valued.

The notion of "learning" through a process involving struggle may lead to such arguments being called "Irenaean." But these arguments about evolutionary development in nonhuman contexts differ from Hick's formulation in that the development is not *moral* learning or, necessarily, maturing *in a given individual*. It is important therefore to clarify that such approaches are not "Irenaean" in any simple sense. What an anthropocentric or biotic developmental defense shares with Hick's "vale of soul-making" defense and Irenaeus's theodicy is the same underlying category of good-harm analysis.

4.3 Constitutive Defenses

As indicated above, the view that the value of human life is seen only in the inevitability of human death would form the basis of a human constitutive defense. Another example would be a view which held strongly to the religious value of martyrdom. Still another, Søren Kierkegaard's sense that we truly understand faith only by looking at a story such as that of Abraham's willingness to sacrifice Isaac in obedience to God's command (Gen 22:1–18).[227]

A variety of anthropocentric constitutive defense might be used to justify the practice of animal sacrifice. In this case the desirability of a certain religious purpose would be regarded as a justification for the suffering of the victim. (Even if the animal suffers no fear or pain, it does suffer the loss of the possibility of future flourishing.) This is a constitutive defense because the death of the animal is intrinsic to, and conceptually inseparable from, the religious purpose. In contrast, the killing of an animal simply for food would require a developmental defense because, although the death of the animal is a necessary effect of its becoming food for another creature, the good is not in the killing itself but in the goal achieved in the form of the nutritional value (and other benefits) of the meal.

A variety of biotic constitutive defense could be developed from consideration of predator-prey relationships. If this form of creaturely interdependence is taken to be a good in and of itself (as is often portrayed in films of wildlife), and suffering seems to us axiomatic within those relationships, then a theist could develop a theodicy based on that good, to which sufferings are intrinsic.

It will be seen that there is an affinity between, on the one hand, developmental-instrumental good-harm analyses (based on sufferings as instrumental to processes that produce goods) and, on the other, constitutive analyses. In both the suffering is logically associated with the realization of the good. But in the constitutive analysis the good and the harm are yet more closely bound up together. The good does not derive from a process

---

[226] E.g., his *Science and Religion* and "Naturalizing and Systematizing Evil" (but see 6.1 for a fuller account of Rolston's position).

[227] Kierkegaard, *Fear and Trembling*.

leading to a goal, with the suffering as instrumental to the process: the good finds its meaning only in relation to the harm. This difference leads to key differences in theological strategy. Where the analysis of goods and harms is developmental, the model is likely to be of a God-created process by which creatures in some sense learn from the world. Where the good-harm analysis is constitutive, the model will not tend to involve any separation between God and the experience that involves the suffering. Constitutive GHAs give rise to a more diverse array of theodicies ranging, for example, from the notion that God the ground of being is not a loving personal agent, but is the ground both of creation and destruction (Wildman's position in this volume), to the affirmation of a God who is personally involved in the world of suffering in such a way that the suffering may, in itself, be the occasion for the creaturely experience of divine goodness. This latter model explicitly informs Kallenberg's Christological position in this volume. We explore divine involvement in processes of suffering further in the sections that follow.

4.3.1 Conceptual and Existential Constitutive Defenses

In an influential paper published in 1955, J. L. Mackie acknowledged the possibility of defenses based on an intrinsic logical connection between good and evil. Mackie pointed out that many theists would wish to hold that God was free to create the laws of logic, in which case, he argued, these defenses fail to solve the problem because God could have created worlds in which the goods in question were not (to use our terminology) constitutively bound to the harms.[228] Nelson Pike, also working within the analytical philosophical tradition, argued in contrast that such defenses might, in principle, offer a coherent form of theodicy.[229] As an illustration he uses an example suggested by Aquinas in chapter 71 of the *Summa Contra Gentiles,* in which a silent pause gives sweetness to a chant. Here the pause is thought of as a harm which is (in our terminology) constitutively related to the goodness of the singing. In terms of Tracy's distinction in this volume, Pike's argument is a thin defense because he attempts only to show the logical compossibility of the existence of good and evil in a world created by an omnipotent benevolent creator.

More recently, several authors have set out what, in our classification, amount to constitutive defenses with a more obviously religious or theological, as opposed to primarily philosophical, emphasis. These more fully developed theological approaches are examples of thick defenses in that, beyond demonstrating the mere logical compossibility of a creator God and the existence of evil, they attempt to offer religiously satisfying responses to the problem of suffering.

An example of one such thick constitutive defense is that elaborated by Diogenes Allen. Allen argues that philosophical discussions of evil frequently overlook the fact that people often find in suffering not evidence against the existence of God but "a medium in and through which his love can be experienced."[230] Drawing on the writing of Simone Weil, Allen

---

[228] Mackie, "Evil and Omnipotence," 28.

[229] Pike, "Hume on Evil," 45–48.

[230] Allen, "Natural Evil," 189.

shows how, at least for some, the obedient acceptance of suffering can be experienced as the occasion for receiving God's gracious presence; "not simply to recognise a gracious presence *through* suffering . . . [but] to find the distress itself as the touch of his love."[231] Weil suggests that we imagine a friend who has been away for a long time who, on returning, grips us very hard: "It feels just as painful as when it is the grip of someone who wants to hurt us, but in this case it is an effect of one who loves us and wants contact with us. Sometimes through the universe of matter God grips us very hard."[232] According to Weil, the theological basis for this is the suffering of Christ, which is brought about by the distancing of Father from Son in the incarnation but is, at the same time, a measure of the extent of their eternal love for one another.[233] "Therefore through Christ it is possible to understand how the Father's love is present in all things, even in suffering. Suffering can be regarded as a mark of our distance from God because we are subject to the cosmos simply by being creatures. Yet, depending on a person's response to suffering, a person can be in contact with God *through* suffering and *in* suffering."[234]

It is not necessary at this stage to assess the theological coherence of this position, or to attempt to show how it might be extended beyond a narrowly human perspective. The point for the moment is to recognize this as an example of a constitutive defense because it rests on an understanding of suffering as *intrinsic to,* rather than instrumental towards (or a byproduct of the production of) the good with which it is associated.

In addition to the distinction between thick and thin constitutive defenses we discern another potential difference of emphasis between different approaches within the constitutive category of good-harm analysis. In Kallenberg's paper in this volume the notion is advanced that suffering after the example of Christ is itself a good. That is what might be termed a "constitutive-conceptual" approach, since adopting the defense would appear (at least on the face of it) to involve accepting as true a particular propositional statement or theoretical construct. In contrast, there is an important strand of thinking which focuses on the personal experience of the sufferer, and which recoils from any attempt to impose a theoretical or conceptual explanatory framework on such experiences. This might be termed a "constitutive-existential" approach.

4.3.2 Logical and Practical Aspects of the Problem of Evil

At first sight the distinction between constitutive-conceptual and constitutive-existential defenses appears to be an example of the distinction made in most post-Enlightenment theodicies between the logical, philosophical problem of evil (why does God allow suffering) and the practical or "existential" problem of evil (how do individuals and communities cope with suffering). Some notable recent approaches to the problem of evil have argued that the hope of offering theoretical responses to suffering is mistaken, and should be replaced by practical theodicies. For example, Ken-

---

[231] Ibid., 201.

[232] Ibid.

[233] Ibid., 201–2.

[234] Ibid., 203.

neth Surin offers a generally negative assessment of post-Enlightenment philosophical theodicies and recommends that we should "evacuate theodicy from the realm of theory in order to relocate it in the realm of practice."[235] Surin's approach involves rejection of property-consequence and developmental analyses in favor of constitutive types of theodicy,[236] underpinned by a resolutely existential emphasis on the importance of listening to the "first order praxis-generating discourse" of suffering.[237] D. Z. Phillips develops a similar position using the Wittgensteinian concepts of "language games" and the proper "grammar" of talk about God.[238] In his view, attempting to formulate theological explanations or justifications for the experience of suffering misunderstands the nature of a genuinely religious response to the problems of existence.

While we recognize the distinctiveness and originality of these anti-theoretical positions, and acknowledge the force of their criticisms of some more traditional defenses, we do not ultimately accept their premise that there is a complete discontinuity between theoretical and practical theodicies. Consider the following example quoted by Phillips: "Suppose there has been an earthquake, and geologists now give an explanation of it. This will not be an answer to the woman who has lost her home and her child and asks 'Why?'. It does not make it easier to understand 'what has befallen us'."[239] In his discussion of this example, Phillips suggests that:

> The question "Why?" in these circumstances does not seek an answer. It does seek reactions or responses to replace the question. Religious responses to the vicissitudes of life are among these. . . . In the religious responses I have in mind, the very notion of God's will arises from and is internally related to these contingencies: things come from God's hands, the God who sends rain on the just and the unjust. We are not sufficient unto ourselves. Everything is ours by the grace of God, something we can be reminded of by both trials and blessings. . . . To think otherwise is to fail to die to the self, to play at being God.[240]

We see here that Phillips's resolutely *existential* approach (his focus on the meaning to the sufferer of the question Why?) cannot help but summon up *conceptual* issues (such as the notions of the grace of God, of undergoing a trial, and of dying to self).

Our rejection of the premise that there is an absolute discontinuity between theoretical and practical theodicies is supported by several recent accounts. Thus, in contrast to those approaches which are hostile to traditional theoretical theodicies (such as Phillips and Surin, mentioned above),

---

[235] Surin, *Theology and the Problem of Evil,* 67. Interestingly, Surin's analysis shares with that of D. Z. Phillips an interest in concepts with a distinctly Wittgensteinian flavor such as the "grammar" of God and "forms of life," along with similar though less specifically Wittgensteinian concepts such as those of "signs" and "narratives" (ibid., 24–28).

[236] Ibid., 112–37.

[237] Ibid., 149.

[238] Phillips, *Wittgenstein and Religion,* 153–70; *The Problem of Evil.*

[239] The quotation is from Rhees, *Without Answers,* 16.

[240] Phillips, Wittgenstein and Religion, 166–67.

others have developed theodicies based on a constitutive understanding that are in greater continuity with the inherited tradition (for example, Allen and Adams[241]). Where such a continuity is accepted, constitutive defenses (more so than property-consequence and developmental approaches) appear to offer a particularly promising way of introducing religious concepts into what might otherwise remain an abstract philosophical debate. John Schneider has remarked upon the "growing support for the view that theology is better equipped to deal with the problem of horrendous evil than is philosophy," but regrets that "a huge rift has grown between Christian thinkers (typically philosophers) who see this fresh theological strategy as complementary to systematic theory, and those (typically theologians) whose stance toward theodicy is entirely adversarial."[242] In his fine discussion of "the wild things" in the book of Job and in Mark's Gospel, Schneider seeks a "convergence" between rational (philosophical) and practical (theological) theodicies.[243] Laura Ekstrom likewise sees a need to combine practical and theoretical approaches to the problem of evil.[244] She discusses the ways in which suffering can, in itself, constitute a religious experience, and on this basis she develops a theodicy based on a constitutive understanding (our terminology) which she calls "the divine presence theodicy." She considers some possible objections to such a theodicy: for example, does it imply cruelty on the part of God, or psychological disturbance on the part of the person inclined to interpret suffering in this way? To these and other objections she offers cogent responses. Significantly, however, she does not regard her approach as anything more than "a partial theodicy," distinct from but not intended to supplant "the traditional soul-making, punishment, and free-will theodicies."[245] This is itself indicative. Constitutive-conceptual understandings are always liable to default to a developmental-instrumental analysis of the soul-making type (e.g., the good of martyrdom is found in the building-up of the church). Conceptual-existential approaches, on the other hand, will always face the possibility of collapsing into meaninglessness—the suffering, in the end, proving to be a theological surd, as in some readings of the book of Job.

5. Some Scriptural and Theological Considerations

5.1 Parallels in Job

None of our three categories of good-harm analysis (property-consequence, developmental, and constitutive) is new. Indeed, all three are put to Job by his first comforter, Eliphaz. The first argument that Eliphaz suggests is a

---

[241] Adams, "Horrendous Evils."

[242] Schneider, "Seeing God," 231.

[243] Ibid., 227. Schneider's discussion also serves to illustrate a point made above, that our classification does not aim to *reduce* theodicies to their underlying category of good-harm analysis. Schneider develops an intriguing new perspective on the book of Job; our classification of his theodicy as principally a constitutive-based one identifies something about the logical structure of the theodicy but does not diminish the specific novelty of its content.

[244] Ekstrom, "Suffering as Religious Experience."

[245] Ibid., 96.

form of property-consequence defense, according to which Job's suffering must be a punishment for some (unrecognized) sin. Job was given freedom, and with it the possibility of sin, and he has committed same. "Think now, who that was innocent ever perished? Or where were the upright cut off?" (Job 4:7 NRSV). A little later Eliphaz attempts to placate Job with a developmental defense, suggesting that his suffering is pedagogical, a means to an end: "How happy is the one whom God reproves; therefore do not despise the discipline of the Almighty" (Job 5:17). In between, Eliphaz hears a voice which offers a constitutive defense: "Can mortals be righteous before God? Can human beings be pure before their Maker? Even in his servants he puts no trust, and his angels he charges with error" (Job 4:17–18). On the face of it this may appear to be another property-consequence theodicy, since Job's suffering is being attributed to his alleged lack of righteousness. However, this must be understood in the context of a view of suffering which was inclined to regard all suffering as in some way due to sin. The distinctive feature of the theodicy suggested in the theophany in Job 4 is that suffering is intrinsic—i.e., constitutively related—to being a creature. Significantly, given the rather mystical quality of many constitutive-based theodicies, Eliphaz hears this voice as part of a dream-like theophany.

5.2 Parallels in the Christian Doctrine of Humanity and Understandings of Atonement

The tension between different forms of theodicy is well illustrated by consideration of the application of the doctrine of the Fall, a motif which has been strong in Western theology. The motif of the Fall of humanity, we would argue, derives its energy from the holding together of two convictions—one as to the goodness of God and of God's creation *ab initio,* and the other as to the fundamental freedom associated with human life. However, these twin convictions can be worked out in very different ways, corresponding to our three categories of good-harm analysis.

There are Christian traditions which emphasize the complete sinfulness of human beings, their total depravity and blameworthiness in the face of the divine offer of freedom. These imply true property-consequence defenses, the fact of freedom squandered in sinfulness. This formulation is deeply reassuring as to the fundamental goodness of God, a goodness which is epitomized in the divine gift of Christ's saving passion.

There are traditions which interpret the Fall as "upwards" into learning, into the development of the capacity of humans to grow and take their place within God's economy. These imply a good-harm analysis based on a process of learning, giving rise to a developmental defense. Christ on this view tends to be the great example of true humanity in its God-consciousness, which has learned obedience to God to the uttermost.

Thirdly there are important strands of Christian writing that emphasize the mystery and inevitability of human fallenness.[246] This strain of writing has strengthened in the modern era, partly as a reaction against the other two proposals, but it taps into an instinct widespread in the Scriptures that harm comes to people both by the evil actions of others *and*

---

[246] A telling example is this from Paul Tillich, "in every individual act the estranged or fallen character of being actualises itself," *Systematic Theology,* 2:43.

by the will of the Lord, and for the furtherance of his glory.

Again, reflection on different foci in thinking of the Passion of Christ helps us. Christ's death can be seen as *transaction,* necessitated by human sin—an approach tied to property-consequence-based arguments based on the primacy of freedom. Christ's protestless self-giving can be seen as the ultimate *example* within life's school of love—a developmental-based approach. Christ's story of betrayal, pain, torture, and defeat can be seen as the *tragedy* of the truly human exposed to inhumanity.[247] This constitutive-existential approach, a refusal to go beyond the experience of the sufferer, has gained increasing currency in the post-Holocaust era. After Auschwitz, even the story of Calvary can only be laid alongside stories of other suffering—Christians must allow those other stories to interrupt their telling of the Passion.[248] It has often been remarked, and remains significant, that no single model of atonement became definitive in the Christian creeds. We associate this with a drawing back from too definite a scheme of theodicy.

5.3 Divine Kenosis

A focus on the sacrificial suffering of Christ on the cross, as in Weil's writings noted above, points us towards the resources available in the ideas of the "creative suffering of God"[249] and of "creation as kenosis."[250] The New Testament casts the notion of (self)-sacrifice in a cosmic perspective of potential importance to our discussion: "the Lamb that was slaughtered from the foundation of the world" (Rev 13:8, NRSV alternative translation). The way of creation is the way of the cross—a paradox which points towards a constitutive understanding of the significance of suffering. That is not to say that a theological concept such as divine kenosis (self-emptying) can *only* find a place within a theodicy stemming from a constitutive understanding of goods and harms. It might be argued, for example, that God refrains from intervening in the actions of free agents in order to permit the exercise of genuine free will (kenosis as augmenting a property-consequence GHA) or that God allows us to make mistakes in order that we may learn from them (kenosis as promoting a developmental GHA). Nevertheless, the idea of God's creative co-suffering with creatures (whether or not this is deemed to stem from kenotic self-limitation[251]) does seem particularly compelling where an intrinsic (constitutive) relation between the suffering and the goods of creation is posited.

---

[247] Cf. Dillistone, *The Christian Understanding.* We note here John Schneider's observations about the narrativist, tragically-oriented approach to theodicy of many contemporary theologians. Schneider is determined to observe—following James Wetzel—that the Christian story can never truly be a tragedy, because the cross of Christ is succeeded by the resurrection. Tragedy is transformed, and ultimately that transformation will operate in all things (Schneider, "Seeing God," 233).

[248] Surin, *Theology and the Problem of Evil,* 161.

[249] E.g., Fiddes, *Creative Suffering of God.*

[250] See the essays in Polkinghorne, ed., *Work of Love.*

[251] See Peters and Hewlett, *Evolution from Creation,* 143, for reservations about the use of this concept.

5.4 Theodicy, Eschatology, and Compensation

In Christian theology the significance of the cross is, of course, inextricably bound up with the theme of resurrection (however the latter is understood), and thus necessarily places the problem of theodicy in an eschatological context (e.g., Rom 8:1–25). As with the concept of kenosis, eschatological elements may be incorporated into several different types of defense. For example, John Hick, working out his developmental GHA, the world as a "vale of soul-making," acknowledges that in *this* world the soul-making process often fails;[252] he therefore introduces an eschatological dimension in order to allow the process to be completed.[253] In a constitutive-based defense, on the other hand, we may expect reference to the eschatological dimension to carry a greater emphasis on the *transformation* or *healing* of suffering, *not simply as a means to an end* (even that of the restoration of health) *but as an intrinsic good.* Such an idea is necessarily elusive. It might be objected. isn't healing always a means to an end? In a sense this is obviously true. However, the fact that some particular suffering has been healed does not, in itself, offer an explanation (or justification) for the existence of that suffering in the first place. What a constitutive-based approach may contribute is a sense that *the very touch of the healing hand,* the opportunity for such a hand to be extended and accepted, is a good *in itself,* a good which is *axiomatically* connected with the existence of the suffering. Again this understanding seems to lurk behind the harrowing story of Abraham and Isaac on Moriah.

The introduction of an eschatological perspective raises the possibility of some form of future compensation for suffering experienced in this world. However, as Phillips has forcefully argued, the notion that suffering may be justified by the promise of some future recompense, especially where horrendous evils are concerned, seems inadequate.[254] In Allen's constitutive-based defense discussed above (section 4.3.1) the role of eschatology is not that of merely offering some form of *compensation* for the sufferings of this world. As Allen puts it, his approach "does not exclude contact with God in a resurrected life"; nevertheless, he emphasizes that "such a future good is not needed or used to make up for present adversities and thus to allow one to maintain God's goodness."[255]

What, then, is the proper role of the notion of compensation in the task of theodicy, and how does it fit into our classification? Two general points may be made. First, the theological problem with any theodicy that rests purely on the promise of some future compensation would be, in effect, to separate the God of creation from the God of redemption. This is the essence of the Gnostic and Manichaean positions, to which the standard theological objection is that such a view undermines the goodness of God as creator. If the creator is not also the redeemer, if some connection cannot be conceived between God's purposes as creator and God's purposes as redeemer, then the theodicist must admit defeat. Therefore any theodicy which does not attempt to show at least some link between the work of God

---

[252] Hick, "Soul-Making and Suffering," 188.

[253] "Without such an eschatological fulfillment, this theodicy would collapse" (Hick, "An Irenaean Theodicy" [2001 ed.], 51).

[254] Phillips, *The Problem of Evil,* 81–90, 247–55.

[255] Allen, "Natural Evil," 206.

as creator and the work of God as redeemer (whether based on a property-consequence, developmental, or constitutive understanding) would appear to be deficient. Second, however, there is no reason why the goodness of God should not be expressed in a glory yet to be revealed to us. We therefore suggest that the notion of compensation may have a role, provided that such recompense can be understood in the context of a reason (such as a good-harm analysis) why creation is ordered so as to include the possibility (or necessity) of the suffering in question. For example, in section 6.5 we shall discuss Southgate's approach to biophysical evil, which combines anthropocentric and biotic developmental GHAs with a conviction that God's love guarantees, ultimately, amends for all creaturely suffering.

6. Good-Harm Analyses Applied to the Suffering of Nonhuman Creatures

To illustrate the distinctions, and also the interplay, between the different categories of understanding of goods and harms we return in this section to the issue of the suffering of nonhuman creatures (anthropocentric or biotic reference, in the terminology of section 3). We show how the same incidents can be looked at in different ways, which correspond to the different categories of GHA in our classification.

6.1 Orcas and Sealions

We draw on the very helpful and telling range of examples offered by the writings of Holmes Rolston III. As mentioned above, in his 2003 article "Naturalizing and Systematizing Evil" Rolston mentions the behavior of certain kinds of orca which, in killing sealions, will toss their victims playfully in the air, prolonging their agony.[256] This type of orca is so feared by its prey animals that dolphins "are known to hurl themselves onto beach rocks in a suicidal frenzy" rather than face their predators.[257]

As we consider this behavior our focus may be on the orcas themselves. The freedom of behavior involved in their lifestyle as predators *can* lead to what seems to human observers like the gratuitous infliction of suffering, but it does not necessarily do so. Other types of orca do not show this behavior, and often predators (unless teaching their young to hunt) kill their prey with the minimum of energy and fuss. Focus on this behavior in orcas, then, would lead to a property-consequence approach to the analysis of goods and harms. Certain properties in created entities can lead—but need not necessarily lead—to suffering in the biotic world.

We could choose to focus instead on the orca's prey, the sealion. We could conclude, with this focus, that the fact of predation progressively develops the abilities of sealions as a species and leads to greater abilities and greater flourishing. Though individuals suffer, the species as a whole becomes better adapted. This type of argument is perfectly illustrated by Rolston's remark that "the cougar's fang has carved the limbs of the fleet-footed deer,"[258] and it points to a developmental approach to goods and

---

[256] Rolston, "Naturalizing and Systematizing."

[257] Chadwick, "Investigating a Killer," 99.

[258] Rolston, *Science and Religion,* 134.

harms, one moreover in which suffering is instrumental to the development of creatures.

We could broaden our focus still further and consider the whole ecosystem of this part of the ocean. It is a system in which eating and being eaten is of the essence. An elaborate chain of interdependent relationships builds up around this dynamic (with the pain that is necessarily caused to various creatures within the process). Focus on this chain of relationships might lead to a constitutive type of understanding of goods and harms. Indeed Rolston also uses this type of language, calling nature "cruciform." "The secret of life," he goes on to say, is that it is a *passion play.*"[259] By this he presumably means that, as at the passion of Christ, the good does not have its meaning without the suffering intrinsic to it.

6.2 The "Insurance" Pelican Chick

Another of Rolston's examples offers more insight into the dynamic between types of argument. He notes the way in which the white pelican, like a number of other predatory birds, hatches a second chick as an "insurance." The insurance chick is normally driven to the edge of the nest by its sibling, and once displaced is ignored by its parents. Its "purpose" is merely to ensure that one viable chick survives. It has only a 10% chance of fledging.[260]

Again, if the focus is on the pelican species as a whole, this strategy, "careless" and "wasteful" of individuals as it might seem, has "worked" for the white pelican, which as Rolston points out has lived successfully on Earth for thirty million years. The process of natural selection has developed in pelicans a strategy which is successful, although in many cases it leads to suffering. A defense based on this analysis would regard the harm as a by-product of the good, and would therefore be a developmental-by-product defense.

But if we shift our focus to the individual that suffers, the "insurance" chick itself, the language of tragedy returns. Rolston talks of "the slaughter of the innocents":[261] the process "sacrifices" the second chick to the good of the whole—it is intrinsic to the system that new life is regenerated out of the chick's death. The victims of the evolutionary process "share the labor of the divinity. In their lives, beautiful, tragic, and perpetually incomplete, they speak for God, they prophesy as they participate in the divine pathos."[262] "Long before humans arrived, the way of nature was already a *via dolorosa.*"[263] Our sense is that there is a tension here. Rolston's under-

---

[259] Ibid., 144, italics in original. Humans' sense of the beauty and importance of these interdependent systems is shown in many films of wildlife. The camera operator does not intervene to rescue the limping impala calf from the hyenas, any more than US National Park officials intervene in cases of nonanthropogenic suffering in animals under their care (see Sideris, *Environmental Ethics,* 179–81, drawing on further examples given by Rolston). Though it operates by way of suffering that on occasion seems tragic, the system is perceived as good and necessary—if a theodicy were being mounted, it would be a constitutive-based one.

[260] Rolston, *Science and Religion,* 137–39.

[261] Ibid., 144.

[262] Ibid., 145.

[263] Rolston, "Kenosis and Nature," 60.

standing of goods and harms, we suggest, remains a developmental one, but the rhetoric by which he elaborates his evolutionary theodicy (here and in the preceding case) implies a constitutive approach.[264]

We see from these examples (sections 6.1 and 6.2) that specific foci within the system can lead to the formulation of property-consequence or developmental arguments, but concentration either on the ills of the individual sufferer, or on the beauty and value of the system as a whole, will take the analysis back in the direction of a constitutive approach.

6.3 McDaniel's Response to Rolston

Jay McDaniel's response to Rolston's analysis is very instructive. He quotes Rolston's remark that "If God watches the sparrows fall, God does so from a great distance."[265] McDaniel is convinced this must be wrong—God's care is present to every sparrow,[266] and it is not enough simply to say of the back-up pelican chick that its suffering benefits the species as a whole. Redemption, McDaniel argues, must be of the creature concerned itself, and must involve a context in which it can respond to God's redeeming initiative. Hence his hope of "pelican heaven," and that "kindred creatures, given their propensities and needs, find fulfillment in life after death too."[267] This is a recourse not only to divine fellow-suffering (as also in Rolston) but also to eschatological compensation for the victims of evolution. As such it differs from the drama of nonhuman redemption that Rolston formulates, in which redemption is understood to occur through regeneration of life in other creatures.[268] By focussing on the suffering creature itself, McDaniel questions whether a developmental defense based on the flourishing of other creatures, or the adaptedness of the species as a whole, can be deemed sufficient. Rolston makes much use of the language of sacrifice, but this only sharpens the point that it is not the evolutionary victim (such as the backup pelican chick) that has *chosen* the good of others over its own. It is the *process* that has "sacrificed" the victim's interests to the interests of the larger whole.

One would expect McDaniel, as a process theologian, to be committed to an approach to theodicy that circumscribes the power of God, diluting proposition P3, that of divine omnipotence (see section 1.1). From the beginning of the universe, he writes, a state existed, apart from God, "possessed of its own ability to actualize possibilities, its own creativity." God offers entities "possibilities for order and novelty,"[269] but "had to take the risk that with the increased capacities for sentience enjoyed by animal life, there would be increased possibilities for pain."[270] The implication of the

---

[264] To say this is not to detract from Rolston's important achievement in drawing attention to the evolutionary dimension of creaturely suffering, but simply to indicate the profoundly difficult character of this branch of theodicy.

[265] Ibid., 140.

[266] Cf. also Edwards, "Every Sparrow."

[267] McDaniel, *Of God,* 45.

[268] A difference already remarked upon by Southgate, "God and Evolutionary Evil," 813.

[269] McDaniel, *Of God,* 36.

[270] Ibid., 41.

latter quotation would seem to be that God could have taken other courses of action, but risked taking the course that led to entities possessing creativity, self-actualization, sentience. Such creaturely freedom to attempt self-fulfillment may lead to harms. Beyond the weakening of P3, then, there is also in McDaniel a property-consequence good-harm analysis with a biotic reference (indeed a reference extending to all experiencing entities, see note 13). Furthermore, McDaniel's invocation of the image of God as "Heart"[271] and his conviction that God suffers with all creaturely suffering recalls A. N. Whitehead's famous description of God as "the fellow-sufferer who understands."[272] Such images for God seem to hint at the further possibility of developing out of process thought a constitutive approach to creaturely suffering, in which that suffering may be the occasion for God to become, in some new and creative sense, present to the creature.

6.4 The Significance of Pain

McDaniel is also much influenced by Arthur Peacocke, and as we explore further the tension between these three categories of theodicy in respect of the nonhuman creation it is important to note a point which Peacocke has particularly emphasized. He writes: "From a purely naturalistic viewpoint, the emergence of pain and its compounding as suffering as consciousness increases seem to be inevitable aspects of any conceivable developmental process that would be characterized by a continuous increase in ability to process and store information coming from the environment. . . . In the context of natural selection, pain has an energizing effect and suffering is a goad to action; they both have survival value for creatures continually faced with new problematic situations challenging their survival."[273] For this reason any consideration of the pain of creatures *must* include a developmental-instrumental element to the understanding of goods and harms, to the extent that it is by the instrument of the pain system that organisms (with sufficiently complex nervous systems to feel pain) learn to avoid harms. (Indeed humans born without pain systems do not long survive.) The theodicist who wishes to advance a developmental argument may well start from the instrumental value of pain.

6.5 Page and Southgate

In illustrating the tensions between different types of good-harm analysis with biotic reference we now turn to the approach of Ruth Page. Part of the vigor and originality of Page's approach to theodicy in her *God and the Web of Creation* (1996) is her rejection of any long-range teleology within her account of God's relation to the creation. She rejects all theologies of creation in which the evolution of freely-choosing self-conscious beings like ourselves is admitted to be any sort of goal (as it clearly is for Ward [see section 4.2] and more implicitly is for most of those writing in this area). The consequences for theodicy of any argument where the development of those faculties in any way justifies the suffering of other creatures, about which

---

[271] Ibid, 48–49

[272] Whitehead, *Process and Reality,* 351.

[273] Peacocke, "Biological Evolution," in *EMB,* 366.

God does nothing, are to Page's mind "scarcely to be borne."[274] She would also reject, by implication, the sort of developmental argument implied by Rolston's comment about the limbs of the deer being carved by the cougar's fang. Page's focus is on "teleology now"—only the goods of the particular creature who experiences the suffering can be of importance to God. She therefore severely limits the scope of developmental arguments (which would be restricted to a narrow focus on the benefits of pain *within that particular creature,* see 6.4 above). Suffering, for Page, can only be instrumental within the creature concerned. Given the extent of evolutionary suffering, it is inappropriate, on her account, to invoke the second type of developmental GHA in which that suffering is a by-product of the process that gives rise to goods like conscious rational creatures.[275]

Where Page leaves her analysis incomplete is in giving no account of her approach to the remaining problem of evolutionary theodicy—individual creatures do suffer, and the God who gives rise to the creation, even if that God has no goals beyond their own good, is in a sense responsible for their suffering by dint of being their creator. Christopher Southgate has called the latter element the "ontological aspect" of evolutionary theodicy.[276] What Page tackles is the *additional* burden for the theodicist of positing longer-term goals for the creation—God apparently using creatures as means to a longer-term end—what Southgate refers to as the "teleological aspect" of evolutionary theodicy.

Where a developmental approach involves a focus beyond that of the individual creature's suffering (the limbs of deer in general, beyond the plight of the individual deer, are honed by co-evolving with cougars; the extinction of dinosaurs was a good in the sense that it permitted the rise of mammals; the retreat and eventual extinction of Neanderthals cleared the way for the full flourishing of Cro-Magnon humans) then suffering is being seen as instrumental towards, or a by-product of, the realizing of long-term divine goals. This is as we have seen what Page rejects. It is not as easy to gauge what Page's own theodicy is as what it is not. It avoids long-range teleology, but does not seem truly to engage with the ontological problem within evolutionary theodicy—the fact of God's having given rise to a creation which is of this form.

Southgate, in contrast, wants to accept the full problematic of evolutionary theodicy in its ontological and teleological aspects, and outlines a scheme in which a developmental-by-product approach—that evolution gives rise to goods in the process of the arising of which harms may (will) occur—is augmented not only by a conviction about divine fellow-suffering and eschatological compensation, but also by a theological anthropology in which humans are accorded a co-redeemerly role in the healing of creation.[277]

Southgate's proposal, then, is multifaceted—he affirms the central im-

---

[274] Page, *God and the Web of Creation,* 104.

[275] Actually the problem of evolutionary theodicy is at its most severe when it is emphasized that the extinction of so many species was *instrumental* in giving rise to humans, as most evidently with the last great extinction at the Cretaceous-Tertiary boundary some 65 million years ago. See Southgate, "God and Evolutionary Evil," on the charge that God used these species merely as means to the divine ends.

[276] Southgate, ibid., 808.

[277] Ibid., 818–19.

portance of humanity in the creation, and that therefore *one of* the goals of God in working with the evolutionary process was the arising of freely-choosing self-conscious creatures like ourselves—thus far this places his analysis of goods and harms within the anthropocentric developmental area alongside Keith Ward. But Southgate also affirms the intrinsic value of all creatures, and the disvalue associated with the suffering of sentient creatures and indeed the extinction of species, and hence of whole ways of being alive.[278] To such harms he responds with a conviction that the triune God pours out love in fellow-suffering solidarity (an approach with constitutive overtones), but also the conviction that fulfillment must come to every creature, if not in this life then in some form of appropriately re-created life (here he is close to McDaniel's position on "pelican heaven"). The anthropocentric developmental element in his argument is further balanced by his conviction that the creation awaits the glorious liberty of the children of God (Rom 8:19). Southgate's endorsement of the idea that humans are in some sense the goal of creation, then, is not an uncompromisingly anthropocentric claim, because humanity's growing into a priestly calling as co-redeemers with God is part of the hope for the healing of all creation. There is therefore an important biotic developmental aspect to Southgate's proposal, in that the suffering incurred in the course of evolution may be regarded as having been necessary to the production of the conditions for the eventual reconciliation of the world to God.

7 The Character of Christian Theodicy

In a sense all theodicies which engage with real situations rather than philosophical abstractions, and endeavor to give an account of the God of the Christian Scriptures, arise out of protest and end in mystery. There are no completely satisfying accounts, only recourses to explorations of God and the world that are bare logical formulations, or systems of partial explanation pointing beyond themselves. What has been of particular interest to us in investigating this way of understanding theodicy has been the extent to which constitutive good-harm analyses are found within the landscape of Christian understandings of the problem of evil. Property-consequence and developmental GHAs, which have often seemed to dominate the philosophical and theological literature on this problem, are special cases derived from viewing the problem with very particular foci of attention. Where there seems to be scope to talk of learning through suffering, be it moral learning in a human individual or group, or the learning that the presence of pain systems makes possible in higher animals, then a developmental shape to the theodicy may come to the fore. If rather the focus is on freedom as an attribute of an entity, then a property-consequence-based theodicy becomes articulated.

But wherever the focus is on the individual in whom there seems to be an overplus of pain, or a powerlessness to learn, or whose life is truncated, there property-consequence and developmental analyses of goods and harms drop away, and the focus shifts back into mystery. The religious theodicist will then be likely to have recourse to talking of tragedy, of goods being intrinsically bound up with harms, and of God's intimate involvement

---

[278] Ibid., 806.

in that relationship. This divine engagement may be in mitigation of the creature's suffering, or in compensation for it, or a combination of the two.

We suggest therefore that there may be a constitutive instinct to the understanding of the relation of goods and harms within Christian theodicy, a sense that though we might not always be able to do the logic there is an intrinsic relationship between suffering and flourishing life. This would be in tune with the central symbol and event of most Christian understanding, the mysterious good represented by the passion and death of the Crucified One. The Resurrection is also central, but Christian reflection has rarely been content to say that all the glory and the victory lies there. As T. S. Eliot put it, Christians insist on calling Good Friday good.[279]

8 Concluding Comments

In summary, we have outlined the elements that underpin some of the options for responses to the problem of evil. The first element in many responses is what we have termed a "good-harm analysis." We offer a classification of these analyses into three *categories* (property-consequence, developmental, and constitutive), each of which may have three modes of *reference* (human, anthropocentric, and biotic). This typology gives rise to nine *types* of defense, each of which may be elaborated in a large number of different *varieties* of argument.

No taxonomy is more than an avenue into a debate. Our hope is that our scheme may help participants in the debate on theodicy to define what their proposals involve. In particular, we hope this scheme may help those engaging with the very difficult area of "bio-physical evil" to be clearer as to their approach. Two key questions in this area are: (1) What category of good-harm analysis is being employed—property-consequence, developmental, or constitutive? (2) In *which* creatures are the "goods" to be found, and by which are the "evils" experienced?[280]

Our remarks in this paper are not so much criticisms of the authors whom we have discussed as observations about the range of arguments being deployed, and a suggestion that future work would benefit from attention to the question of what exactly is the category of good-harm analysis, and what the reference of any defense, that is being proposed. We regard our scheme as a way of analyzing more clearly the problematic behind the varieties of defense and theodicy, and consider that it may be particularly useful in the developing area of evolutionary theodicy. We commend it to those working in this area, and in theodicy in general.[281]

---

[279] Cf. Eliot, "East Coker," IV.25, *Four Quartets,* 182.

[280] Consideration of the results of the 3.8-billion-year history of evolution on Earth may incline one to a biotic developmental approach—participation in an evolving biosphere is the process by which species "learn" the strategies of being that are often so fascinating and admirable. That biotic developmental approaches remain problematic is shown by the way a whole range of thinkers seek to balance and supplement them by reference to divine fellow-suffering and eschatological fulfillment. As we have seen, this may lead to a rhetoric more akin to a biotic constitutive understanding of goods and harms—Rolston's talk of a "passion play"—or to a more anthropocentric developmental scheme such as Southgate develops.

[281] We thank many colleagues for feedback at the conference, and also Dr. Mark Wynn for his generous help in sharpening our thinking.

DELETE THIS "P.91" AND "P.92" FOLLOWING FROM PDF

II. SCIENTIFIC AND PHILOSOPHICAL RESPONSES

William R. Stoeger, SJ

Robert John Russell

Nancey Murphy

Thomas F. Tracy

Philip Clayton and Steven Knapp

Terrence W. Tilley

ENTROPY, EMERGENCE, AND THE PHYSICAL ROOTS OF NATURAL EVIL

William R. Stoeger, SJ

The natural evils of transience, dissolution, death, and the pain, suffering, and loss they induce have their roots in the underlying characteristics of nature, particularly in matter and its interactions, including gravity, and in the increase of entropy—the second law of thermodynamics—which is always in force. The interplay of the various physical interactions—via the exchange of matter and gravitational entropy—is however also crucial for the emergence and development of structure and complexity. Thus, transience, dissolution, and death are the entropic price demanded for the exploration of new possibilities, the generation of novelty and the support of highly organized systems in our evolving universe. A further meta-question, of course, is: Are there other possible universes in which the laws of nature are such that highly complex systems could emerge and thrive without these natural evils? Could our universe therefore have been different in this regard? I shall briefly consider this speculative issue towards the end of the paper, concluding that a finite self-evolving universe in which more and more complex levels of organization gradually emerge avoiding significant transience, fragility, and dissolution is not a coherent idea. This would be a possibility only in a universe which is carefully designed and micromanaged from the outside. Thus, throughout this paper I shall assume that the regularities, processes, and relationships within nature are as we find them.

1 Introduction

In approaching the issue of natural evil as a theological problem, establishing an adequate context is crucial. It is only in reference to this context—including the assumptions and perspectives that constitute it—that we either make progress in resolving the problem, or fail to do so.[282]

The components for constructing such a context are all those which bear on adequately representing the relevant aspects of the realities involved. Among these are nature, its structure and dynamics, and God and God's creative and salvific action. Included must be an anthropology, scientifically based, but also with philosophical and theological dimensions, including that of eschatology. Along with this must be a careful analysis of the concepts we employ in framing the question itself—those of "evil," "suffering," "God," "God's activity as Creator, Redeemer," etc., and the reference they have to the way things are in the world. That must nuance and guide theological reflection.

Such an approach would take much more than is possible in a single paper. Here I propose to focus on just one component of this context—the structural and dynamic elements of nature at the level of physics, particu-

---

[282] For further discussion of an adequate context for resolving the problem of natural evil, see Southgate, "God and Evolutionary Evil," 808–24; Kropf, *Evil and Evolution*; Stoeger, "The Problem of Evil: The Context of a Resolution."

larly thermal entropy and gravitational entropy and their connection with dissolution, transience, and death as well as their crucial role in emergence and complexification.

In section 2, I shall describe the evolutionary character of the universe and of nature in general. In section 3 we look at the concept of entropy and the second law of thermodynamics, and their connection with "natural evil." In section 4, we turn to the discussion of gravitation and gravitational entropy, their relationship to classical non-gravitational entropy, their importance for the evolution of the universe, and their connection with both emergence and natural evil. In section 5 we conclude with an elucidation of emergence, and the role of the dissolution of emergents.

2 The Evolution of the Universe and of Nature

We live in an evolving universe. It began very simply as an expanding, hot, nearly homogeneous configuration of matter and energy. As it expanded it also cooled. And, as it cooled, environments changed—both microscopically and macroscopically—allowing new entities to emerge and to organize at every stage through the dominant physical and chemical processes obtaining at specific temperatures and under specific conditions. Those new entities in turn constituted new environments, with new possibilities of further organization and emergent systems. This general pattern is repeated over and over again in different ways, depending upon the particular phase of development the system under consideration is in and the details of its previous evolutionary history.

There are several general characteristics which are key to this process. The first is that, as the universe evolves, structures gradually develop within it. The nearly homogeneous, undifferentiated expanding universe breaks up into galaxies and clusters of galaxies, and, within these, stars and clusters of stars form. Around these stars there are planets, comets, and asteroids. Each of these systems constitutes an evolving context within which new systems can emerge—microscopically, macroscopically, and mesoscopically.[283] Secondly, the ways in which these structures and entities develop depends on fundamental relationships or ways of interacting at the physical level—on the basic interactions of gravity, electromagnetism, the strong and weak nuclear interactions—or on the more basic unified interactions at higher temperatures leading to these—and on the conditions within which these interactions operate. Without gravity, for example, there would be no possibility of any macroscopic structures forming at all. But, equally, without vacuum energy to drive the cosmic expansion gravitationally, there would be no cooling and diffusion of available mass-energy to enable middle- and low-temperature phenomena to emerge. At higher levels of organization, other emergent relationships become important, supported by these underlying basic physical interactions. In general, then, differentiated relationality is key.

---

[283] The mesoscopic realm is in between the microscopic and the macroscopic ones. In particular it is the realm in which the peculiar characteristics of the quantum world are evident and directly observable—where the transitions from the microscopic to the macroscopic realm are accessible to observation.

Thirdly, there is the closely related characteristic that the universe we inhabit possesses a large variety of systems which manifest many nested levels of organization and complexity—some cosmic venues being much more notable in this regard than others. This is the result of the operation of the interactions and relationalities that have developed over time in very different or even just slightly different contexts. The vast majority of these do not rise to the level of life—or even to that of complex chemical molecules. The conditions are not right. But at least a few do so. Nearly all, however, are affected by the formation of elements heavier than hydrogen within stars—which then spread them throughout the universe, to provide the building blocks for a large variety of molecules and more evolved systems. We shall revisit these first three characteristics in a more thorough way in section 5.

Fourthly, and most relevantly for this paper, there is the transience and fragility of emergent systems. This is an essential feature—a condition, as well as a consequence—of evolution. What are consequences of earlier evolutionary stages become the conditions for further evolutionary explorations. New, more complex systems can only emerge at the expense of the alteration or demise of old ones. That is simply because in an integral and interconnected universe, what emerges does so from its environment—from what is already present. That is, it needs to draw material and energy from the environment within which it develops. The entities and systems which form the building blocks or components of more complex ones derive in turn from older ones, or from their remnants. Furthermore, more complex systems are usually much more vulnerable to being reduced to their components and inevitably over time suffer deterioration and dissolution. They themselves, or their components, are often subsequently incorporated into even more complex systems, or provide the necessary environment for their emergence. If this were not the case, there would be severe evolutionary bottlenecks, and the overall evolutionary process would stagnate. The permanence, and strong resistance of earlier systems to incorporation and replacement, would prevent any further modification or change. In fact, these systems would never have emerged in the first place. As environments change, systems must change as well. What flourished in an earlier environment, may not flourish in a later environment. What was impossible in an earlier environment becomes likely in a later one. Along with this transience and fragility is also a certain contingency and chance, which is born from the simplicity and universality of the basic regularities and processes. If that were not the case, there would be little openness to change or novelty in the overall evolutionary process. It is crucial at the same time to recognize, however, that this contingency and chance is always functioning within the larger framework of "law and order."

In sections 3 and 4, we shall delve more deeply into the detailed physical properties underlying these characteristics, particularly that of the transience and fragility we have just discussed, but also the complexity, the relational dependence and the environmental sensitivity of these systems. We shall find, confirming what we have just pointed out, that the basic materiality and finitude of the world, which finds its concrete expressions in the conservation of mass-energy and in the persistent importance of thermal and gravitational entropy (the second law of thermodynamics), requires transience and fragility for evolution and emergence—if there is to be formational and functional integrity and relative autonomy of nature's innate dynamisms, and if it is to be an evolving cosmic ecological complex, which obviously it is. These two features, of course, also imply very stringent theological constraints on how we can conceive of God and God's relationship to us as Creator and as the source of Salvation. The theological perspective consonant with this is God present and active in and through the regularities, relationships, and freedoms of nature and of history, but deeply respecting them and empowering them to operate on their own—without becoming another "secondary cause" within creation.

An important and easily understandable example of how nature operates at its most basic level is in the formation and evolution of stars, and their ultimate demise. Without stars and their eventual dissolution the universe would be neither complex nor life-bearing. Stars, first of all, produce all the elements heavier than helium and lithium, thus enabling chemistry and chemical complexity. Secondly, stars in their death-throes produce the heaviest elements and spread all that they produce throughout the universe—enriching the material out of which succeeding generations of stars and planets will form. Thirdly, they thus provide the venues near which cool chemically rich environments can explore complexity. Fourthly, at the very same time they also provide the energy by which such complexity can emerge, flourish and evolve. We shall return to this simple but important example in section 5.

We now discuss in more detail several of the mysteries underlying this remarkable and pervasive behavior of the physical world—especially thermal and gravitational entropy, and the evolution they enable.

3 Entropy and the Second Law of Thermodynamics

Many[284] have recognized that the principal underlying physical reason for the transience and fragility of any physically or chemically based system—any material entity—is the second law of thermodynamics. It states that the entropy, the measure of disorder, in any thermally isolated system must either remain the same or increase in any process or change. Thus, without the continual injection of energy or material—or nourishment—any system tends to run down or undergo eventual dissolution.

Organization, complexity, and evolution towards greater order and complexity, require special conditions—and the input of material and energy to form or maintain those conditions as well as the relationships which enable the system to function as a whole. The components of a given complex system are always in very special dynamic relationships with one another. If those relationships and interactions, and the conditions necessary for them, falter, the system fails and deteriorates. But, as we have already seen, all of this means that a successful system must *not* be isolated from its environment. For it is from its environment that it draws its nourish-

---

[284] See, for instance, Russell, "Entropy and Evil." Russell explores the notion of entropy as a metaphor for evil—"prefiguring evil on the physical level" (457). However, he stresses that at the same time entropy underlies disorder and decay at one level, it is also an agent of change and growth and emergence of new order at other levels. Russell, to my knowledge, was the first to make this point explicitly. Here I briefly elaborate on these two aspects of entropy in a systematic and purely scientific way.

ment in material and energy. This also means that the overall larger system of which it is a part must, as an isolated system, increase in its overall level of disorder, to maintain the healthy functioning of all its more complex subsystems.

Why is that? The basic point is that any ordered system which functions well has its components relating to one another—interacting with one another—in very precise and specific ways, in terms of energy and information transfer, and the careful maintenance of the conditions for its network of continuing processes. There is just one, or at most a few, configurations involving those components which "work" to maintain its proper functioning. The higher the degree of order in a system, the more dependent it is upon the intricately nested relationships and processes which enable to it to function. These inevitably require layers of feedback control loops, and often also involve the reliable transfer and precise implementation of vast quantities of information (as in living systems). But there are very many other possible configurations of those same components—many other states—which do not "work." There are very few states of the system in which everything works in harmony as a whole, and many other states in which it does not work. Thus, it is relatively easy for any ordered or complex system to shift into one of those nonfunctional states. Just a relatively slight rise in temperature, a failure of a key component, or some toxic contaminant can completely undo a system or a functioning organism. As we have already emphasized, it takes work, or the input of energy, as well as the maintenance of internal and external conditions, to keep a system functioning in a given environment.

But why aren't such systems more robust and sturdy? Why isn't it possible to construct them so that they are invulnerable to changes of environment and independent of energy and nutrients? A significant part of the answer is that any system which is complex and well-organized depends, as we have already stressed, on maintaining a large number of highly differentiated precise relationships among its components. Those relationships depend inevitably upon carefully maintained conditions within the system, as well as on special conditions in its immediate environment. They, together with the necessary conditions upon which they depend, are easily upset, and therefore require continual support. Systems that do not depend on such precise relationships are rarely complex or well-organized and thus relatively insensitive to their environment. The relationships among their components are highly undifferentiated and incapable of preserving or transferring information reliably.

Thus, in the course of evolution, the more complex a system is, and the more interactive with and responsive to its environment, the more vulnerable and fragile it will be to disturbances and hostile conditions, and more it will need to be nurtured and maintained in its integrity. Basic physical systems, such the atom of a stable isotope, or a star, are relatively—but not completely—stable against changes in environment. It would take an nuclear reaction—usually at high temperature—in the case of the isotope, and an internal or external catastrophe in the case of the star, to disrupt these systems. Fortunately, then, the basic underlying physical organization of nature is relatively robust. A bacterium or a plant, in contrast, is much more fragile, and can be destroyed quickly in any number of ways, involving relatively small exchanges of energy or material.

What we have just been describing, without use of the term itself, is the second law of thermodynamics as applied to ordered systems. It is the inevitable tendency of these systems—of any system, actually—to become more disordered, to return to a more probable, simpler, less intricate or less ordered state. A highly ordered system is always high maintenance!

What then is entropy, as a measure of disorder? As described above, a system in a highly ordered (low entropy) state is in a very special or un-usual configuration, relative to all its possible configurations. There will be only a few configurations or ways in which its components are related to one another which will enable it to function in this precise, highly ordered way. There are innumerable other ways in which its components can be configured with respect to one another for which the system would not function with the same precision or order. Thus, the number of microscopic states or configurations which give an equivalent macroscopic state is a measure of its entropy. Technically, we define the entropy as

$$S = k \ln W$$

where k is the Boltzmann constant and W is the density of microscopic states for a given macroscopic configuration. ln is the natural logarithm (of the density of states)—that is, relative to the base of the special number e, which equals 2.7183.... A macroscopic state that possesses a large number of equivalent microscopic configurations will be of higher entropy (more disorder) than one with a fewer number of microscopic configurations.

Thus, if we look at the space of all possible microscopic configurations of a system (phase space), we shall see that there are very few correspond-ing to some macroscopic states, and an enormous number corresponding to other macroscopic configurations. As the system evolves, it will naturally evolve toward regions of phase space where there are more microstates—unless it is provided with the energy it needs to stay in the very small region corresponding to a highly ordered macrostate. Another way of saying this, is that the probability is much higher that the system will be found in a macrostate corresponding to a large number of possible microstates, rather than in a macrostate corresponding to a relatively few microstates. The second law of thermodynamics is equivalent to saying that if a system is in a low entropy state and evolves, it is very hard for it to stay in that low entropy state without the input of energy—without doing work on it. It will naturally tend to shift to a macrostate that corresponds to more microstates, and therefore to occupy a much larger region of phase space. As matter of fact, it will gradually tend to migrate towards an *equilibrium* macrostate, which is the macrostate corresponding to the largest number of microstates. This is also, generally, the configuration where there is no free energy within the system for doing any other further work. Equilibrium within a system is a "death state," which is thermally static and unproduc-tive.[285]

For nongravitating systems—or systems in which the influence of gravity is negligible—the macroscopic high entropy states are normally those representing higher-temperature, more homogeneous, unclustered or dissociated equilibrium states of matter. In order to organize such a sys-

---

[285] Interestingly enough, it is only in states close to equilibrium that bulk macro-scopic parameters such as temperature, pressure, density, and even entropy are rigorously definable.

tem, work must be done on the system from the outside—to cool it, compress it, etc.

Thus, from this discussion we begin to see that, because of the basic physical structure of our universe, complexity only arises and can be maintained in nonequilibrium conditions which encourage, support, and cater to the special highly ordered relationships among the components of systems, and between the systems and their environments. This generally involves highly differentiated relationships, and the input of energy and information from outside—in interaction with their environment. Realizing this, however, how it is that such systems can ever emerge in our universe? Where does the capacity of ordering come from—and from where does the energy necessary for keeping entropy at bay derive?

4 Gravity and Gravitational Entropy

In large part, the answer to these questions is "gravity" and "gravitational entropy." In our discussion so far, we have limited our considerations to systems in which gravity is not important. But gravitating systems are radically and fascinatingly different from boring nongravitating ones! In a way entropy applies to them, too, but not in the same way—and it cannot be conceptualized precisely. This is because a gravitating system does not, in general, exhibit the possibility of thermal equilibrium. As gravity acts, the system collapses, and heats up; and there is no classical limit to that gravithermal catastrophe. Insofar as we can assign an entropy to a gravitating system, the more gravitational clustering there is, the more entropy the system possesses—the upper limit for the system being a black hole. That is, for a gravitating system of a given total mass, its maximum entropy is reached when it has collapsed to a black hole. Its lowest entropy state is when the gravitating particles are all homogeneously distributed throughout space. This is just the opposite of a nongravitating system![286]

The rationale for this striking difference is the following. The more a gravitating system condenses, the less gravitational degrees of freedom it has, that is, the more microstates there are representing *each* macrostate, and the more "thermalized" those gravitational degrees of freedom are. Once the whole system has become a black hole, there are an enormous number of equivalent microstates for that very simple black-hole macrostate. Furthermore, it is obvious that there is no available energy that can be extracted from the black hole for anything!

However, for a gravitating system that is relatively spread out, but partially clustered and still evolving towards becoming a black hole, there are relatively few gravitational microstates for each of the many possible gravitational macrostates the system is exhibiting—and there is a lot of energy still available. As the system evolves its condensations grow under the action of the gravitational fields, heating up those condensations to high temperatures (stars!)—providing energy for work, which can help maintain nonequilibrium systems elsewhere. The energy within the stars is also used to create more complex atoms, which then can be employed as building blocks for more complex molecules, and eventually for living systems. It is

---

[286] See Penrose, *The Emperor's New Mind*, 337–39 (see also pp. 317–22), and *The Road to Reality,* 702–12.

precisely in this gravitating process that the system moves from a low entropy state to a higher entropy state—thus providing energy to the larger system. That energy can be used to keep entropy at bay on planets or on spaceships—to power the emergence and maintenance of more highly ordered systems, if the conditions are right.

Thus, the expanding, relatively homogeneous universe has over time provided a relatively low entropy gravitating system which has gradually evolved into a richly differentiated, hierarchically structured system of systems, containing regions where there are highly complex entities and organisms. Essentially gravity has provided for the manufacture of a rich variety of chemical elements, and also for nonequilibrium systems based upon them, from the relatively low entropy radiative energy generated by stars— high-energy photons, in contrast to the low energy, high-entropy photons which pervade most of space.[287] Without gravity, none of this would have been possible—no new chemical elements, no energy for powering growth and complexity, no sites or environments in which complex life forms could flourish.

Of course, before all that could have occurred, it was crucial that the universe expanded and cooled. Thus, the big bang and whatever triggered it—i.e., cosmic inflation, which not only spread out the matter of the universe in its low-entropy configuration, but also cooled it and generated density fluctuations in it, so that gravitational clustering could begin—was also essential to the whole process. But gravitational clustering only proceeded when the universe, after what must have been a very brief period of exponentially rapid expansion (inflation), finally settled down into a relatively slow expansion phase, allowing gravity to act effectively. There has been a relatively fertile balance between expansion and gravitational contraction. Otherwise we would not be here to enjoy it!

5 Dissolution and Emergence

From what we have already seen, entropy, both thermal and gravitational—and the transience and fragility they induce—are natural consequences, and probably unavoidable requirements, for a universe like ours. It is finite, dynamic, and evolving, organizationally multilayered and intricately structured, relying on networks of evolving internal and external constitutive relationships, and therefore open to the emergence of ever more complex systems and behaviors. Such a universe must, it seems, be marked by change, limitation, contingency, extinctions, local catastrophes and tragedies, demises to provide the energy, material and opportunity for the emergence of new and more complex systems—for the exploration of the full range of possibilities. It is very difficult to imagine how it could be otherwise, given those characteristics.

But, returning to the meta-question we briefly introduced at the beginning, are there other possible universes in which the laws of nature are such that complex systems could emerge, thrive and evolve towards life and consciousness without being enslaved by entropy and natural evil? As has been indicated by a number of cosmologists recently, it is likely that whatever primordial process generated our universe would have also spawned

---

[287] Penrose, *The Emperor's New Mind*, 318–19.

many trillions of others, each with different values of the fundamental constants and parameters, and possibly even with very different laws of nature. In order to begin to examine this idea critically, Ellis, Kirchner, and Stoeger have suggested how a preliminary parameter space of such a collection of universes can be described.[288] There is compelling evidence from a number of considerations that the vast majority of possible universes would have characteristics which would render them unsuitable for hosting complexity and life. This is just a consequence of "anthropic principle" arguments. Our universe is "fine-tuned" in many ways for complexity and life—one or two very slight changes in the underlying physics would make it completely sterile.

What we are really looking for is a possible universe in which complexity emerges and thrives, leading to levels of ever higher organization, but also in which there is very little cost in terms of entropy, disorder, fragility, transience, dissolution. Is that possible? If so, then God could choose to initiate or create a universe like that. However, we really do not know enough about all the requirements and possibilities for a consistent set of laws of nature—or about God—to say. However, I think we can argue in a preliminary way that any universe constituted by limited, finite, material entities developing successively more complex relationships with one another, characterized by its own dynamical autonomy and requiring something like cosmic and biological evolution for the realization of highly complex systems, will inevitably be subject to natural evils, and to something similar to entropy. In fact, it seems that any universe which is finite and which at the same time possesses it own integrity and inner dynamisms automatically needs something like energy and material for its ongoing maintenance and development. That has to come from somewhere—which means that the objects within that universe will at each stage have to undergo continual dismantling and rearrangement.

The only way of avoiding this is if there is no autonomy, with the universe and its contents rigidly designed, micromanaged, and maintained from the outside—or, alternatively, if the entities which constitute nature, though finite, are immaterial. In either case, there would be no evolution as such. In other words, this would be a universe in which a "heavenly," relatively static state is fashioned and maintained by a creator or demiurge from the beginning. This is not a possible universe we can say much about—it falls outside any multiverse scheme about which we can critically speculate, scientifically and philosophically.

From a theological point of view, we might wonder why the Creator would want to initiate and maintain a universe like ours. We might either refuse to answer this question, simply pointing out that we do not know, but that this is the way the universe is. Or we might speculate that these characteristics are in some way reflective of God's priorities and purposes. I shall not discuss where such speculations would lead. It is worth mentioning, however, that others, as well as I,[289] have suggested that the

---

[288] Ellis, Kirchner, and Stoeger, "Multiverses and Physical Cosmology"; Stoeger, Ellis, and Kirchner, "Multiverses and Cosmology: Philosophical Issues."

[289] Ellis, "Theology of the Anthropic Principle," in QCLN, 377–99; Stoeger, "Conceiving Divine Action in a Dynamic Universe"; and Stoeger, "Divine Action in a Broken World."

relationality, autonomy and freedom, predictability, divine hiddenness, and subtle inviting immanence of God in this world are precisely the characteristics for a creation whose ultimate priority and goal is loving free communion with the divine and among its constituents, commensurate with their different capabilities.

We have referred often to "emergence" as a central feature of our evolving universe at every stage—from the simple fragmentation of collapsing perturbations into galaxies and stars to the much more intricate and complex initiation of life in reproducing megamolecules, or the dawning of self-consciousness in our primate ancestors. In doing so I have emphasized the inevitable and necessary cost of emergence in the fragility, transience, and dissolution of entities, systems, and organisms, tracing that to the finiteness, multilayered, intricately relational character of nature and the universe, its flexibility and its evolutionary dynamism, and its need for the continual influx of low entropy energy and material to drive that evolutionary propensity. Now I shall set that hastily sketched portrait within a more integrated description of emergence, returning afterwards to situate transience, fragility, and the "natural evils" within that context in a more detailed way.

What is "emergence"? Generally speaking, it is the appearance of a new, qualitatively different entity, structure, property or behavior from more basic entities, structures, properties or processes, whether or not it is causally or ontologically reducible to the underlying processes or entities. The properties and behavior of the emergent entity or system are qualitatively different from the sum of the properties and the behaviors of its components.[290] Though we cannot delve deeply into the fundamental philosophical and scientific aspects of emergence, and the controversies surrounding them, here,[291] careful reflection on a few key features and relating them to the points we have emphasized in our earlier discussions will bring the overall picture into focus.

As we have already briefly mentioned, one of the common features of emergence is its reliance on structures with different levels of order, with constitutive relationships both on each level, and linking the levels with one another.[292] At each level there are structures and properties specific to that level, which have necessary conditions in the levels below their level, and constraints (boundary conditions and initial conditions) from levels above. But there will typically also be essential, same-level relationships which are important to what happens at that level. The higher levels cannot be understood or even described simply in terms of lower-level language and descriptions. Thus, there are always bottom-up, same-level, and top-down causal factors involved.[293]

A second key feature directly enabling emergence is that the structures at each level of organization are *modular,* that is, constituted by "combina-

[290] Stoeger, "The Emergence of Novelty in the Universe and Divine Action."

[291] For a survey of the philosophical issues, see Silberstein, "Reduction, Emergence, and Explanation," 80–107; for summary of the scientific aspects of emergence, see Clayton and Davies, *Re-emergence of Emergence.*

[292] Ellis, "On the Nature of Emergent Reality," and references therein.

[293] Ibid. See also Peacocke, "God's Interaction with the World," in *CC,* and references therein.

tions of semi-autonomous components with their own internal state variables, each carrying out specific functions."[294] These semi-autonomous functional components or modules, occupy levels of organization just below that of direct interest. In each module, typically, many lower-level states of the submodules correspond to a single modular state, since each modular semi-autonomous state is obtained by some effective averaging over lower-level states of its components, thus throwing away large amounts of lower-level information.[295] Thus, what goes on within the submodules is relatively hidden or isolated from the higher-level organization functional states of the module itself. No part of the system depends on the internal workings of any other part. This is sometimes referred to as *encapsulation*.[296] In the human body, for instance, we have the various organs (heart, kidneys, etc.) as the higher-level modules, with cells as the submodules within the organs. The cells themselves, of course, are also constituted by their own submodules (the nucleus, ribosomes, mitochondria, plasma membrane, cytoskeleton microtubules and microfilaments, etc.), and so on to the lowest levels of organization. As Ellis stresses, the success of such hierarchical structuring is directly due to enlisting separate submodules to perform lower-level processes, and then integrating these into higher-level structures whose overall functionality depends only indirectly on the internal process within the submodules.[297] This modularity supplies a certain robustness and redundancy to complex systems—a way to thrive, even when entropy and deterioration affect the modules and their submodules. The modular boundaries partially insulate the system itself from such degradation, so that the system's overall function is not compromised.

Thirdly, and finally, there are the constitutive relationships themselves. These are the complex of connections and interactions among the components of an entity or organism, and with its environment and its forbears, which endow it with a definite unity of structure and behavior, distinctive characteristics, and a persistence and consistency of action.[298] Depending on the levels of organization involved, these constitutive relationships may be metaphysical, physical, chemical, biological, or social in character. For a human being, for instance, they involve all of these categories. Essentially, then they are the foundation for the complex unity an entity or organism manifests and for the functions it fulfils[299]—that network of relationships which all together unites the lowest-level material constituents of a system into a unified functioning whole.

In pursuing this discussion, I include the metaphysical connections—which within scientific investigations are usually left out, or rather presumed. I include them because they are the ones that link the whole system, the basic components, and especially the other constitutive relationships themselves, with the Ground of their being, with the Creator. Though we do not adequately understand them, and cannot adequately

---

[294] Ellis, "On the Nature of Emergent Reality."

[295] Ibid.

[296] Ibid.

[297] Ibid.

[298] Stoeger, "Mind-Brain Problem," in *NP*, 129–46, see especially pp. 136–37.

[299] Ibid.

model them philosophically or theologically, they determine the existence of all that is, and the underlying order which supports and renders all other constitutive relationships effective. Those metaphysical relationships are the answers to the basic questions: Why is there something rather than nothing? And why is there this type of order in reality rather than some other kind?[300]

An aspect of emergence which is closely related to the three key characteristics we have just explored is the selection, preservation, transfer, and generation of information, both syntactic, semantic, and functional (pragmatic).[301] Living systems select, store, replicate, and use vast amounts of information. This always involves pattern recognition of some sort and feedback control loops. At the self-conscious level, abstraction, symbolic representation, and the implementation of predictive models are prevalent and crucial.[302] We do not have time to develop these ideas here, but it is important to mention them, as they will be referred to below when we briefly discuss the different types of emergence. Finally, in this regard, it is helpful to point out that emergence occurs in very different general categories of processes: cosmic evolution, uninstructed prebiotic chemical evolution, instructed prebiotic chemical evolution, biological evolution, biological development (e.g., of the fertilized egg to the adult organism), functional behavior of the individual, social, and technological evolution and development.[303]

At this point it is instructive to emphasize, however briefly, that the phenomenon of emergence is not uniform. Every example has its special character and specific underlying processes and network of organizing relationships. In order to appreciate some of the differences that can be involved, it is helpful to outline two classifications of emergence which have recently appeared.

Terrence Deacon in a recent paper[304] has identified three categories of emergence, which are really distinguished by the character and complexity of the relationships between the emergent level and its immediate lower level components. Deacon's first (and lowest) order of emergence in characterized by distributed relationships among the microelements of the lower level (for instance the molecules of a gas) determining statistical dynamics which lead directly to emergent level collective properties (such as temperature and pressure of the gas). Another example would be the emergence of surface tension in a liquid.

In Deacon's scheme, second-order emergence involves "spatially distributed re-entrant causality allow[ing] microstate variation to amplify and influence macrostate development."[305] The macro-level (emergent level) relationships constrain and bias the micro-level relationships, leading to amplification of the selected microstates. The very helpful example Deacon gives is the growth of a snow crystal.

---

[300] Ibid.

[301] See Küppers, *Information and the Origin of Life,* 31–56.

[302] Ellis, "On the Nature of Emergent Reality," 4.

[303] Ibid., 2; Küppers, *Molecular Theory of Evolution.*

[304] Deacon, "Three Levels of Emergent Phenomena."

[305] Ibid., 102–5.

Deacon's third-order emergence, of which biological evolution is the primary exemplar, adds the preservation of selected information (memory) and therefore the distribution of causality between the levels over time. Information from the environment is received at the emergent level and used to select and amplify certain lower-level characteristics, which in turn lead to adaptation with increasing divergence, complexity and self-organization at the emergent level.[306]

George Ellis[307] has suggested a somewhat different but very helpful five-level (or five-order) categorization of emergent phenomena. He speaks of levels of emergence, but here, so as to avoid confusion with "levels of organization," I shall continue speaking of orders of emergence. Ellis's focus in his categorization in not so much on the character and complexity of relationships between the levels involved, as on the character and capabilities of the emergents themselves. Ellis's first and lowest order of emergence includes all those cases where bottom-up action leads to emergent level properties, but not to essentially new emergent level structures. Examples would be, as with Deacon, the properties of gases and liquids.[308]

In his second-order category Ellis includes all those situations in which bottom-up action plus boundary conditions lead to emergent level structures not directly implied by lower-level behavior nor by the boundary conditions themselves. Examples are convection patterns in fluids, stars and galaxies, inorganic and organic molecules. At this level there is as yet no true complexity, nor any "goal-seeking" behavior.[309]

Third-order emergent phenomena are distinguished by the fact that bottom-up action occurs in highly structured systems, leading to the existence of feedback control at various levels. This, in turn, leads to coordinated responses to the environment and top-down action of the emergent structures on the lower-level components. Thus, we have coherent, non-reducible emergent-level action directed by implicit inbuilt goals. Examples would be viruses and living cells. At this level there is no capacity for individual learning.[310]

In the fourth order of emergence, we have for the first time the capability for such individual learning. In addition to the bottom-up and top-down action distinctive of third-order emergence, we have here feedback control systems directed by specific events in each individual's history. The best example here is learning in animals.[311]

What characterizes fifth-order emergent phenomena is that some of the operative goals are explicitly expressed in language and/or determined by symbolic understanding or complex modeling of the physical and social environment. Examples are human self-conscious reflection, intentional action, and social and cultural systems, which are marked by intersubjective, consciously intended actions and projects.[312]

---

[306] Ibid., 106–7.

[307] Ellis, "On the Nature of Emergent Reality."

[308] Ibid., 13.

[309] Ibid.

[310] Ibid.

[311] Ibid., p. 14.

[312] Ibid.

This in-depth description of emergence is the context now within which to place our further discussion of thermal and gravitational entropy, and the resulting transience, fragility and fertility of nature and the universe. We can look at this from two different complementary points of view, that of the evolutionary history of the universe and of the Earth, and that of the present requirements for maintaining the structures we have as they are, with attention to the prospects for the future.

The first perspective is especially illuminating in appreciating the essential role transience, fragility, dissolution, and death play in emergence of structure, complexity, life, and consciousness. Over enormously long periods of time layer upon layer of organization developed, each new layer crucially supported and informed by the layers below it, and nourished and challenged by mutually evolving environments. At the physical levels there is the overall directionality provided by the expanding and cooling universe, and the universal attractive action of gravity, which provides the basic macroscopic differentiation and organization of nature, as well as its localized energy sources and reservoirs of chemically enriched material in stars. At the nuclear and chemical levels there are the nuclear fusion interactions and the various processes of chemical bonding which under many different conditions led to the production of all elements heavier than hydrogen, and later the rich array of molecules. Natural selection, symbiosis, and sexual reproduction dominate the advanced chemical and the biological spheres. And now culture, education, and technology drive the evolution of personal and social consciousness and relationality.

At all these levels, and in all the innumerable stages of development falling within each level, the possibility of change and emergence depends upon not only the stability of basic lower-level relationships, but also on the flexibility and vulnerability of systems to incorporation into more complex systems or to being dismantled and rearranged into new entities or organisms. Furthermore, as discussed earlier, such incorporation and rearrangement into new and more highly organized systems requires free energy and material. All those processes needed for emergence, then, demand the stability of systems at all levels within certain limits, but at the same time their transience and fragility outside those limits, including their sensitivity and adaptability to the evolving environments they inhabit. That, as we have stressed earlier, inevitably involves the dissolution or demise of earlier systems in order to make way and provide the components for later systems. Without such a paradoxical combination of qualities, emergence of the nested levels of organization that characterize nature, and the novel relationships and behaviors which accompany that, would simply not have occurred.

To be more specific, what is it about an emergent system which depends upon entropy production and transfer, and energy and material flow? We have already indicated the general characteristics. But it is helpful to be more precise. Genuinely emergent systems and phenomena—Deacon's second- and third-order emergence, and Ellis's second- to fifth-order emergence—always involve nonlinearity, strong couplings with the environment and with the emergent structures themselves. This obviously means that the systems must be open. Both to trigger the emergent subsystem and to maintain it also requires keeping the system dynamic and out of—but usually not too far from—equilibrium, and at the same time stable.[313] Oftentimes the emergent structure is the result of a random fluctuation which stabilizes, persists, and grows in the nonequilibrium environment under the influence of the nonlinear dynamics. This always means supplying a continual flow of material and/or energy, which is used to power or nourish the system, keeping it active, developing, and vibrant. A good simple example is a hurricane. Just to keep it in a nonequilibrium state requires such input. Equilibrium, as we have already seen, is equivalent to death and dissolution. Use of the energy and material flowing into the system for increasing or maintaining its organization generates entropy—waste products and disorder, which must be transferred out of the system to the surrounding environment, whose entropy then must increase. That process, too, requires work, energy. That energy and material for maintaining the emergent system and keeping its entropy relatively low must come from the environment—ultimately in our case from the Sun and from the terrestrial energy sources which store solar energy. Obviously, as we have already discussed earlier, all the processes involved inevitably require the transience, fragility, and eventual dissolution of other entities and systems. Without those features we would have a boring, changeless, dead world in perfect equilibrium, with no possibility of the emergence of anything. There has been a great deal of research done in the last few decades filling in the innumerable details of this picture for many different examples of such emergent dissipative structures and systems. This brief summary simply provides the basic connection between emergence and entropy.

The second point of view, that of the present need for maintaining the structures we have, and our prospects for the future, complements the first perspective, but cannot be fully appreciated without it. As we come to understand more about ourselves and our environment, we realize more and more how intricately interconnected and delicately balanced our world is— and how fragile we and our context must be. And it is always changing, subject to both well-known and to unforeseen natural and cultural influences. Furthermore, we experience ever more forcefully how much we need energy, food, and other resources in order to thrive and prosper, and more disturbingly how utterly dependent we are upon both our cosmic and terrestrial environments. Not only are our resources more limited than we imagined a hundred years ago, but also we are reminded much more often how helpless we are in the face of eventually inevitable cosmic and terrestrial catastrophes. We are beginning to encounter and recognize our physical and biological limits and to see beyond them to what could emerge here or elsewhere in the universe.

This is the way the universe is—the way nature is. And, as we have seen, it is very difficult to see how reality could be any different and still be a reality that is dynamic, relational, interconnected, integral, evolving, open to new possibilities, predictable and understandable, personal and engaging. Transience and fragility are unavoidable consequences of these characteristics, and—along with entropy and emergence—turn out to be

---

[313] See, for instance, Peacocke, *Introduction to the Physical Chemistry of Biological Organization,* 17–72; and Weber, *Entropy, Information, and Evolution*; see especially the articles by Lionel Johnson, Jeffrey S. Wicken, and E. O. Wiley.

necessary conditions for evolutionary fruitfulness.[314] There is hidden in that strange counterpoint an intimation of possibilities beyond what seems destined at first sight to remain unfinished.

From a theological point of view, what is the possible destiny of such a universe, and of the beings in it? Is there any room for "the eschaton"? Earlier, I have hinted at this issue in mentioning the importance of the metaphysical constitutive relationships with the Creator, which transcend scientific investigation, and in speculating why the Creator would initiate a universe like ours. All the characteristics of nature and of the universe are consistent with what is necessary for eventual full realization of loving free communion with one another and with the Creator. This is not possible this side of death—but we already witness partial realizations of it. Though we do not understand how this would occur, we can see that there is a movement towards this in our world. Thus, the eschaton, or completion, of the world would be the entry of nature and the universe as it has developed into communion with the life of God, God's self. The processes of evolution and the transience and fragility implicated in it would find their completion and meaning is such communion. The life, death, and resurrection of Jesus—followed by the sending of the Spirit and the life and celebration of the Eucharistic community of the church in Christ—provides the basic pattern of and witness to, this ultimate destiny, and to the significance of the context of evolutionary and natural evil. In some definite sense, this is the fulfillment and completion of evolution and the process of emergence. Though obviously beyond what the natural sciences can recognize or adjudicate, it is not beyond what is accessible to human experience and understanding enlightened by discerning and committed faith.[315]

---

[314] As should be clear from what I have said at the beginning of this paper, I am not here constructing either a defense of religious belief nor a theodicy (see Terrence W. Tilley's paper in this volume for an explanation of this distinction). Instead I am laying out the details of some of the relevant scientific aspects of the natural world which can be used as a component in developing arguments in either strategy.

[315] I am very grateful to all those who have interacting with me before, during, and after the conference in Castel Gandolfo, and from whom I have learned much—especially to Wesley Wildman, Tom Tracy, Phil Clayton, and Don Howard, and to my coeditors Nancey Murphy and Bob Russell.

PHYSICS, COSMOLOGY, AND THE CHALLENGE TO CONSEQUENTIALIST NATURAL THEODICY

Robert John Russell

1 Introduction

The task of this paper is to assess the possibility of constructing a robust "natural theodicy," that is, a theodicy[316] whose focus is not on moral evil[317] but on what is frequently called natural evil.[318] For the purposes of this paper I will work with a consequentialist natural theodicy in which natural evils are an unintended consequence of God's choice to create life through natural means, namely, through the biological processes of evolution (section 2).[319] I believe this approach is vastly preferable, both for theological, ethical, and pastoral reasons, to theodicies in which natural evil is seen as a result of intentional divine action. Moreover, with our focus on theology in dialogue with the natural sciences, consequentialist theodicies are, in my opinion, reasonably robust when considering evolutionary biology.[320]

---

[316] Terrence Tilley makes a very important distinction between a theodicy and a defense in "The Problems of Theodicy." As I read Tilley, a theodicy is an argument that attempts to show that there are rational reasons for believing in why God, whose existence is presupposed, allows evil to exist in the world, whereas a defense is an attempt to show that there are rational reasons for believing in God and that such belief is not incompatible with the existence of evil. Because of this distinction, the modes of reasoning and the kinds of evidence employed by a theodicy and a defense are strikingly different. In this paper, I will continue to use the term "theodicy" for what Tilley calls "defense" as that seems to fit with many of my personal theological sources. The reader, therefore, should recognize that I am agreeing with Tilley about the difference in these terms and methods.

[317] This term generally includes both personal sin (e.g., choosing a lesser good than God) and social sin (e.g., patriarchy, institutional racism, economic domination, political oppression, environmental devastation, etc.)

[318] I am using the term "natural evil" as a generic term for both biological natural evil and physical natural evil. Biological natural evil includes pain, suffering, disease, death, and extinction. Physical natural evil includes geological phenomena (e.g., earthquakes), meteorological phenomena (e.g., hurricanes), oceanographic phenomena (e.g., tsunamis), astrophysical phenomena (e.g., impact of asteroids), radiative phenomena (e.g., radioactive decay), etc. Both biological and physical natural evil affect human life and prehuman life. In addition, physical natural evil is present in nature before life began and serves as a prerequisite for the biological evolution of life, and it is constitutive of much of biological natural evil. Of course much of biological and physical evil is the result of human moral evil, but that is not the specific focus on this paper (as it would be when considering environmental ethics or eco-justice, etc.). Note that Christopher Southgate uses these terms in a slightly different way: he refers to what I am calling physical and biological natural evil as "evolutionary evil" when they affect nonhuman nature and "natural evil" when they affect humanity. See Southgate's paper in this volume.

[319] Examples of consequentialist theodicies in this volume include the chapters by William Stoeger, Nancey Murphy, Tom Tracy, and Terrence Tilley.

[320] I have addressed the issue of theodicy in a variety of previous publications. They tend to fall into three groups: (1) Theodicy as such. See for example Russell,

But biology presupposes physics even while building upon it, and yet, surprisingly, the implications brought to theodicy by physics are seldom made explicit in the literature on natural theodicy. Hence in this paper I will move below biology to the physics that makes biology possible and outwardly in time and space to the physical cosmology which describes our universe (sections 3–6). It is here that we discover how the fundamental laws of physics and the fundamental constants of nature make our universe fine-tuned for the possibility of life, and it is here, therefore, that a robust consequentialist natural theodicy will have to be constructed through a fully-developed research program in theology and science. I will refer to a natural theodicy whose domain extends beyond biology and physics to physical cosmology as "cosmic theodicy." This short paper, then, is an initial exploration of some of the directions for more detailed research on the strengths and weaknesses of a consequentialist form of cosmic theodicy.

My tentative conclusion, which is subject to revision after more research is completed, is that a consequentialist cosmic theodicy faces two severe challenges. To understand the first challenge one must remember that a consequentialist natural theodicy depends on the laws of physics being a given, as they are in biologically-based theodicies. Here one can argue persuasively that, given these physical laws, God had no choice but to permit *biological* natural evil because God's intention is to create life through the processes of evolution. But when we expand our perspective downward to physics and outward to cosmology, consequentialist theodicy runs up against the fact that the fine-tuning of the universe that makes life possible is viewed theologically in terms of the doctrine of creation: the laws and constants were created *ex nihilo* by a free act of God unconstrained by any ontologically or temporally prior conditions. Hence the first challenge is stark: Why did God choose to create *this* universe with *these* laws and constants? Are there *no other possibilities open to God* to create a universe in which life could arise by natural processes and yet without the physical natural evils that characterize our universe?

Here an *a priori* scientific issue arises as to whether we can even frame the question of other possible universes in which life could arise in light of contemporary physics, for with this issue we arrive at the frontiers of fundamental physics and cosmology, and we must admit up front that a unique and well-supported scientific answer may not be possible at the present. Having said this, it is an intriguing fact that some of the research surrounding the anthropic principle might be helpful for our concerns in constructing a consequentialist cosmic theodicy. Specifically, a variety of "many worlds" type responses to the anthropic principle are being developed. These theories, although highly speculative, suggest that our universe is one "domain" of a multidomain mega-universe, one topological branch in an endlessly branching, eternally inflating universe, or one layer

---

"Natural Theodicy in an Evolutionary Context." (2) Theodicy as exacerbated by the claims of noninterventionist objective divine action, particularly based on quantum mechanics (i.e., QM-based NIODA). See for example Russell, "Divine Action and Quantum Mechanics," in *QM*. (3) Theodicy as a rationale for developing an eschatology that responds to the challenges raised by theodicy as well as by scientific cosmology, an eschatology based on the bodily resurrection of Jesus as a transformation of creation into new creation. See for example, Russell, "Bodily Resurrection, Eschatology and Scientific Cosmology."

within a brane-structured multiverse, and so on. The general implication of these theories is that the fine-tuning of the laws and constants of our universe is merely the random result of the existence of many universes embodying different laws of physics and different values of the constants. While this tends to mitigate against the anthropic principle as a design argument for the existence of God,[321] some of the underlying arguments in these many worlds theories might help us in our quest for a cosmic theodicy—in a limited way—by implying that God's choice is not at the level of the laws and constants of *our* universe but at some megalevel of universes, laws, and constants.

Yet this in turn just reopens the question of cosmic theodicy at a higher level of abstraction: (1) Would slightly different values of the constants have produced a universe in which life evolved and in which the extent of natural evil was lessened? (2) Would completely different sets of these values have produced a universe in which life evolved and in which the extent of natural evil was lessened? (For brevity I will focus in this paper on variations in the constants.) If we can show to any reasonable extent that there are *no* other possible universes in which life could arise by natural processes but with significantly less suffering, then a consequentialist natural theodicy succeeds in the sense that the suffering involved in the evolution of life is an unintended consequence of God's choice to create our universe, and God created our universe because these particular laws and constants are *uniquely* necessary for life. I shall explore these questions briefly in section 7 as a way of pointing out directions for future research.

Even if a consequentialist cosmic theodicy at this level of abstraction can be supported, there is, as mentioned above, a second and even deeper challenge to consequentialist cosmic theodicy: *is life worth the price of such extensive suffering?*[322] In my opinion this second challenge actually points us in a different direction and leads us to a much different conclusion about the entire project of natural, versus moral, theodicy. I base this on the fact that natural theodicy is framed within the doctrine of creation. Thus the challenge here is not due to current scientific theories as such, nor to their vulnerability to historical relativity (i.e., the inevitable changes in scientific paradigms), nor—need I say it!—to our using science within the project of theology. Instead I believe the challenge is due to the fact that the Christian tradition's response to evil has always been to set the present world in a wider, eschatological context: the new creation proleptically begun in Christ and forming the ultimate future of our—and every—universe. It is only when the new creation is the starting point for reflecting on evil that we can hope to give a response to its origin and meaning in this present, broken world. I will close, then, by turning to an eschatological framework and by hinting at ways in which it might offer a fruitful interaction with

---

[321] In a previous publication I suggested that a design argument for the existence of God can still be formulated even if we consider not only variations in the values of the constants of nature but also variations in the laws of nature, and even variations in the kinds of logic that govern the structure of these laws. See Russell, "Cosmology, Creation, and Contingency."

[322] I am grateful to Nancey Murphy for her helpful suggestion that I distinguish between whether a consequentialist theodicy fails because of the suffering incurred or because it cannot show that the evolution of life is worth the cost of such suffering.

science, in particular with scientific questions about the laws and constants of nature (section 8).

Before proceeding, however, I need to make several preliminary comments. First, I accept Tom Tracy's distinction between logical and physical necessities. "Only the former constrain God's choices in creation. So when you say . . . that God had 'no choice,' the necessity is merely conditional. If God wants adaptive complexity within a system of natural law, then God must employ natural selection, and this in turn requires biology and physics and the rest. But what are the goods that require, as a logically necessary condition for their realization, the existence of a world like this? If we can answer this question, then the next issue that arises (as you say) is whether this good is worth having at this price."[323] It is when I move to the level of physical and mathematical abstraction dealt with in section 8 that physical and logical necessities seem to merge, leading here at least, if not even sooner, to the "no choice" formulation.

Second, we have no reason to believe that our present-day understanding of these laws and constants is the final word on what are, ultimately, the truly fundamental laws and constants of nature—quite the opposite, given the brisk pace of scientific research in physics and cosmology during the past century. Nevertheless I will use what we have at present to make this case for two reasons: (1) I will choose a strategy that does not rely directly on current theories in physics but more specifically on the constants of nature since I believe these are more likely to survive further paradigm shifts in theoretical science than the theories themselves. (2) I do not believe that future scientific paradigms will lead us to view natural evil as somehow less pervasive and real than we now believe it to be, and thus future scientific paradigms will not circumvent the need to shift our theological response to evil from creation to redemption and new creation.

Third, let me add that I am aware of and sensitive to the concern that in responding to the problem of theodicy one must never be, in effect, justifying evil, minimizing the horror of evil, or ignoring the responsibility we all share for perpetrating evil. At the same time I do not believe we can ignore the challenge of theodicy to Christian faith today. To the extent that contemporary theology has done so it has contributed to the mass exodus from a Christian faith perceived to be worshiping a god who permits, let alone performs, "horrendous evil," to use Marilyn Adams's blistering phrase.

Finally, the challenge posed by moral evil to Christian faith has been exacerbated by atheists who coopt the vast canvas of suffering, disease, death, and extinction that characterize the evolution of life to argue against the existence of the Christian God. Hence a serious rethinking of the problem of natural theodicy is clearly needed. I hope that this chapter can offer a few suggestions towards a much longer research project focused on this goal.

2 Theodicy Posed in the Context of Evolutionary Biology

Michael Ruse, never an unambiguous friend of Christianity, nevertheless offers a quite helpful response to the way the problem of biological natural evil is used by Richard Dawkins to support atheism. Following a particularly stark quotation from Dawkins in support of the meaninglessness of the world given the astonishing adaptation of both predator and prey to the

---

[323] Private communication written in response to my initial paper.

predator-prey cycle, Ruse writes that God is not really to blame for this kind of evil:

> Dawkins ... argues strenuously that selection and only selection can (produce adaptedness). No one—and presumably this includes God—could have got adaptive complexity without going the route of natural selection. ... The Christian positively welcomes Dawkins's understanding of Darwinism. Physical evil exists, and Darwinism explains why God had no choice but to allow it to occur. He wanted to produce designlike effects (including humankind) and natural selection is the only option open.[324]

Here in attempting to defend God from being blamed for the pain associated with natural selection, Ruse apparently takes for granted the underlying chemical and physical regularities that serve as preconditions for the possibility of biological evolution. A more careful response to natural evil, however, will require that we not make the same mistake. Instead we must press below the domain of biology and reach downwards in complexity to the fundamental laws of physics and the fundamental constants of nature, and then by using these laws and constants reach backwards in time before biological evolution arose on planet Earth ~3.85 billion years (Byrs) ago, back before our Sun and solar system formed ~5 Byrs ago, back before our galaxy formed ~10 Byrs ago, until we come finally to the earliest phases of the universe some 13.7 Byrs ago and circa the Planck epoch, $t_{Planck} = 10^{-43}$ seconds. From this perspective on the universe as a whole and its characterization in terms of the fundamental laws and constants we can finally pose the question of natural evil in relation to the universe as such using the language of physics.[325]

3 Evolutionary Biology Requires the Macroscopic World of Nature as Described by Classical Physics

Our search for the fundamental laws and constants of nature begins with the everyday world of nature in which evolution takes place, the macroscopic physical world of oceans and atmosphere, sunlight and weather, the orbits of the Earth and its moon, earthquakes, volcanoes, and tsunamis, the frozen Antarctic and the sweltering African savanna. The physical sciences that describe this world were originally based in classical physics: classical mechanics, electromagnetism, Newton's theory of gravity and thermodynamics and their application to such fields as meteorology, geology, oceanography, and nonrelativistic astrophysics. Biology presupposes these physical sciences in order to discuss such key topics as variation, natural selection, adaptation, ecology, and so forth. Yet even here more is required than classical science could provide.

---

[324] Ruse, *Can a Darwinian Be a Christian?* 136–37.

[325] Whether we can probe back prior to the Planck time to the essential singularity t = 0 as described by standard big bang cosmology, or whether our universe emerged from a previous universe described in diverse ways by inflationary big bang and by quantum cosmology, or whether ours is the only universe or only one of a possibly infinite series of universes, as multiverse theories suggest, will require further exploration (see below, section 5.2). For an earlier technical discussion see Russell, "Finite Creation without a Beginning," in *QCLN,* 468.

Take variation as an example. One source in the macroscopic world is the changing environment in which organisms live, reproduce, and compete over finite resources, and these changes are routinely described by the physical sciences just mentioned. Other sources, however, lie deeper within the microscopic and submicroscopic world probed by molecular biologists in the twentieth century. Here the DNA macromolecule, which carries the genetic code for the organism, is subject to random mutations brought about by such processes as replication and recombination during cell division (mitosis and meiosis), as well as by exposure to radiation, and so forth. All of these processes are ultimately rooted in quantum mechanics. A third source of variation involves the chance juxtaposition of random environmental variation and random genetic mutations, and thus a mixture of classical and quantum processes.

4 The Classical World, in Turn, Requires Quantum Mechanics and Special Relativity

We now know that the theories of classical physics are not truly fundamental, although their epistemologies often warrant nonreducible (i.e., emergent) properties and processes ranging from the simple viscosity of water to the complex features of self-organizing systems. (Whether strongly emergent processes can exert top-down causality on the processes underlying them is a crucial and highly debated question, but it is not relevant here.) Instead, their true foundations are the two pillars of twentieth-century physics: special relativity (SR) and quantum mechanics (QM). These in turn have been combined during the past century into increasingly complex theories that deal with physical phenomena from elementary particles to the inflationary universe.

A promising way to explore the revolution in physics that led to relativity and quantum mechanics is to study the role of three constants of nature during that revolution: the gravitational constant, G, the speed of light, c, and Planck's constant, h. Each marks the revolution in theoretical physics in its own way (table 1).[326] Planck's constant has no corollary in classical physics; it was posited *sui generis* by Max Planck in 1900 and marks the birth of quantum mechanics. The speed of light has roots in classical physics, emerging out of two previously unrelated fields of study (electricity and magnetism), which were combined by James Clerk Maxwell in his theory of electromagnetism (1864). It was Maxwell who recognized that light is a wave in the electromagnetic field, but the true significance of the speed of light as a universal and invariant constant of nature required the framework of Einstein's special relativity (1905). The gravitational constant was posited by Newton in 1687 to scale the force of gravity between masses. It crossed the divide between classical and contemporary physics unscathed, finding a new home in Einstein's relativistic treatment of gravity, the general theory of relativity (1915). It is worth noting that there are no constants of nature in classical mechanics or in thermodynamics, clearly suggesting their merely phenomenological status.

---

[326] For a helpful resource on the constants of nature see "The NIST Reference on Constants, Units, and Uncertainty"; see also "Table of Physical Constants."

The diversity of paths taken by these three constants—the *de novo* origination of h, the true meaning of c, and a continuous transition for G—

signify the complex and multilayered meaning of the revolution in physics from classical to twentieth-century theories. It is often stated that twentieth-century physics reduces to, or contains as a limiting case, the theories of classical physics: quantum mechanics reduces to classical mechanics for h → 0, special relativity reduces to classical mechanics, and general relativity to Newtonian gravity, for h → ∞.[327]

But the story is more complicated. As we shall see, both quantum mechanics and special relativity continue to play key roles even in the world of ordinary phenomena described by classical physics. The transition between classical physics and contemporary physics is subtle indeed.[328]

$h = 6.626 \times 10^{-34}$ J-sec   Planck's constant

$c = 2.99 \times 10^8$ m/sec   speed of light in vacuum

$G = 6.6742 \times 10^{-11}$ m³/kg-sec²   gravitational constant

Table 1: The Current Values of Three Crucial Constants of Nature
(See n. 12 for a current estimate of the uncertainties in these values.)

4.1 Quantum Mechanics (QM)

One of the key fundamental constants of nature is Planck's constant, $h = 6.626 \times 10^{-34}$ Joule-seconds. It arises most famously in the Heisenberg uncertainty principle which states that our knowledge of both the position x and the momentum p of a physical system can never be exact: If Dx is the uncertainty in our knowledge of position x, and Dp the uncertainty in momentum p, then Heisenberg's principle is expressed by $Dx \cdot Dp \geq h/2p$. Its greatest significance, however, may lie in the discovery that energy, a continuous variable in classical physics, is in fact quantized, leading to such

---

[327] For a recent and illuminating presentation of these and other limiting conditions in diagrammatic form, see Barrow, *Constants of Nature*, chap. 3, esp. fig. 4.2, p. 59.

[328] In 1899, Planck proposed that these three constants be combined into what he considered to be a preferable format, namely, into "natural units" of mass, of length, and of time as follows:

$m_{pl} = (hc/G)^{1/2} = 5.56 \times 10^{-5}$ grams

$l_{pl} = (Gh/c^3)^{1/2} = 4.13 \times 10^{-33}$ centimeters

$t_{pl} = (Gh/c^5)^{1/2} = 1.38 \times 10^{-43}$ seconds (the "Planck time" referred to below)

(Actually Planck also included Boltzmann's constant in a fourth term.) The enormous difference in the sizes of the Planck constants—28 to 38 orders of magnitude difference between the unit of mass and that of length and time—led Planck to speculate that the value of the natural units would be discovered by any form of intelligence with any kind of scientific method. For a discussion, see Barrow, *Constants of Nature*, 23–30.

remarkable phenomena as the particle-like characteristic of light (corpuscles of light now referred to as photons.)[329]

Quantum mechanics relates to the macroscopic world in at least three distinct ways:

(a) Within certain qualifications,[330] QM reduces to classical mechanics as represented by the correspondence principle in the limit that $h \rightarrow 0$. Quantum mechanics provides a theoretical explanation of the routine phenomena of the macroscopic world which previously had simply been posited in classical physics. These include the properties of the chemical elements as represented by the periodic table and the properties of solids, including impenetrability, specific heat, and electrical conductivity. Quantum mechanics accounts for the color of the sky, the color of dyes, the viscosity of water and why water expands when it freezes, why candles and fires glow red rather than blue as they cool, and on and on.

(b) Quantum mechanics also accounts for certain phenomena, such as superfluidity and laser light, which cannot fit easily into the classical account of the world since they arise as bulk macroscopic quantum states and retain much of their quantum mechanical character in the otherwise classical world of everyday life. This fact contradicts the often-repeated claim that quantum mechanics is irrelevant to the macroscopic world because the size of Planck's constant, h, which serves as a measure of the length scale/energy of atomic processes is so minute.

(c) Specific states in the macroscopic world can arise from the amplification of individual quantum events. The (in)famous example is Schrödinger's cat, where a single quantum event, such as the radioactive decay of an atom, is detected by a Geiger counter and amplified, leading to one or another distinct outcomes in the world around us. Such amplification of quantum events does *not* require chaotic processes (*pace* John Polkinghorne), although chaos can provide such amplification.[331] These amplification-induced states are another example of the direct relevance of quantum mechanics to the classical, ordinary world.

Quantum mechanics is relevant to biological evolution in several ways. It underlies all of organic and inorganic chemistry in fields ranging from cell physiology to neuroscience. It plays a central role in genetic mutations as described above. Biology treats mutations as a given and describes them as "chance" events (or, with Dawkins, "blind chance"), but mutations, in fact, require the making or breaking of a hydrogen bond within the DNA

---

[329] It was first written down by Max Planck in 1900 in his work on black-body radiation. The classical equipartition theorem gave erroneous results when used to predict the spectrum of black-body radiation. Planck solved the problem by assuming that the energy E of the radiation is a discrete variable instead of a continuous variable as classical physics presupposed. The result is his famous equation relating E to the frequency n of the radiation: $E = hn$. Planck's constant occurred again in Einstein's explanation of the photo-electric effect, in which Einstein assumed that light has particle-like properties. There are many other illuminating examples of the crucial role played by Planck's constant in the thirty-year development of quantum mechanics.

[330] According to Michael Berry, singularities in the semiclassical limit play a key role in the applicability of the correspondence principle. See Berry, "Chaos and the Semiclassical Limit of Quantum Mechanics," in *QM*.

[331] Russell, "Divine Action and Quantum Mechanics," in *QM*, 299.

molecule, and these processes are essentially quantum mechanical in character. Since genetic mutations are a necessary source of variation—but certainly not the only source—it is clear then that neo-Darwinian evolution requires, at a foundational level at least, QM as part of its explanation of the history of life on Earth. Indeed the expression of genetic mutations in variations in the characteristics of phenotypes, and their relative capacity to contribute to reproductive fitness, can be considered a "biological Schrödinger's cat"—although the time scales involved between genetic mutation to phenotypic expression are enormous compared to the "instantaneous" consequences at the macro level of a specific quantum event such as described under (c) above.

4.2 Special Relativity (SR)

In 1905 Einstein published his now famous paper, "On the Electrodynamics of Moving Bodies."[332] Here he dealt with a fundamental challenge facing classical physics: the irreconcilability of Newton's principle of relativity (PR) with Maxwell's equations of electromagnetism. Foundational to Newton's entire system of mechanics is his principle of relativity: absolute rest and absolute motion are physically meaningless, and thus the form of the laws of physics must be the same for all observers in uniform relative motion ("inertial observers"[333]). These are expressed most clearly in the second of his three laws, $\mathbf{f} = \mathbf{ma}$, where the acceleration, $\mathbf{a}$, of a physical system of mass, m, is related to the total external forces, $\mathbf{f}$, applied to that system. This equation holds for all inertial observers when their coordinates are related by the Galilean transformations.[334] The problem is that Maxwell's field equations of electromagnetism do not remain the same (they are not invariant) when the Galilean transformation is applied to them. Hence they seem to violate Newton's principle of the relativity of motion.

The constant, c, appears prominently in Maxwell's field equations and denotes the speed of propagation of the electromagnetic waves in the field. It was only then that physicists realized that light must be composed of waves in the electromagnetic field and that c denotes their velocity. Interestingly, the constant c is actually a combination of two unrelated constants arising separately in the study of electricity, tracing back to Thales of Miletus in ancient Greece (ca. 600 BCE) and of magnetism, with roots in the experiments of the Roman Pliny the Elder (ca. 50 CE). The former constant is the "dielectric constant of free space," e_0; the latter is the "permeability of free space," m_0. The dielectric constant has genuine physical significance: it provides a measure of the electrical insulation of various

---

[332] A. Einstein, "Zur Elektrodynamik bewegter Körper," *Annalen der Physik* 17 (1905): 891–921.

[333] The specification of which observers are inertial and which are accelerating led Newton to introduce the metaphysical idea of absolute space and absolute time.

[334] For observers in uniform relative motion along the x-axis, these equations are: $x' = x - vt$, $y' = y$, $z' = z$, and $t' = t$, where the primed and unprimed coordinates are those of the two observers whose relative velocity is v.

materials.[335] Appearing together in Maxwell's equations they yield the speed of light, $c = (e_0 m_0)^{-\frac{1}{2}}$. This provided convincing evidence for physicists to reject Newton's corpuscular theory of light and accept light as a wave in the electromagnetic field.

As with other kinds of waves (e.g., sound, ocean waves), the speed of light should depend in part on the velocity of its source through the aether. The Michelson-Morley "null experiment" is famous for having not detected the Earth's motion through the aether. It is often cited as the empirical reason for Einstein's construction of special relativity, but Einstein's primary motivation was to find a way to make Maxwell's equations consistent with Newton's principle the relativity. To do so, Einstein made a radical proposal: the speed of light in vacuum, c, must have the same value for all observers in uniform relative motion. This means that unlike all other phenomena in nature—including both particles and waves—the speed of light is *independent* of the velocity of an observer. To make his proposal work, Einstein showed that the laws for transforming coordinates between observers cannot be the classical Galilean transformation but instead must be the Lorentz transformation.[336] The result is Einstein's special theory of relativity (SR).

It was only with SR that physicists could appreciate the truly profound significance of c: although its roots lie in combined strata from electric and magnetic domains, it can now be seen as a universal invariant linked to the shift in worldview from Newtonian space + time to Minkowskian spacetime geometry, to the counterintuitive behavior of clocks (such as time dilation), the downfall of the assumption of a universal present globally dividing past from future, the strange paradoxes of SR (such as the twin paradox), and so on. In passing we should note that special relativity also explains Michelson-Morley's null result.

Special relativity relates to the macroscopic world in at least three distinct ways:

(a) SR reduces to classical mechanics when the speed of light, c, is set to infinity ($c \to \infty$). According to SR, each momentary event, P, in spacetime is associated with three domains of events which are defined by their causal relationship to P: the event's causal future, causal past, and acausal "elsewhere." These domains are demarcated by the worldlines of photons arriving at and being emitted by P and forming the "lightcone" surrounding P. In the limit $c \to \infty$, the acausal "elsewhere" collapses to a surface, leaving only the causal future and the causal past which are shared by all other events like P in the collapsed surface which we call "the present." The present is a global property of the macroscopic world, bifurcating time into the global past and the global future. Our intuitive ideas about "flowing time" and "history" take for granted the idea of a global present. According to SR, however, the global present is not a physical feature of nature but an an-

---

[335] When an insulating material such as glass is placed between two electric charges, it reduces the force between them, and the reduction is inversely proportionate to the value of the dielectric constant of that material. Magnetic permeability, on the other hand, is merely a constant of proportionality between the magnetic flux density and the magnetic field strength for a given material.

[336] The Lorentz transformations are: $x' = g(x - vt)$, $y' = y$, $z' = z$, $t' = g(t - vx/c^2)$, where $g = (1 - v^2/c^2)^{-1/2}$. Within two years of their publication, Hermann Minkowski had given SR a geometric interpretation as a four-geometry called spacetime.

thropomorphic relic produced by the fact that most of human experience is confined to relative velocities far less than the speed of light. Thus in an ironic way SR accounts for what we think of as one of the most basic features of the world, the present, even while reducing it away as an illusion.[337]

(b) SR provides a theoretical explanation of the (in)famous equation for the interchangeability of two of matter's fundamental properties, energy e and mass m: $e = mc^2$. In classical physics, energy and mass are unrelated and separately conserved properties. Like the bulk effects of QM in the macroscopic world, the transformation of mass to energy that makes sunlight possible contradicts the claim that SR is irrelevant to the ordinary, macroscopic world.

(c) Physical phenomena involving electromagnetism occur throughout the ordinary, macroscopic world. This in turn means that SR is of crucial relevance to ordinary phenomena even when they are characterized by velocities $v \ll c$, contradicting the claim that SR is irrelevant to our everyday experience. SR, through its intrinsic relation to electromagnetism, is also crucial to biological evolution in a variety of ways. Electromagnetism accounts for such phenomena as the Earth's magnetic field, lightening, the electroflourescence of primitive fish in the Earth's deepest oceans, the magnetic homing patterns of migrating birds, and of course vision from its primitive forms in light-sensitive multicellular creatures to the human eye.

Special relativity in combination with QM provides a theoretical explanation of nuclear fission and fusion. These are crucial for biological evolution but inexplicable in the framework of classical physics. The splitting of the atom and the fusion of nuclei are quantum mechanical processes, and SR accounts for the conversion of mass into energy during these processes. Through nuclear fusion the Sun transforms 4 billion tons of hydrogen into helium every second; the mass lost in the process is transformed into photons and carried away as sunlight. Sunlight, in turn, heats the Earth and is a basic source of energy for the biosphere as it is transformed into sugars by the photosynthetic biochemistry of green plants. The differential between the temperature of sunlight flooding the Earth and the temperature of the microwave radiation being irradiated away by the Earth helps make possible the evolution of complex biosystems. Here organisms, as open thermodynamic systems, can be driven to states of higher complexity by dumping entropy into the environment. Nuclear fusion in stars that undergo (super) novae explains the origin of all of the heavy elements (starting with carbon).[338] Elements such as iron are required for the formation of planetary systems like the solar system, while carbon, oxygen, and nitrogen, together with hydrogen, are essential to the organic chemistry of life on Earth. Nuclear fission, together with the residual gravitational energy left over from the formation of the Earth, continues to heat the core of our planet, producing the magnum that generates

---

[337] Put more precisely, the relativistic topology of the lightcone and the classical topology of time as past-present-future are entirely distinct; it makes no sense to see the latter as a "topologically limiting case" of the former.

[338] In the first 3 minutes following the big bang, hydrogen was fused into 25% of the helium now found in the universe. The remaining 75% was produced by nuclear fusion in stars.

the Earth's magnetic field and that erupts as molten lava in volcanoes to terraform the Earth's surface. All of these contribute to natural selection by the changes they produce in the environment.

4.3 Newton's Theory of Gravity

Newton's theory of gravity involves a force proportionate to the product of the masses of gravitating objects and falling off as the square of the distance between them. It is scaled by the natural constant G: $f_g = Gm_1m_2/r^2$ where m_1 and m_2 are the interacting masses separated by a distance r. G is one of the fundamental constants of nature. Newtonian gravity is the basis of the classical model of the solar system, including the Keplerian motion of the planets and their moons, the asteroid belt, and the comets, all of which gave final vindication to Copernicus's paradigm over that of Ptolemy. It is a ubiquitous factor in the physical and biological history of the Earth and the evolution of life. A striking example of the intrusion of a catastrophic factor into the evolution of life that is specifically due to gravity is the extinction of the dinosaurs at the K/T Boundary—the transition from the Cretaceous (K) period of the Mesozioc era to the Tertiary (T) period of the Cenozoic era—65 million years ago due to an enormous asteroid that collided with the Earth. On a more routine level the force of gravity accounts for the flow of rivers to the ocean, the aerodynamics of the flight of birds and insects, the indirect proportion between atmospheric density and altitude, the tides of the ocean, meteorological phenomena such as tornadoes and hurricanes produced by "fictitious" gravitational Coriolis forces, and tsunamis, which result from undersea earthquakes. The value of G determines the intensity and resulting implications of all these phenomena.

5 Further Developments in Twentieth-Century Physics and Cosmology

5.1 General Relativity and Cosmology

In 1915 Einstein published his General Theory of Relativity (GR), which was quickly shown to account for several anomalies in Newton's theory of gravity.[339] In GR, the flat pseudo-Euclidean spacetime of SR is curved by the presence of matter, and matter then follows a curved (geodesic) path through spacetime—thus accounting for the orbits of the planets without introducing the concept of gravitational force. General relativity is based explicitly on SR (it is a relativistically correct theory of gravity), and thus it incorporates the natural constant c as well as the natural constant G.

When applied to astronomical data from distant galaxies in the first half of the twentieth century, it became clear to most cosmologists that the universe itself is expanding and that GR can provide the most successful models for this expansion. By the mid 1960s, the resulting big bang model had come to dominate contemporary cosmology with its prediction that the

---

[339] David Hilbert published the same field equations at approximately the same time, and there has been some controversy over who actually should be credited with their discovery. See for example Corry, Renn, and Stachel, "Belated Decision in the Hilbert-Einstein Priority Dispute," 278.

universe began in an "essential singularity" referred to as the beginning of
time, t = 0 (currently estimated to be some 13.7 billion years ago).

5.2 Inflation and Quantum Cosmology

Since the 1980s, inflationary and quantum cosmologies have been developed to solve some of the lingering technical issues in standard big bang
cosmology, including the problem of the essential singularity. Today these
and other scenarios depict the universe as being one of an infinite set of
domains in a mega-universe, and all of this as emerging not out of an initial singularity but out of a prior superspace of quantum universes subject
to eternal inflation, or perhaps as one of many in universes in a multiverse
of parallel worlds, and so on.

Of course these recent models are highly speculative. What is crucial
for our purposes is that whatever the origin of our universe, the values of
the natural constants seem to apply to it, and the form of the fundamental
laws of physics as we know them certainly apply to our universe after its
initial inflationary period of 10^{-36} to 10^{-32} seconds.[340] If we move further
back in time "behind" the era of inflation we come up against a time which
appears to represent a fundamental limitation in our ability to make physically meaningful statements about the early universe, the Planck time,
$t_{Planck} = (Gh/c^5)^{-1/2} \sim 10^{-43}s$. We simply cannot project back further using
GR, since it is here that GR breaks down and a quantum theory of gravity
is required. In the opinion of many scientists, the question of whether there
was an essential singularity, t = 0, as predicted by GR, is formally
"undecideable."[341] Nevertheless, even in the speculative versions of
quantum gravity that are currently being explored, the natural constants,
at least, seem still to apply and thus these—and to a more limited extent
the fundamental laws of physics as currently understood—shape our
scientific description of the universe and, in turn, our theological formulation of the physical and cosmological preconditions for the possibility
of natural evil, or what I have called "cosmic theodicy."

5.3 Unification of the Fundamental Forces and Its Relation to Cosmology

Along with the development of big bang, inflation and quantum cosmologies, research in atomic and subatomic physics during the twentieth century has led to fundamental theories of particle physics that are now seen
as directly involved in our understanding of the very early universe. Quantum field theory (QFT) was created by Paul Dirac in 1927 as a relativistically correct treatment of QM. It eventually resulted in quantum electrodynamics (QED) in the 1950s. The weak nuclear force that accounts for
radioactivity was unified with the electromagnetic force (1961–1967) by
Sheldon Glashow, Steven Weinberg, and Abdus Salam, and the electroweak interaction was unified with the strong nuclear force that binds the

---

[340] During this incredibly short period in the very early history of the universe,
the size of the universe may have increased by a factor of between 10^{20} and 10^{30}—
remarkably, a much larger increase than the increase in size from then to now!

[341] See Barrow, *Impossibility*.

nuclei, resulting in the standard model of particle physics in the early 1970s. To date, however, the unification of the strong and electroweak force with gravity has proved to be an outstanding challenge; string theory is just one current approach to this goal. This final unification, and its application to cosmology, remains at the edge of current theoretical research in fundamental physics.

5.4 Additional Natural Constants

In addition to c, h, and G, there are other natural constants which we might eventually consider. One attractive candidate involves the lowest masses in each class of elementary particles. Here we can define the ratio of the mass of the electron, m_e, to the mass of the proton, m_p: $b = m_e/m_p \sim 1/1836$. Of course these masses may in fact not be fundamental but instead derivable from theories such as string theory. Another frequently discussed constant is the fine structure constant a which is actually a combination of h, c, and the charge of the electron, e: $a = e^2/hc \sim 1/137$. It gauges the coupling strength between photons and electrons. We will see below (section 7) how these constants also play a role in characterizing our universe.

5.5 "Surplus" Ramifications Regarding Natural Evil

The fine-tuning of the constants and the form of the fundamental laws that make the evolution of life and moral agents possible lead to additional or "surplus" examples of natural evils (and moral evils) as unintended consequences of God's choice. The following is a partial list of these surplus evils:

(a) h together with c: fusion is the basis for the hydrogen bomb and fission the basis for the atomic bomb. Their horrific consequences were visited sixty years ago on the cities of Hiroshima and Nagasaki, the beginning of the age of nuclear weapons.[342]

(b) h: QM can be interpreted in terms of ontological indeterminism: according to the Copenhagen school, the world is not seen as fully deterministic. This, in turn, is a key part of the argument that the choices of sentient creatures capable of some degree of (incompatibilist) free will can be acted out somatically—something which is not possible in a Laplacian world of strict ontological determinism. For humankind, the possibility of freedom provides one dimension of the basis needed for ethical and moral responsibility.

(c) h makes noninterventionist objective divine action (NIODA) possible and thus a robust form of theistic evolution[343]—yet this again circles

---

[342] I have given an extended treatment in a recent article for an ORC conference in Kyoto remembering the 60th anniversary of the bombing of Hiroshima and Nagasaki. See Russell, "Religion and Peace in the Nuclear Age."

[343] NIODA-based divine action is robust in comparison with Arthur Peacocke's early version of theistic evolution that could easily collapse into statistical deism: God was said to work through both law and chance, but chance was really a shorthand for epistemic ignorance, i.e., "an unknown deterministic law." See for example, Peacocke, *Creation and the World of Science,* and see my comments above about theodicy and QM-based NIODA, n. 5.

back to raise the issue of natural theodicy: if God creates life by acting within the processes of genetic mutations (no need for ID!), what is God's relation to the estimated 4000–6000 gene-based human diseases, given that God acts within the processes that result in genetic mutations? These diseases include cancer, heart disease, diabetes, Alzheimer's, Parkinson's, HIV/AIDS, and multiple sclerosis.

6 Summary

The preceding discussion leads us at last to "ground zero": natural theodicy shaped in terms not of evolutionary biology but of fundamental physics, big bang cosmology, and the constants of nature. The Ruse-Dawkins exchange illustrates the kind of challenge/response to natural evil that one might give if one were considering the level of biology. Since God's only choice in creating life by natural means is to create through the requirements of evolutionary biology, that is, variation and natural selection, the consequences of this choice—natural evil—are unavoidable. This means that biological and physical evils will be constitutive of life and *not* the consequences of a primordial human choice (see Genesis 2–3 and its appropriation by Augustine in the traditional free-will defense theodicy).

But neo-Darwinian evolutionary biology is not a fundamental theory. Although it is not epistemologically reducible to classical physics, it nevertheless presupposes classical physics, as well as quantum mechanics—e.g., genetic mutations—while going beyond the categories of both classical and quantum physics as an emergent discipline. This means that *in principle* any biologically framed response to natural evil must ultimately be an insufficient response to natural evil. In short, God's choice is not grounded in the stand-alone requirements of evolutionary biology. Instead these requirements themselves presuppose the underlying laws of classical physics, and so, then, must natural theodicy.

It is ironic that, while classical physics does seem to provide a portion of the foundations for evolutionary biology in terms of its account of the ordinary macroscopic world, classical physics, and the macroscopic world with which it works, are ultimately an epiphenomenon of the underlying microscopic world described by QM and SR and shaped by the cosmology described by GR. More to the point, QM and SR produce the classical world while snatching it away. They give as patently phenomenological what they then undermine as merely epiphenomenal. Instead of the understandable world of the macroscopic, with its populations of animals, its environment of forests, mountains and streams, its changing weather and its predator-prey cycle, we are led inevitably to consider the almost unimaginable world of QM and SR and the twentieth-century expanding cosmology of GR.

Here we find candidates for the status of truly fundamental laws and constants of nature. Yet this comes with a price: gone is the ordinary world of animals and plants, planets and suns, universal and sufficient efficient causality[344] and the unambiguous, global present demarcating past from future. Instead the ontology is one of wave-particle contradictions and nonlocal, nonseparable, and nonpicturable structures, perhaps even strings in multiple dimensions beyond mere 4-spacetime. This, however, is the world-

---

[344] Here I am referring to one version of the Copenhagen interpretation of QM.

view we are finally driven to in our search for the foundations in nature for the problem of natural evil. This is the worldview in which the posing of the problem of theodicy must ultimately take place. It involves the language and worldview not of the ordinary world and its biological realities such as suffering, disease, death, and extinction, but the austere and almost unfathomable language of fundamental physics. Here theodicy takes on a starkly inanimate and cosmological perspective which I am calling "cosmic theodicy."[345] How do we respond to what in my introduction I called the first challenge posed by cosmic theodicy, which we can now sharpen as follows:

> Did God have a choice in the values of the constants of nature (c, h, G, etc.) and the form of the fundamental laws of physics (QM, SR, GR, etc.) if God's intention in creating the universe was to create a universe in which life could arise by divine action in and through natural processes?

7 A Bit of Light to Shed on the Problem, and Then Darkness

There may be a way to shed a bit of light on the seemingly intractable problem of cosmic theodicy framed, as has been done above, in terms of the natural constants and laws of nature. The idea is to start by considering possible universes with different values of the constants of nature. Next we might consider universes in which the "constants" of nature vary in time, or universes with different laws of nature, or even more exotic possibilities: universes with multiple temporal dimensions, and so on. Here I will focus mainly on the first option and only briefly touch on the others.[346] The purpose is to use this discussion to ask two closely related questions: (1) How much leeway did God have in choosing the fundamental constants? That is, would slightly different values of the constants have lessened the problem of natural evil? (2) Are there completely different sets of these values that would both allow for the biological evolution of life and that would have also lessened the extent of natural evil?

---

[345] To capture the enormity of this transition from the ordinary framework in which we think of theodicy to the new framework, I would cite a somewhat similar question, but far less challenging in the difference in mindset it requires than cosmic theodicy: Suppose we were transported back in time to the K/T Boundary 65 million years ago and were witnesses first to the predator-prey world of the dinosaurs, and then to their demise resulting from the collision with the asteroid that caused the first true "nuclear winter." Would we have been able to formulate a doctrine of creation that affirmed that "all that is is good" and thus a theodicy that needed to account for the contradiction that is natural evil? Would we have been able to avoid a thoroughly Manichaean view of the world that Augustine overthrew with the help of neo-Platonism and finally of Genesis? Our situation is much harder to imagine and much more stark: what would our doctrine of creation be, and thus our view of natural evil as a surd against its template, if we take the doctrine of creation to be a theological interpretation of the most fundamental laws of nature and its natural constants, and the ontology of elementary particles and fields as a gauze over a seething multidimensional universe of strings and possibly an endless multiverse?

[346] For an excellent overview and detailed references see Barrow and Tipler, *Anthropic Cosmological Principle,* chap. 4. For a readable and recent introduction, see Barrow, *Constants of Nature.*

7.1 Question 1: The Implications of Variations in the Values of the Constants of Nature on Cosmic Theodicy

A helpful way to consider the first option—variations in the values of the constants of nature on cosmic theodicy—is to draw in an unusual way on the anthropic principle (AP). In order to do so it will be helpful to present a brief overview of the AP.

7.1.1 Brief Overview of the Anthropic Principle (AP)

As is well known, there are two forms of the anthropic principle (AP) as first published by Brandon Carter in 1974.[347] The "weak AP" focuses on why we exist in *this* time and *this* place in the history of the universe. The usual answer is that the biological evolution of life, spelled out in detail, could clearly not have occurred in an earlier epoch of the universe. But the "strong AP" is much more pointed: it states that our universe must be such that the evolution of life can occur. In particular, the values of the constants of nature must be "fine-tuned" for life: they must have precisely the values they actually have within a very small margin of error (typically one part in a million or even a billion) for it to be possible for life to evolve in the universe.

This fact then raises the obvious question, why is our universe fine-tuned for life? One answer is then God: that is, the fine-tuning leads to a design argument for the existence of God as Creator of the universe *ex nihilo* with the purpose of creating life through natural means (i.e., through the secondary causes described by neo-Darwinian evolution). Nancey Murphy and George Ellis have expanded this argument in a brilliant move to claim that we should seek to explain not only the fact that life has evolved in the universe but that intelligent life, capable of moral behavior, has evolved, and that this fact provides even further understanding about the nature of the God who is our Creator.[348]

The operative assumption here is that our universe is the only universe. In response, the proponents of the many worlds argument (MW) claim to offer evidence that ours is not the only universe; instead there are many universes, perhaps an infinity of universes, each with a different set of values of the natural constants. If this were proven, then the fine-tuning of our universe would be a mere tautology: we have evolved in that particular universe which is capable of our evolution. No need for God, at least not to explain the "design" of the natural constants. More complex MW arguments can be constructed in terms of the laws of nature, but they are more difficult to mount and will not be discussed here.[349]

Purported evidence for MW comes from various recent approaches to research on "multiverses." One approach sees our universe as one of an in-

---

[347] Carter, "Large Number Coincidences and the Anthropic Principle in Cosmology."

[348] Murphy and Ellis, *On the Moral Nature of the Universe.*

[349] In 1989 I argued that the entire framework which includes both design arguments and MW counter arguments is a specific form of the juxtaposition of elements of necessity and of contingency that characterizes cosmology as such, and that this fact, too, calls for a design explanation of sorts. Russell, "Cosmology, Creation, and Contingency."

finite set of "parallel universes." Another looks at our universe as one of an
infinite set of domains in the vastly inflated universes. Still another possibility being explored is that the four spacetime dimensions of our universe
are part of a much larger set of dimensions (perhaps 11). If one or more of
these approaches gains significant support, one might conclude that the design argument for God is defeated. Given this, I take the recent article
"Multiverses and Cosmology: Philosophical Issues" by William Stoeger,
George Ellis, and Ulrich Kirchner as offering a promising counterargument
to all MW approaches. These authors give what is potentially an in-
principle argument for why an actual infinity of universes is impossible.[350]

7.1.2 Leslie's Contribution to the God/MW Debate

In order to pursue the two questions posed above I suggest we turn to the
extensive analysis of the Design vs. MW debate given by John Leslie.[351]
One of Leslie's most helpful arguments here involves an attempt to defeat
the claim that, if there are many universes consistent with the possibility
of life, then the fine-tuning of our universe is not surprising. Leslie starts
by considering variations in the values of the natural constants around
their actual values such that universes characterized by these values would
still be life-bearing. For example, for constants c_1 and c_2, we consider a
region surrounding them marked out by Dc_1 and Dc_2 such that any universe
characterized by a value of c_1 and c_2 within Dc_1 and Dc_2 is life-bearing. He
then considers the possibility that there are other quite different regions
containing values of the constants c_1' and c_2', which also characterize life-
bearing universes and which include their own small regions with allowed
variations Dc_1' and Dc_2'. He then compares the distance between the first
set, c_1 and c_2, and the second set, c_1' and c_2', with the sizes of Dc_1, Dc_2, Dc_1',
and Dc_2'. Leslie claims that all it takes to defeat the MW counterargument
is that the allowed variation in the natural constants (Dc_1, Dc_2, Dc_1', and
Dc_2') be much smaller than the distance to any other set of values of the
natural constants that would also allow for the evolution of life.[352] In other
words, if the variation in the value of the constants is small compared to
the distance between regions of constants, then the fact that other such
regions exist is irrelevant; all we need to make the design argument work
is that the region of life-bearing values of the natural constants surrounding those values for our universe is very small compared to the distance to
the next such region. Leslie then asserts that this condition does indeed
hold for the values of the natural constants for our universe, thus supporting the design argument.

---

[350] Stoeger, Ellis, and Kirchner, "Multiverses and Cosmology." In the end, however, I disagree with their argument for rejecting an actual infinity, and thus their
case against multiverses, insofar as it depends on this argument, is not convincing.

[351] Perhaps the most useful place to start is with John Leslie, *Universes*. The
first chapter is a reworking of John Leslie, "How to Draw Conclusions from a Fine-
Tuned Universe," in *PPT*. Note that based on his analysis, Leslie concludes that the
Design and the MW arguments are equally persuasive; at the same time he does not
follow the design argument to the biblical God but to a form of neo-Platonism.

[352] Leslie is known for communicating his arguments through stories and images. This one is called "The Story of the Fly on the Wall." Leslie, "How to Draw
Conclusions," 17–18.

7.1.3 Returning to Question 1

Here I will use insights from the above discussion not for the purpose of a design argument but to point to a nascent response to, or at least a line of reasoning about, cosmic theodicy. Let us return to the first of the two questions posed above: (1) How much leeway did God have in choosing the fundamental constants? That is, would slightly different values of the constants have lessened the problem of natural evil?

Based on Leslie's reasoning I suggest the following response: If as Leslie claims the domains within which c_1 and c_2 fall are much smaller than the distance to the next set of domains, then God had little choice in determining the values of c_1 and c_2. This in turn would lessen the challenge posed by cosmic theodicy (somewhat analogously to the Dawkins/Ruse "no choice" debate). But this is problematic as it implies that there is something given a priori about the mathematics governing the space of the values of the constants which God must reckon with, and this quickly leads to a Platonic interpretation of God as a demiurge, a view which Christian theology long ago rejected. I will postpone further discussion of these ideas for a latter time.[353]

7.2 Question 2: The Implications of Entirely Different Values in the Constants of Nature and the Form of the Laws of Nature for Cosmic Theodicy

Our second question involves an even further generalization of the AP-type reasoning: (2) Are there completely different sets of these values that would both allow for the biological evolution of life and that would have also lessened the extent of natural evil?

Here we consider universes in which the "constants of nature" vary in time in our universe, universes with different laws of nature, and even more exotic possibilities: universes with multiple temporal and/or spatial dimensions, and so on. Specifically, we might pursue the second question using Leslie's argument in the following way: is the amount of natural evil found in a universe characterized by c_1' and c_2' less than that found in our universe? And what about universes characterized by slight variations in the values of c_1' and c_2': are they universes in which the biological evolution of life is possible and the amount of natural evil is lessened?[354]

---

[353] In 1998 Max Tegmark published a lengthy paper that draws together the extensive research over the past two decades on the question of possible universes governed by different values of the constants of nature (and even different laws of nature, see 7.2 below). Tegmark studied variations in two constants, which I have only mentioned in passing above: the ratio of the electron to proton mass, ß, and the fine structure constant, a. According to Tegmark's analysis, an increase in ß would give us a universe in which molecules are impossible. Small increases in a would preclude nonrelativistic atoms. I believe his analysis will add in very fruitful ways to our future research on cosmic theodicy along lines similar to the implications of Leslie's work just described. See Tegmark, "Is 'the Theory of Everything' Merely the Ultimate Ensemble Theory?" particularly figs. 4 and 5.

[354] Leslie gives a brief but tantalizing reference here to a "pioneering paper by Rozental, I. Novikov, and A. Polnarev." According to Leslie, by considering various strengths for the forces of gravity and electromagnetism, they find a second tiny window of possible life-encouraging universes in addition to "our" window, the one sur-

In the late 1960s, George Gamow first suggested using astronomical observations to determine whether the constants of nature vary in time in our universe. A variety of approaches to this possibility have been pursued over the past four decades. Notably, a recent publication by John Barrow's research team has finally offered substantial evidence that the fine structure constant may be growing slightly in time, perhaps to the extent of 5 parts per ten million per billion years.[355] Moreover, in a 1998 paper, Tegmark provided a detailed analysis of the basic mathematical structures underlying the laws of physics as we know them, particular GR and QFT.[356] His analysis might help in considering other laws of physics and their implications for life in the universes characterized by them. Tegmark also studied the possibility of life in a universe characterized by different numbers of temporal dimensions (T) and spatial dimensions (D) but with the same laws of physics. Barrow summarizes his conclusions succinctly: Universes with $T \neq 1$ and/or $D \neq 3$ "are too simple, too unstable, or too unpredictable for complex observers to evolve and persist within them. As a result we should not be surprised to find ourselves living in three spacious dimensions subject to the ravages of a single time. There is no alternative."[357]

As with the first question (section 7.1), I hope to pursue these and other leads in future work on cosmic theodicy. For now we have come up against current limitations in fundamental scientific theories in physics and cosmology. Until this research is further along, a definitive conclusion is not possible, although the door is still open for a consequentialist cosmic theodicy. However, even if a consequentialist cosmic theodicy succeeds in the future, we still face the second kind of challenge to any theodicy: *is life worth the price of such extensive suffering?* It is to this second challenge that I turn briefly in the conclusions to this paper.

8 The Shift from Creation to New Creation: Conclusions and Future Research

While it is certainly worth pursuing the kinds of research questions outlined in section 7, the second challenge leads in a very different and exciting direction: namely, eschatology. This challenge requires us to relocate the problem of suffering in nature, as we always have regarding human suffering, to a new theological doctrine, one that is not confined by the universe as it is now according to science and thus the theological context of the doctrine of creation but one expanded to the universe as it will become

---

rounding the strengths as measured by us. See Rozental, "Physical Laws and the Numerical Values of Fundamental Constants"; Leslie, "How to Draw Conclusions," 209.

[355] Webb, "Further Evidence for Cosmological Evolution of the Fine Structure Constant"; for a more accessible version, see Barrow and Webb, "Inconstant Constants"; and Barrow, *Constants of Nature,* chap. 12.

[356] For example, QFT depends on three structural flows: (1) differential operators; (2) formal systems > Boolean algebra > lower predicate calculus > semi-groups > rings > fields > Abelian fields > vector spaces > linear operators; (3) sets > topological spaces > metric spaces > Banach spaces > Hilbert spaces. Here arrows indicate mathematical structures with additional axioms and/or symbols.

[357] Barrow, *Constants of Nature,* 223–24. See Tegmark, "Is 'the Theory of Everything' Merely the Ultimate Ensemble Theory?" figs. 6 and 7.

as it is transformed eschatologically into the new creation. It is redemption, not creation, that saves. It is to this project of beginning a redemptive view of nature that I now turn for the remainder of this paper.[358]

8.1 New Creation: Eschatology as Transformation

If an eschatological response to natural evil is truly required, as I believe it is, it means we should look at the universe through the lens not just of creation theology (as is routinely done in "theology and science," focusing on t = 0, the anthropic principle, molecular and evolutionary biology, etc.) and with it the problem of theodicy as posed above, but more importantly through the lens of eschatology, hints of which come from the New Testament accounts of the bodily resurrection of Jesus and the moral teachings of the early Christian communities; from the life and witness of Christian mystics, saints, and martyrs; from Christian art; from liturgy, worship, and the sacraments; from the requirements for Christian community and a life of discipleship today, and so on. In earlier papers[359] I have argued that the best way to understand the bodily resurrection of Jesus is in terms of transformation, with its elements of continuity lodged amidst massive discontinuity. The elements of continuity argue against resurrection accounts that are limited to strict discontinuity (other-worldly, spiritualized, and Gnostic views of the resurrection of Jesus as mere appearances with the body left in the tomb; many of the discussions of the Jesus Seminar), while the elements of discontinuity contradict arguments for mere continuity (inner-worldly views of the resurrection, such as the "swoon" theory or resuscitation with the risen Jesus still fully human and destined eventually to die; the "physical eschatology" of Frank Tipler, Freeman Dyson). This view of the resurrection of Jesus in turn implies an eschatology based on the transformation of the creation—the universe as it is now, with the existing laws of nature and natural constants—into the new creation.[360]

8.3 A Tantalizing Possibility: Eschatology as Illuminating Contemporary Science

I will close this brief introduction to the long-range research program ahead with one tantalizing possibility. Once we have shifted the framework for responding to the problem of natural evil to eschatology, we might find new insights that could simply not have arisen otherwise. To search for these insights, however, we first need to take one additional step by asking a highly unusual question: what implications might theology have *for* science? In previous writings I have approached this question by suggesting

---

[358] Having distinguished between God's choice as a conditional versus logical necessity (see n. 3 above), Tom Tracy then wrote: "I agree that we're not likely to be able to give an adequate answer to this last question (i.e., whether life is a good worth having at the price of extensive natural evil) without discussing eschatology."

[359] Russell, "Eschatology and Physical Cosmology"; Russell, "Bodily Resurrection, Eschatology and Scientific Cosmology."

[360] John Polkinghorne has used the term "creation *ex vetere*" to signify such a transformational view of eschatology. See for example Polkinghorne, *The Faith of a Physicist: Reflections of a Bottom-up Thinker.*

that if we place theology and science in a relation of what I call "creative mutual interaction" there can be genuine benefits for both fields from the interchange. Not only can an eschatology of transformation be constructed more fruitfully in light of contemporary science, but such an eschatology might lead to fruitful suggestions for new directions in scientific research.

In these publications I focused on the relation between time as understood in physics and a Trinitarian understanding of divine eternity. Here in the context of cosmic theodicy, my suggestion is that an eschatology of transformation might yield some interesting questions to pose to the scientific discussion of cosmic fine-tuning and to scientific reflection on the role of the constants of nature. In particular, since transformation implies the existence, in the present universe, of elements of continuity, it is possible to expect that some of the laws of physics and some of the constants of nature can be identified with, or at least related to, those elements of continuity. If so, their role as elements of continuity in an eschatology of transformation might provide a new starting point for a response to cosmic theodicy, one in which their role in nature would be "benign" and not a contributing factor to the emergence of the features of natural evil that pose the problem of cosmic theodicy. It might also provide scientific insight into the way these constants of nature serve as a truly fundamental feature of the universe, since they will be part of the universe even through its transformation into the new creation.

Similarly for discontinuity: do some of the constants (or combinations of them) play a more constitutive role in the emergence of natural evil than others? Would they therefore be elements of discontinuity to be transformed away in the dawn of the new creation? And would this then be part of what we mean by the healing and transfiguration of the world so that the biology of suffering, disease, death, and extinction, with its roots in the physics of this world, will be no more in the new creation? Finally, what light might this shed for current scientific research if we focus on the relation between these constants and those which are a more permanent characteristic of the creation as it is to be transformed into the new creation?

I hope in future research involving eschatology to pursue the potential insights that these questions about the fundamental laws and constants of nature might reveal regarding what is truly eternal compared to what are tragic, but only passing, features of our universe. Such an eschatological response to cosmic theodicy might contain additional insights for scientific research into the nature and meaning of the fundamental laws of physics and the constants of nature that are derived from the eschatological perspective in which they are framed and without they which cannot be obtained—thus reflecting the interaction between theology and science which I believe provides a valuable method for future research in both fields.

SCIENCE AND THE PROBLEM OF EVIL: SUFFERING AS A BY-PRODUCT OF A FINELY TUNED COSMOS

Nancey Murphy

1 Introduction

After the tsunami of December 26, 2004, *Newsweek* magazine devoted a page to religious responses to the event. Kenneth Woodward asked scholars how adherents of various faiths would be likely to respond to the questions: Why us? Why here? Why now? Hindus in poor fishing communities were hardest hit. "Along the coast of south India, Hindus tend to worship local deities, most of them female and far down the Hindu hierarchy of divinities. But like Shiva and other classic gods and goddesses, these local deities are ambivalent: they have the power to destroy as well as to create." Richard Davis is quoted as saying: "Relating to the local deity and cooling her anger through propitiation is more important than thinking about personal or collective guilt for what has happened."

Along the coast of Thailand and Sri Lanka, Buddhists have many weather gods to both blame and propitiate. "But when the time comes to make sense of it all," says Donald Lopez, "Buddhists will look to the idea of karma and ask what they did, individually and collectively, that a tragedy like this happened." Muslims recognize no natural laws independent of God's will, so all that happens in nature is Allah's doing. Even the destructive tsunami may have some hidden purpose. Akbar Ahmed says, "On the individual level, they have also this notion that God is testing them by taking away a child or a spouse. . . ."

Woodward writes that the example of Jesus' crucifixion has made the acceptance of suffering a deeply embedded aspect of the Christian worldview, yet the death of so many innocent children "was an excruciating test of the Christian belief that their God is a God of love." He offers no Christian answer, and concludes: "Little wonder that from Sumatra to Madagascar, innumerable voices cry out to God. The miracle, if there is one, may be that so many still believe."[361]

These positions incorporate two standard responses to suffering. One is to deny the (total) goodness of God; the other is to attribute suffering in some way or other to human sin. Indeed, the standard response throughout much of Christian history has been to count all suffering, both human and animal, as either caused by or punishment for sin. However, this strategy has become less and less credible in the modern era due to rejection of the idea of a historic Fall, whether angelic or human. The few resources developed in response to these changes, such as John Hick's account of "soul making," strike this author as still too anthropocentric, focusing as they do on good for us coming as a consequence of suffering.

This paper offers an alternative account, in which humans still play an essential part, but suffering is seen not so much as a means to good for humans but as an unwanted but unavoidable by-product of conditions in the natural world that have to obtain in order that there be intelligent life at all. This pattern of reasoning can be traced to Gottfried Wilhelm

---

[361] Woodward, "Countless Souls Cry Out to God," 37.

Leibniz's argument for "the best of all possible worlds" but enhanced by considering current understandings of the interconnectedness of natural systems, especially as seen in arguments for the "fine-tuning" of the laws of physics and the constants of nature.

In the following section I sketch my version of this argument. However, this will parallel a number of other arguments in this book, so the center-piece of my paper will rather be to attempt to set out and address objections and questions that *undermine* its usefulness for addressing the problem of evil. The questions fall across a range from being readily answerable to being utterly imponderable. As I lay out the questions I shall indicate the replies that I believe to be available.

Next I take up the claim that all that is logically possible is possible for God, concluding that the distinction between logical possibility and physical possibility cannot be drawn clearly. This discussion provides some vocabulary for classifying objections to my argument as: (1) sensible and in need of an answer, (2) meaningful, but such that we do not know how we could go about answering them—these I call "imponderable," or (3) unanswerable but essentially meaningless.

2 The Argument

On January 11, the day after the *Newsweek* piece described above, the *New York Times* Science Section ran an article by William J. Broad explaining that earthquakes like the one on December 26 are a crucial part of the constant recycling of planetary crust, which has the effect of producing a lush, habitable planet. "The advantages began billions of years ago, when this crustal recycling made the oceans and atmosphere and formed the continents. Today, it builds mountains, enriches soils, regulates the planet's temperature, concentrates gold and other rare metals and maintains the sea's chemical balance. . . . The tragic downside is that waves of quakes and volcanic eruptions can devastate human populations."[362]

So what about Woodward's questions: Why here, why now, and why us? The here and now are matters for geologists to explain. "Why us?" leads to questions of economics in two ways: The recycling of the crust is what makes fishing and coastal farming lucrative. But poverty explains why such large populations live in regions most likely to suffer natural calamities.

And where was God in all of this? God is creator of a universe intended to support life, particularly life with the capacity to respond to him in love. The tsunami was an unwanted by-product of the natural conditions that make life possible on Earth. This is but one illustration of the basic assumption on which this paper is based: The better we understand the interconnectedness among natural systems in the universe, and especially their bearing on complex life, the clearer it becomes that it would be impossible to have a world that allowed for a free and loving human response to God, yet one without natural evil.[363] This suffering is due primarily to the

---

[362] Broad, "Deadly and Yet Necessary," 1.

[363] I say "human" here, but the arguments that follow do not refer specifically to *Homo sapiens* but to any creatures with the intelligence and emotional capacities to know and love God.

ordinary working of natural causes. (For shorthand I'll say the ordinary working of "the laws of nature," although I understand the universe to behave in a law-like manner rather than as governed by natural laws.[364]) Work associated with the anthropic principle, in particular, shows the impossibility of a world with life but with natural laws and constants significantly different from our own.[365] Note that I am using "natural evil" as it is often used in the Christian tradition to refer to the suffering of animals and the suffering that nature causes for humans. Another traditional category is "metaphysical evil," which refers to limitations and imperfections of creatures. Both of these are distinguished from moral evil (sin), yet in any particular instance of suffering it is often the case that all three sorts of evil are involved.

George Ellis and I began to work on the problem of evil in our jointly authored book, *On the Moral Nature of the Universe*.[366] Our focus in that section of the book was on the nature of divine action, but the book as a whole was an attempt to argue for an anabaptist theological account of cosmological fine-tuning. So this paper is a continuation of that project. I summarize here our position as developed so far. Our purpose in particular was to argue for an account of evil consistent with a kenotic understanding of God, and to show that the major alternative, dependent largely on Augustine's work, was incoherent.

The centerpiece of an Augustinian theodicy (at least as reconstructed by Hick[367]) is the free-will defense. God created all things good, but mutable. First the angels rebelled and then Adam and Eve. Inasmuch as the whole human race was present in Adam's loins, all future humans inherited Adam's guilt. Suffering of all sorts, death, and ultimately consignment to Hell are fit punishment for so great a sin. Thus, moral evil, freely chosen, provides an explanation not only of the suffering humans cause one another, but also of suffering at the hands of nature. Hick writes:

> The theodicy tradition, which has descended from Augustine through Aquinas to the more tradition-governed Catholic theologians of today, and equally as we find it in the Reformers and in Protestant orthodoxy, teaches that all the evil that indwells or afflicts mankind is, in Augustine's phrase, "either sin or punishment for sin" (*De Genesi Ad Litteram,* Imperfectus liber, chap. i, para. 3).[368]

So it is sin that *justifies* death and all sorts of human suffering. But the Ancients and Medievals also had a causal account of how the once perfect universe came to have its "malevolent" features. In the fall of the angels, the great chain of command from God to lesser beings was broken. In that the angels were thought to play a crucial role in the governance of nature, those aspects of nature over which they have control have become disor-

---

[364] See Murphy, "Divine Action in the Natural Order," in *CC*.

[365] See the chapter in this volume by Russell and references therein.

[366] Murphy and Ellis, *On the Moral Nature of the Universe*, 243–49.

[367] I note that many have criticized Hick's account. See Tilley, *Evils of Theodicy*, chap. 5.

[368] Hick, *Evil and the God of Love* (rev. ed.), 172–73.

dered.[369] This disorder provides an account of animal suffering as well as of human misery.

In the modern period, developments in science and accompanying metaphysical and theological changes have reconfigured Christians' responses to evil and suffering. An evolutionary account of human origins contradicts the Augustinian view of sin as a heinous fall from original perfection. Hick has constructed a useful replacement for this view, with roots in the writings of Irenaeus and Friedrich Schleiermacher. In place of Augustine's account of biologically inherited original sin, Hick emphasizes the need for human development: sin is not so much the effect of mature, informed rebellion against God, but more a manifestation of the race's child-like state of ignorance and weakness. But human sin contributes to a social environment that tends to perpetuate sin. Natural evils are permitted by God in order that the world be a vale of "soul making"; moral development is only possible in a world filled with the challenges of pain, temptation, uncertainty.

Why does God require the race to go through this painful learning and developmental process? The answer is subtle but important. Virtue, moral character, cannot be instilled or implanted; by its very nature it must be acquired through a process of learning and testing. In addition, God's ultimate goal for human life is a relationship with him, and such a relationship cannot be coerced or created unilaterally by God.

There is much in Hick's defense with which Ellis and I agree, and it is consistent with an evolutionary view of the development of the human species. No biological account of the transmission of original sin is necessary, and the idea of the creation *de novo* of morally perfect free beings is dismissed as incoherent. What is lacking is a replacement for Augustine's account of disorder in nature. Hick has a partial answer: God permits these things in order to create a morally challenging environment for human development. Yet the disorders of nature seem to go far beyond what is needed for human learning. In fact, if the amount and distribution of evils corresponded with our need for challenges and moral lessons, the problem of evil would not have arisen in the first place.

Biologists in the past century have added to the intellectual problem by pointing out that animal suffering and death preceded humans' coming on the scene by tens of millions of years. Thus, Augustine's account of their suffering as a result of our sin is implausible. Their suffering cannot be justified as leading to moral development for themselves or for the human race.

Ellis and I believe that the two most important changes regarding the problem of natural evil are, first, the change in conception of the governance of the universe—natural laws in place of angels—and the consequences of this change for understanding cosmic disorder and human and animal suffering at the hand of nature. Second, there has been a widespread change in conceptions of God. In place of the impassible God of classical thought (to be affected by something outside oneself is an imperfection) we have the various kenotic accounts of God's relations to creatures. Some (such as Moltmann) emphasize the notion that God suffers with and in creatures. Others (Peacocke, Polkinghorne, Diogenes Allen) emphasize that suffering results from God's voluntary self-limitation, not only with

---

[369] Ibid., 331–32.

regard to human freedom (the free-will defense), but also with regard to other creatures.

Here the modern account of the governance of the universe and accounts of the character of God come together. The regularities in nature are necessary in order to have a cosmos (an order) at all. They are also a necessary basis for meaningful human action—we must be able to predict the effects of our acts. Yet, as by-products of those inexorable laws, both humans and animals suffer: illness, starvation, natural disasters. The problem of evil thus shifts to the question why God does not cause exceptions to the (usually necessary) laws—why not occasional interventions, overruling natural processes when greater good will come from the exception than from following the rule?

It is the general consensus among liberal theologians, as well as among scholars working in the area of theology and science, that God's action must be understood in a noninterventionist manner. I have argued that God has apparently decided not to violate the "natural rights" of created entities to be what they are.[370] This means in part that God voluntarily withholds divine power out of respect for the freedom and integrity of creatures. This means, as well, that God takes the risk and suffers the cost of cooperating with creatures whose activity violates or fails to measure up to God's purposes. This cost is accepted in order to achieve a higher goal: the free and intelligent cooperation of the creature in divine activity. This relation between God and creatures is one of God's highest purposes in creating. This mode of divine activity extends all the way through the hierarchy of complexity within the created world. Hence God cooperates with, but does not overrule, natural entities. For this reason it is useful to conceive of God as acting at the quantum level, for here direct divine action is possible without violating determinate patterns of behavior of the entities in question.[371]

The crucial consequence of this view of noncoercive divine action at all levels of reality is that natural processes will be expressions not only of God's will, but of the limitations imposed by the creaturely natures of the entities with which God cooperates. At the human level, God's action is limited by human limitations but also by free choices in rebellion against God. At the lower levels of complexity the issue is not sin but simply the limitations imposed by the fact that the creature is only what it is, and is not God.

Earlier dualistic accounts of evil have identified the material aspects of reality with evil. While Ellis and I reject this claim, we recognize that it was based on accurate observations that material beings are resistant to divine action. I emphasize that this is not evil in itself—it simply reflects the fact that the material is not divine—it is over against the divine. Yet this over-against-ness (what Simone Weil describes by a metaphorical extension of the term "gravity"[372]) is in the process of being transformed, slowly and painfully, into the "image of God." The metaphorical expression in Genesis that God creates out of the dust of the earth creatures in God's

---

[370] See Murphy, "Divine Action in the Natural Order," in *CC*.

[371] See Ibid.; and see especially Russell, "Divine Action and Quantum Mechanics," in *QM*.

[372] See Weil, *Gravity and Grace*; and for a summary account see Allen, *Traces of God,* chap. 3.

own image, creatures enough like him to relate to him, has turned out to be closer to the literal truth than many have imagined.

So the aims of God, the creation of God-relating beings out of recalcitrant matter, is achieved slowly and indirectly and painfully because the quality of relationship toward which God aims is not to be achieved by a process that violates that very quality. Therefore, the process, too, must reflect noncoercive, persuasive, painstaking love all the way from the beginning to the end, from the least of God's creatures to the most splendid. Just as sin is a by-product of the creation of free and intelligent beings, suffering and disorder are necessary by-products of a noncoercive creative process that aims at the development of free and intelligent beings.

Figure 1 is an attempt to extend the position developed here by reflecting on the necessary preconditions for creation of creatures capable of a loving response to God. It is a decision tree showing that the only path leading to loving response to God is through options that have evil as an unwanted but unavoidable by-product.

The first set of options is to create, in biblical terms, either a cosmos or a chaos; or in contemporary terms, to create a universe that is or is not law-like in its operation. The reasons why a non-law-like universe would fail to result in creatures able to relate to God are too many and too obvious to consider. The dashed arrow from "law-like" to "metaphysical evil" is based on an argument of Austin Farrer's. In considering whether any physical universe would involve evil, he writes: "let us speak of the mutual interference of systems as being the grand cause of physical evil. . . . [I]n the many and various interactions of the world there are innumerable misfits, vast damage to systems, huge destruction and waste."[373] To the question of why God permits this he answers first by asking whether a world without conflict would be better and argues that if the universe were "not a jungle of forces but a magically self-arranged garden" there would be enormous loss of vitality and splendor.[374]

More important is his argument that such a universe would not be possible. It would not be a physical universe at all. This raises the question of what is meant by "physical." From a theological perspective:

> God's personal creatures share his spirituality and answer his speech. His physical creatures express his actuality, and mirror his vital force. They are action-systems, for to act is to be; they are what they do, or what they are apt to do. . . . Physical things are physical agents. When God creates physical creatures he lets loose physical forces; and until he dis-creates them again, they will do what they will do.[375]

The interference among elementary particles is not to be lamented. However, these elementary particles combine to constitute "richer systems of being," which act "to build and perfect and maintain their own organization, seizing on the matter which suits them, and resisting interferences. It is impossible to see how strife between them is to be avoided. . . ."[376]

---

[373] Farrer, *Love Almighty,* 48–49.

[374] Ibid., 50.

[375] Ibid.

[376] Ibid., 53–54.

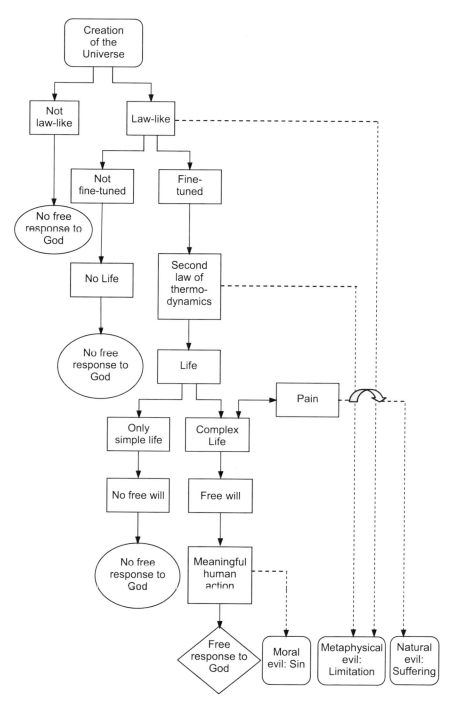

Figure 1: Reasons for Evil
Solid arrows represent choices for God. Dashed arrows represent unwanted but necessary by-products of those choices.

So, to summarize: Farrer's point is that any rich and complex world will be one in which there will be waste, damage, destruction. Should we call this evil? It certainly seems to fit the category of metaphysical evil—the competing systems will inevitably limit and damage one another.

The next set of options is based on arguments for the (apparent) fine-tuning of the cosmological constants. The value for explaining natural and metaphysical evil is worked out in detail in Russell's chapter. Prior to this sort of work it was possible to imagine that the world could have been vastly different in ways that would ameliorate suffering. It is now possible to show why gravity, for instance, *needs* to be as strong as it is relative to the forces that hold the tissues of human and animal bodies together.

As an intermediate step between "fine-tuning" and "life" I have placed the second law of thermodynamics. Russell has pointed out that it is more appropriate to include the laws of thermodynamics as an aspect of fine-tuning and has detailed a number of reasons why the second law is essential for the existence of life. One example is the Earth-Sun system. Entropy production within biological organisms is fueled by solar energy, driving the systems toward greater complexity. In fact, thermodynamics "is involved in every physical process of energy transfer in the universe, from the initial quantum mechanical production of helium from hydrogen in the first three minutes after the big bang to the astrophysical development of the sun and the planets of our solar system. . . ."[377]

Nonetheless, I have set the second law of thermodynamics apart from the other elements of fine-tuning for visual reasons. It is valuable to be able to show its particular relevance to the issue of metaphysical evil. In short, the second law "represents an inevitable limitation on the varieties of processes which could occur according to the laws of physics."[378] Thus, the effects of entropy limiting human (and animal) life are everywhere: the need for food; the need for clothing and shelter to conserve energy; fatigue; aging; and ultimately death. These limitations in human life are not moral evils but certainly provide much of the motive for sin, from instances simply of being too tired to do a good deed, to robbery, murder, and many wars. Entropy plays a major role in causing suffering as well: hunger pangs, certain forms of disease, predation.

The next set of options is a universe with only simple organisms versus one in which higher forms develop as well. I have represented the capacity to feel pain as both a consequence of complex neural systems and also as a necessary condition for the existence of complex organisms. Very simple organisms can be hardwired to withdraw from harmful stimuli, but this only allows for a simple repertoire of responses that are adequate in a narrow range of environments. For great flexibility in behavior we need *motivation* to avoid self-destructive situations. Pain (or something equally noxious, such as fear) is necessary for higher organisms' survival. And of course it is just these capacities that allow for suffering. Farrer says:

> we may boldly say that pain, and the remedial action which normally springs from it, are as vital as any functions of animal consciousness. Without them no living species above the most elemen-

---

[377] Russell, "Theological Consequences of the Thermodynamics of a Moral Universe," 20; see also the chapter in this volume by Stoeger.

[378] Russell, "Entropy and Evil," 458.

tary would have the faintest chance of survival. Pleasure is a serviceable lure, fastening attention on the continued pursuit of a wholesome gratification. Yet we can conceive a creature capable of survival, which knew no positive pleasure, only the escape from pain. We cannot view as viable a creature knowing no pain but the lack of pleasure. It would perish in a thousand deaths.

The animal capacity for flexible action, in turn, is one of the building blocks for human free will. "Free will" is inserted in the diagram in deference to the important place the free-will defense has traditionally played in Christian discussions of the problem of evil. I agree with Wildman (in response to proposal) and Tracy (this vol. sec. 3.1) that human freedom in itself is not a great enough good to justify permitting the magnitude of suffering it generates. I also agree with Hick and others, though, who argue that it is a necessary condition for a *loving* response to God.

The very mention of "free will" raises a host of philosophical problems. It is not possible or appropriate to attempt to address them all here, but it may be worthwhile to say a little about my position. The philosophical debates in the literature are stalled over the libertarian-compatibilist issue: is free will compatible with determinism or must free actions be in some sense undetermined? I have argued that determinism versus indeterminism is too clumsy a set of categories. One has to ask, determined by what? And in the case of each possibility one has essentially a different free-will problem. The most realistic philosophical worry is neurobiological determinism, yet to speak of neurobiological *determinism* is subtly to misdescribe the problem. It is not a question of whether neurobiological processes are deterministic or not, but rather whether neurobiological *reductionism* is true. And of course if the work of Arthur Peacocke and others on downward causation or whole-part constraint is valid, then all organisms can be expected to exert downward constraint on their own parts, including their neural systems.

What needs to be added to this animal agency for a serviceable concept of free will is the capacity to evaluate one's own behavior and cognitive processes, particularly in light of moral concepts, enabled by the development of sophisticated language and cultural systems.[379]

If space permitted in figure 1 I would show social structures as both a necessary condition for and a product of meaningful human action. The formation of communities was an essential prerequisite for the evolution of humans; the formation of complex societies is necessary for full human development. Social structures are the "principalities and powers" of the New Testament, necessary for human life, yet fallen in that they do not serve God's purposes adequately, and in that humans find themselves in bondage to these powers. This has a bearing on one of the theological problems with regard to free will—the extent to which humans are in bondage to sin. Thus, the diagram ought to show structural evil as an additional by-product of the necessary conditions for free and loving response to God.

---

[379] For an extended treatments, see Murphy and Brown, *Did My Neurons Make Me Do It?* chap. 7.

3 Objections

The gist of the foregoing argument is the claim that if God is to have living, intelligent, free, loving partners for relationship, then the universe God created had to be almost exactly as it is with respect to the ways in which nature produces human and animal suffering, and in which creatures are limited and subject to imperfections. Natural and metaphysical evil are un-avoidable by-products of choices God had to make in order to achieve his purposes in having creatures who could know, love, and serve him.

My purpose in this section is to consider objections to the foregoing argument of the sort: "But why could the universe not be different (and possibly better) in *this* way?" While all of the questions or objections can be phrased in roughly this form, they will turn out to be questions of a variety of sorts. So at the end I shall attempt to catalogue the kinds of questions involved and comment on their importance.

3.1 Question 1: Why Not More Special Divine Action?

One question that arises, especially given the historical context of this project,[380] is this: even if the natural world must operate in the law-like way it does, why does God not perform more special divine acts in order to allevi-ate or prevent suffering? This question can be posed in two forms: Why can God not so act as to remedy all suffering; and Why can God not act to rem-edy more suffering? The first is the easier of the two; several lines can be taken in response to it.

3.1.1 Remedies for Suffering?

(a) Hick and others' arguments known as the soul-making defense are rele-vant here. The perfection of human creatures does not take place at the species level by a natural and inevitable evolution, "but through a hazard-ous adventure in individual freedom."[381] Hick rightly regards human good-ness slowly built up through personal histories of moral effort as more valuable in God's eyes than humans with a nature created good.

I would go much further here and say that human nature created *de novo* with good moral character is an incoherent idea. Certainly children are born with more or less suitable temperaments for developing moral goodness. But the concepts of character and virtue are "past-entailing predicates"; that is, they cannot apply now if certain things have not been true in the past. Virtues are *acquired* human characteristics; the virtue of courage is to be distinguished from innate fearlessness because to be cou-rageous is to have the capacity to feel fear and yet to have developed the capacity to act in face of that fear as one knows one ought to act. Thus, Hick is correct in saying that dangers, hardships, pain, and other kinds of suffering are necessary conditions for development of the moral character prized by God.

(b) The universe must be orderly enough to allow for human responsi-bility. If the world were completely unpredictable then there would be no

---

[380] That is, following on the project regarding divine action sponsored by the Vatican Observatory and CTNS.

[381] Hick, *Evil and the God of Love* (rev. ed.), 256.

meaningful human action. Hick says that the effect of God's preventing every event that would cause pain would result in a world in which the laws of nature would not operate consistently: "sometimes gravity would operate, sometimes not; sometimes an object would be hard and solid, sometimes not. . . ."[382] In such a world there would be no point in trying to do science, and prediction of the consequences of our actions would be impossible. Furthermore, if God acted in all cases to prevent suffering (e.g., putting out abandoned campfires to prevent fawns from being burned in forest fires) then humans would have no sense of responsibility for their actions.

(c) The two foregoing replies have to do with necessary conditions for human character development and responsibility. God cannot act, consistently with his purposes in creating human life, to prevent or remedy all causes of suffering (and limitation). These are both anthropocentric arguments if we think that all of the suffering of animals, even millions of years before the appearance of hominids, is permitted in order to give us a chance to develop our full potential. I want to draw here on previous work of mine and others that takes a broader view of God's purposes in relating to the whole of nature.

The Vatican Observatory and Center for Theology and the Natural Sciences have sponsored a series of conferences on the problem of divine action: how to explain in a way consistent with science how God might act in the natural world—against the backdrop of modern liberal theology, which tended to deny any place for special divine action over and above the conservation of natural processes. Many of the participants' focus was on the question of where is there "room" for special divine action in nature. Those of us who argue for the locus of divine action at the quantum level have been accused of allowing too little scope for special divine acts. This criticism, though, becomes a strength when confronting the problem of evil. The question here is why God appears to be so *inactive* in combating evil and suffering. I believe that the question is still open as to what can be done by an omniscient ordering of quantum-level outcomes. It is important to add here whatever can be achieved by using quantum-level events to trigger or guide chaotic processes.[383] Another important factor, I hypothesize, is that God refrains from imposing unnatural behavior on higher-order, more complex creatures. It is an interesting fact that Christians (in the mainstream) have rather set ideas about what it is and is not appropriate to pray for. We pray for healings, for good weather, for divine guidance, but not for the re-growth of a severed limb or the resuscitation of a corpse. Will it turn out that the sorts of special acts we have come to expect (or not) correspond to the sorts of events that could (or could not) be orchestrated at the quantum level?

---

[382] Ibid., 306.

[383] Russell has argued that one instance in which a quantum-level action can be magnified is in genetic mutation. This raises the question, though, of why it is not the case that God prevents all harmful mutations, or the parallel question falling under the next heading, why not just a few (more) of them. See Russell, "Special Providence and Genetic Mutation," in *EMB*.

For these various reasons it is understandable that God does not act so as to prevent or remedy every case of natural evil by means of special divine action.

3.1.2 A Little Less Suffering?

The more difficult question (and perhaps the most difficult question of all) is whether God could do a *little* more to remedy suffering. This question receives attention in a variety of ways throughout the present volume. Wildman argues that if there were a personal God, God would need to do a *great deal* more for the suffering. Clayton and Knapp argue that aiding the suffering is an all or nothing affair. If God aids one, then God is morally bound to aid all and this would destroy the necessary regularity of the world. Tracy argues that we are simply not in an epistemic position to judge whether or not God is doing all that can be done. Brad Kallenberg complicates the issue by invoking the role of chaotic systems in amplifying small changes. We may think that God surely could have created one less mosquito in Minnesota without otherwise disturbing the overall plan of the universe. But "[c]haos theory at least indicates that we are bumping up against the limits of human knowing when we ask questions such as [this]" (response to proposal). Thus, I consider this an extremely important question, but one that we do not know how to answer—imponderable.

3.2 Question 2: Why Not Life of a Different Sort?

Fine-tuning arguments only show that the universe had to be this way for there to be life as we know it. Why not a different sort of life? There are actually a number of questions here. They are arranged in order of difficulty.

3.2.1 Life without Pain?

I have already argued that only the simplest of life forms could exist without some noxious signal system to deter them from injury.

3.2.2 Life without Carnivory?

Nature programs on television show us countless scenes of large carnivores pursuing and tearing apart prey. "Nature red in tooth and claw" was Alfred Lord Tennyson's answer to William Paley's account of the biological world as "miriads of happy beings." Could there not be an ecosystem without carnivores?

(a) While carnivores often attack healthy prey they also bring a swift end to the suffering of diseased and injured animals. In the case of healthy prey they are part of the system that drives the evolutionary process, removing the weaker members of the population.

(b) The brain is the most metabolically expensive organ in the body. When the body is at rest it consumes up to ten times the glucose and oxygen, for its mass, than any other organ.[384] It has been argued that humans, with their disproportionately large brains, could not have evolved without

---

[384] Deacon, *Symbolic Species,* 157.

the concentrated nutrients in meat or bone marrow from animals already dead. We can now live on vegetarian diets but it is due to improvements in agriculture and knowledge of which amino acids from vegetables to combine in order to produce complete proteins.

(c) Frans de Waal notes that in the animal kingdom as a whole, sharing of food is most prevalent in species that feed on meat. While sharing within a small family unit can be explained on the basis of kin selection, group-wide sharing is likely the result of carnivory: the food is easier to catch if group members cooperate; it is highly valued because of its concentrated nutrients but prone to decay. He writes:

> If carnivory was indeed the catalyst for the evolution of sharing, it is hard to escape the conclusion that human morality is steeped in animal blood. When we give money to begging strangers, ship food to starving people, or vote for measures that benefit the poor, we follow impulses shaped since the time our ancestors began to cluster around meat possessors. [385]

3.2.3 Biological Life of an Entirely Different Sort?

Many of the arguments for fine-tuning are arguments regarding the necessary conditions for the evolution of carbon-based life. Could there not be life of a different sort, such as silicone-based life? Barrow and Tipler note that biologists believe carbon-based life to be the only sort that could arise spontaneously. [386] John Leslie responds with the point that small changes in the fundamental constants such as force strengths, particle masses, and Plank's constant "would have meant the total absence of 'nuclei, atoms, stars, and galaxies': not merely slight changes in the cosmic picture but rather 'the destruction of its foundations.'" [387]

3.2.4 Nonbiological Life?

Leslie considers life based not on chemistry but on the nuclear strong force or gravity and concludes that neither could produce elements as precisely positioned as the electrons whose positioning is crucial to our genetic code; furthermore, if the basic constants had been much altered the cosmos either "collapses in a thousandth of a second or flies to pieces so quickly that there is soon nothing but gas too dilute to become gravitationally bound." [388]

Other sorts of speculation are Freeman Dyson's suggestion that intelligence could survive indefinitely in an enormously dilute universe, [389] and Tipler's hypothesis of very fast ticking intelligent life in the final instants of a collapsing universe. [390] Russell replies that such accounts reduce life to

---

[385] de Waal, *Good Natured,* 146.

[386] Barrow and Tipler, *Anthropic Cosmological Principle,* 3.

[387] Leslie, *Universes,* 52.

[388] Ibid., 53.

[389] Dyson, "Time without End."

[390] Tipler, "The Omega Point Theory," in *PPT.*

nothing more than intelligence.[391] This would be a defeat of God's intention to produce creatures who could respond in love. In addition, both proposals have been questioned on scientific grounds.[392]

3.2.5 Why Not Life of an Entirely Nonmaterial Sort?

This question includes two possibilities. The first is: could we not be bodiless souls, as many dualists suppose?

(a) Soul-making arguments are relevant here.

(b) Brian Hebblethwaite, in a paper on Austin Farrer's theodicy, argues that it is not clear that we can even imagine the creation of finite persons through anything other than "rooting them in, and drawing them out of, a physical universe of law-governed energies and forces that can and do at the same time cause so much harm."[393] Hebblethwaite notes that Farrer gave different answers, sometimes saying that we are unable even to ask such a question; other times categorically rejecting the notion of persons not rooted in physicality.

A second question is: could we not have been created angels?

(a) If this were the case, then it would mean that there simply are no such things as human beings. So it is a question of whether it would have been better to create a universe with angels but without humans. Farrer replies as follows:

> The problem of evil in any form only arises if we are inclined to believe in God, and in his goodness. For it is only then that we are moved to ask why, being good, he allows evils to multiply in his creation. Now we should not believe in the goodness of God, unless we were ready to acknowledge our existence as a blessed gift; and our existence is inseparable from its context, the world in which it is physically rooted. Believers must be glad to be, and to be in the world; they cannot, therefore, ask why God has done so ill as to make a world essentially of this kind. We could only wish the world had been made otherwise, if we could wish to be creatures of another sort. But we cannot; we want to be ourselves; better men, no doubt, and happier, but still men. We love our physical being: we do not want to be angels; and to be human is to be active in the world we know.[394]

(b) It is important to counter with the question of whether angels are in fact conceivable. Are there really such things? Religious traditions have long described a world full of "spirits"; there is the Greek notion of the Great Chain of Being, with gradations from God to lesser spiritual beings, to human souls and then to material entities. The problem of how such beings could interact with matter is well known. A more important question is how they could interact with one another. How to make sense of their having location? What are the criteria of identity to distinguish one spirit

---

[391] Russell, response to paper proposal. See also Russell, "Cosmology and Eschatology."

[392] Russell, response.

[393] Hebblethwaite, "God and the World as Known to Science."

[394] Farrer, *Love Almighty,* 57.

from another and to re-identify one later? This is an instance of the question of whether a nonmaterial being other than God is intelligible. (See also Question 3.)

Farrer raises comparable questions:

> We shall begin by attacking the assumption that it is possible to begin with archangels. It may be right to think that God would gladly create beings as near as possible to his own perfection; but then they must be *beings,* with, as we say, a soul of their own, and an action which they exert of themselves. . . . What can their minds be, but mirrors of the divine thought; or their actions, but executions of the sovereign will? There is no real creaturely core to such a manner of being. Is not the idea of an archangel, thus conceived, the idea of a contradiction?[395]

3.3 Could God Not Have Created an Entirely Nonmaterial Universe?

This is a question far out on the margins of answerability. On the one hand, the intuitions of countless generations point to a positive answer. A negative one goes very much against the grain of most religious traditions, early Hebraic thought being an exception. On the other, William Stoeger has said that the two metaphysical categories Christians need are God and *non-God.* Hebblethwaite (again reporting Farrer's position) says:

> In order to have finite personal creatures, with a being and nature of their own, the Creator has to place a kind of screen between his infinite glory and their creaturely selves. The material, evolving, universe is just such a screen, begun with the most elementary organisation of energy, and gradually built up, level by level, till rational and personal beings emerge, thoroughly rooted and grounded in what is fundamentally other than God.[396]

I take a radically dualistic view of God and creation—rejecting the Great Chain of Being. God is spirit and nothing created is spirit. The world, at the beginning, is in most ways the antithesis of God: chaotic, dumb, incapable of relating to God. Is it possible that *non-God* is necessarily matter? This is a question that is not only unanswerable, but what I shall call an imponderable question in that we do not even know what it would take to answer it. Could there be any grounds for choosing between these two sets of philosophical intuitions, and if so, what would they be?

3.4 Question 4: Could There Be a Material Universe of a Radically Different Sort?

The fine-tuning arguments assume that matter is (roughly) as current physics tells us. Is this the only possibility? For example, might there be a universe made of atoms as Plato conceived them? This possibility may already be ruled out by fine-tuning research—insofar as the possibilities are

---

[395] Ibid., 64.

[396] Hebblethwaite, "God and the World as Known to Science."

imaginable—but if one asks about some as yet unimagined sort of matter, then the question becomes imponderable.

3.5 Question 5: Why Not the Eschaton Now?

If Christians are looking forward to a new creation, the eschaton, understood as a radically transformed cosmos no longer subject to the laws of nature, why not create it this way in the first place?

(a) Soul-making arguments are relevant here. Earlier I made use of the concept of *past-entailing predicates,* and argued that accounts of moral goodness in terms of virtue and character are past-entailing. One can only be virtuous by having become virtuous. Thus our moral status in the eschaton is dependent on a previous life in which it is earned. Hebblethwaite makes a related point regarding the supposition that a human being could have been created five minutes ago with a "remembered" unreal past:

> 'human beings' posited in being five minutes ago with built-in 'memory' traces would not be human beings. The suggestion is logically incoherent, if we take logic to include factual necessities as well as analytic ones. The point being made by both Farrer and [Saul] Kripke is that a real history of formation and growth, and a real history of interpersonal relations are internal to what it is to be a human being.[397]

3.6 Question 6: If This Is the Best of All Possible Worlds, Then Better Not to Create at All?

If the foregoing responses to objections are valid, and if this is indeed (in the relevant sense) the best world possible, then the question arises of whether it would have been better for God not to create a world at all. This is the question that takes center stage in the latter part of Russell's chapter.

(a) An answer to this question requires attention not only to the amount of evil and suffering in the world, but also to God's responses to suffering and evil. I suggest that nearly the whole of Christian doctrine can be understood under the heading of God's responses to evil. Understandings of salvation that focus on forgiveness of sin and blessedness in the next life are much too narrow. Figure 2 lays out the shape of a theology that understands salvation broadly as including all of God's responses to evil. So I concur with Russell that a robust eschatology is a critical component but even that is too narrow an approach.

(b) Finally, even after all of this is taken into account, we are left with an imponderable question: Where could one stand epistemologically to judge that such a world is better or worse than no world at all? Can anyone ever judge even for a single individual that no life is better than a miserable life? Here the distinction between theodicy and defense is relevant. The only source of confidence we can have here is to trust that a good God can and has judged it better to create than not. Such an answer would be circular if used in a theodicy, but makes a reasonable part of a defense.

---

[397] Ibid.

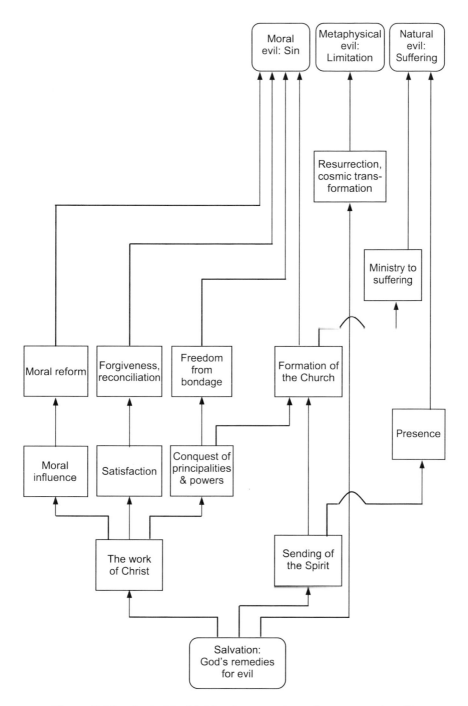

Figure 2: Theological loci laid out as a system of responses to evil.
In contrast to Figure 1, this is not a decision tree; arrows are additive
rather than options.

4 In What Sense "Necessary"?

The argument of this paper is to the effect that if God were to have creatures with whom to relate, then the typical causes of suffering (apart from human actions) are unwanted but necessary by-products. Kallenberg raises the question (in response to proposal): unwanted by whom? and unforeseen? Presumably unwanted by creatures, but also, presumably, foreseen and unwanted by God. I do not believe that it is unwarranted anthropomorphism to attribute sorrow to God for the foreseen suffering of creatures. This is an aspect of the biblical God that can be recovered with a kenotic doctrine of God.

Tracy claims that the project of this paper requires sorting out physical versus logical necessities, at least in a provisional way, since only the latter bind God (response to proposal). My response is to say that physical and logical necessities do not so much need sorting out as "fuzzy-ing up." Richard Mason has done just that; in *Before Logic* he devotes one chapter to the distinction between logical and physical (ordinary) possibility, and another to necessity. I shall report on his work here both for present purposes and because it will offer resources for sorting the questions in my foregoing section into types.

The general purpose of Mason's book is to show that logic has a historical dimension. Some of the problems internal to contemporary logic have their origins in philosophical moves made centuries earlier.

> When we talk about concepts of inference or necessity we tend to mean *our* concepts, and when we talk about our concepts we tend to mean our concepts *now*. . . . A central strand in the argument will be that some—apparently purely logical—notions acquire their senses within specific frameworks, and that some of the apparently purely logical problems surrounding them tend to arise when those frameworks are removed or denied. Both logical possibility and necessary truth will be approached this way.[398]

Mason argues that the concept of logical possibility owes its force to a succession of intellectual contexts, and has, in fact, outlived its history. The early modern context was theological. Descartes's whole system depended on the premise that what is conceivable (for humans) is possible, but this premise depends on two theological assumptions: that God created our minds so that our power of conception *fits* in this way, and that possibility-in-principle is equivalent to possibility-in-principle for God.[399] Conceivability (mental picturability), without Descartes's theology, continued throughout much of the modern era to serve as the criterion for logical possibility. The link between representation and possibility reached its clearest extreme in Ludwig Wittgenstein's *Tractatus,* but here the focus shifted to representation in language.[400] Mason describes Wittgenstein's later position as ambivalent, sometimes blurring the distinction between logical and ordinary possibility and other times saying that the essence of what is possible is laid down in language. Mason acknowledges that there

---

[398] Mason, *Before Logic,* 5.

[399] Ibid., 10.

[400] Wittgenstein, *Tractatus.*

is no inconsistency here because, for Wittgenstein, the logical possibilities in language are there because things are like that[401]—language and world are internally related.[402]

Mason concludes that the concept of logical possibility can only be characterized unambiguously in logic. If this only means within some specified logical system it is unproblematic but uninteresting. If what is meant is any *possible* logic, "we are heading for circularity."[403] Thus, he concludes that "logical possibility is . . . helpless when God disappears."[404]

Of course God has not disappeared, but Mason's reflections raise worries nonetheless. Tracy implies that God can do whatever is logically possible, but does this mean what is possible in a two-valued logic, or three-valued . . . ?

Mason shows that the distinction between logical and physical necessity is also problematic. Is necessity a characteristic of sentences or of things, events? Consider the possibility that it is a characteristic of things. Spinoza and Hume look at the same world; Spinoza sees it all as necessary, Hume as contingent. Contingency is a problem for one system; necessity for the other. How to choose?

Leibniz attempted to address the problems of necessity and contingency by applying necessity to sentences, and his is the route most have followed since. This allows for the creation of logics and modal logics, and makes the concept of necessity quite manageable, but raises the problem of the relation between necessarily true sentences and the world. Can it be that "It is necessarily true that A" (logical necessity) has nothing to do with "Necessarily A" (physical necessity)?

Two points at which these problems bear on my argument (and on others in this volume, especially Russell's and Clayton and Knapp's) are with regard to the status of the laws of nature and the applicability of mathematics to reality. The point at which this book has something to offer that was not available from Leibniz through Farrer and Hick is the use of arguments for fine-tuning. These arguments begin with physics and cosmology (as we know them so far); they involve changing numbers in the equations and then calculating the effect such a change would have on a universe otherwise the same as this. So a facile answer to Tracy's question would be that mathematical necessity is logical necessity and therefore these calculations do have a bearing on what God can and cannot do.

The attempt to reduce mathematics to logic has been abandoned. It seems more reasonable (to this nonspecialist in the philosophy of mathematics) to see mathematical systems as sets of human conventions, but conventions developed because they work in the real world. But *must* the world be this way?

There are problems enough with widely accepted views of the universe as *governed* by laws. Consider, for example, the difficulty of deciding whether the deterministic form of the logistic equation entails that chaotic systems are deterministic.[405] Many have called the ontological status of the

---

[401] Ibid., 24.

[402] Kallenberg, *Ethics as Grammar,* 218–19.

[403] Mason, *Before Logic,* 29.

[404] Ibid., 31.

[405] See *CC,* and particularly the introduction by Russell.

laws of nature into question, claiming that they are merely descriptions of regularities produced by the joint operation of God and nature.[406] This takes us out of the realm of (pure) logical necessity altogether, but does not mean that there are no limits on what God can do. It means instead that there is no neat distinction to be drawn between logical and physical impossibility.

5 On the Status of the Objections

My purpose in this section is to use some of the (messy) philosophical resources from the previous section to reflect on the varying status of the objections I have raised (in section 3) to my own argument. It seems clear that the questions pose challenges of different strengths, but I also suspect that they do not simply lay along a spectrum from easier to more difficult. I hope to show here that some are legitimate and others not.

Is the problem of suffering about the world or a problem with how we speak (and think) about the world? Suffering is real and immense, and I have provided my particular statement of why it is a necessary part of life in this world. Suffering itself is a problem in the world; we must do as much as we can to combat it. Suffering becomes an *intellectual* problem only when we think and talk about it, and then only against the background of a number of things we hold to be true, such as the existence, goodness, and power of God. How much of the rest of our knowledge needs to be in place to get the problem of suffering off the ground? My reason for asking this is to see whether some of the objections considered in section 3 above are so contrary to fact as to be not only unanswerable but meaningless.

Here, in abbreviated form are the objections:
1. More special divine action?
2. Life of a different sort?
    2.1 Without pain?
    2.2 Without carnivores?
    2.3 Different biology?
    2.4 Nonbiological?
    2.5 Nonmaterial?
        2.5.1 Souls?
        2.5.2 Angels?
3. A nonmaterial universe?
4. A universe of radically different material?
5. The eschaton from the beginning?
6. Better no universe at all?

Questions 1 through 2.3 leave enough of our knowledge of the world intact so that we can have some sense of the state of affairs in question and evaluate its possibility (and preferability). Questions 2.4 through 4, however, call for a blank slate with regard to our general knowledge of reality. Recall that concepts of logical possibility went through stages: early moderns required that the situation be imaginable; the early Wittgenstein required that a proposition describing the situation be say-able—a well-

---

[406] E.g., Murphy, "Divine Action in the Natural Order," in *CC*; Russell, "Divine Action and Quantum Mechanics," in *QM*.

formed formula. This set of questions comprises well-formed formulas, yet (purport to) describe situations that are so different from anything we know that they are beyond our capacity fully to imagine them. Without even being able to picture the situation, we are entirely unable to form a judgment as to whether it is possible at all (or preferable to the state of this world). So the questions are not just unanswerable but what I have called imponderable because we do not even know what would be involved in trying to answer them. In short, our ability to form sentences describing a different state of affairs outstrips our ability even to imagine the state of affairs and a fortiori to evaluate it.

In fact, we believe that we can attach significance to these questions because they are in proper grammatical form and we know the meanings of the words. But if the world were as different as is proposed, then all of the referential and semantic relations would be different in ways that we cannot specify. We cannot make any more full-blown description of the state of affairs in question. Thus, I conclude that the sentences are in fact meaningless.

My conclusion, then, is that the intellectual problem of suffering can easily grow beyond the boundaries of sense, and it is important to attempt to locate that boundary so that empty words do not create an unnecessary stumbling block to faith in a loving God.

6 Postscript

The story of Adam and Eve depicts pain in childbirth, harsh conditions of labor, and death as consequences of the first sin (Gen 3:16–19). While pain, toil, and death are not caused by sin, they are not causally unrelated. Toil and death are the consequences of the finely tuned laws of physics that allow us to be here. The pain of childbirth is due to the large heads of human infants, which are necessary to house brains large enough to learn language, and ultimately to pray, and write theology—but also, tragically, to sin.

THE LAWFULNESS OF NATURE AND THE PROBLEM OF EVIL

Thomas F. Tracy

Injury, disease, pain, and death befall living things simply by virtue of their participation in the system of nature. If one thinks that the universe has been called into being by a perfectly good intentional agent who freely chooses what sort of world to make, then the ubiquity of hardship and death presents a compelling moral challenge. Outside a theological context of this kind, the generation and destruction of lives is just a striking feature of the universe, and there is no "natural evil." But theists who affirm the classical divine attributes of omniscience, omnipotence, and perfect goodness cannot avoid the question of why such a being would create a world in which living and dying are so intimately intertwined. The problem concerns not just pain and suffering (though I will often refer to it by this shorthand), but all the forms of damage done to creatures that have a good of their own, and therefore can flourish or fail within the system of nature.[407] What we are now learning about the universe suggests that an astonishingly restrictive set of conditions must be satisfied if life is to arise at all. But these conditions bring with them the necessity that the history of life will unfold by trial and error, with vast numbers of individuals being generated and destroyed. We live in a universe characterized by improbable fecundity and ever-present perishing, and as we rejoice in the former and mourn the latter, we cannot help but wonder whether it has to be this way. It will not be enough to observe that the laws of our universe necessarily make it so. If God is the source of these laws, rather than being subject to them, then it is fair to ask why a loving God chooses to create a world that has this terribly costly structure. This question plunges us into the turbulent crosscurrents of debate over the problem of evil, so I will begin with a brief explanation of my approach to this controversy.

1 Setting the Context: Explanation and Salvation

In *Paradise Lost,* John Milton suggests that theoretical discussion of the problem of evil is one of the pastimes of the damned in hell. While Satan is away on a brief but gratifying visit to the newly created Garden of Eden,

---

[407] There are fascinating debates in ethics about what it is required in order to have "a good of ones own." Peter Singer argues that sentience is a necessary condition for having interests that can be given equal consideration in utilitarian weighing of alternatives (e.g., *Animal Liberation*). Tom Regan, on the other hand, extends an analog of Kantian respect for persons to all living things that are conscious subjects of a life (e.g., *Case for Animal Rights*). Paul Taylor contends that every living thing, whether sentient or not, has a good of its own that we are obligated to respect (e.g., *Respect for Nature*). Other environmental philosophers break with what they see as the "individualism" of the received moral tradition, and argue that the basic unit that commands moral respect should not be particular living things but rather the ecosystems to which they belong (e.g., J. Baird Callicott, *Companion to a Sand County Almanac*). These debates have important implications for the shape and scope of the problem of natural evil.

some of the other rebellious angels wile away the hours discussing the great issues of philosophical theology.

> Others apart sat on a Hill retir'd,
> In thoughts more elevate, and reason'd high
> Of Providence, Foreknowledge, Will and Fate,
> Fixt Fate, free will, foreknowledge absolute,
> And found no end, in wand'ring mazes lost.
> Of good and evil much they argu'd then,
> Of happiness and final misery,
> Passion and Apathy, and glory and shame,
> Vain wisdom all, and false Philosophy. (bk. 2, lines 557–565)

Milton's sardonic wit in constructing this scene makes use of the comedy of misplaced priorities; the fallen angels energetically debate theoretical puzzles about God's nature, though their recent actions have forever excluded them from enjoying the immediate presence of God. Clearly, Milton had read enough philosophical theology to recognize the limits of what such arguments can accomplish, and his depiction of the diabolic discussion group points to the importance of not losing sight of practice in our preoccupation with theory. I what follows I will join their circle of conversation, but it is important initially to note a difference in emphasis between philosophical and theological approaches to the problem of evil, a difference that is linked to this distinction between theoretical and practical concerns.

1.1 Philosophical and Theological Approaches to the Problem

Among philosophers of religion, the problem of evil is typically discussed as an abstract general difficulty for any form of theism that affirms God's perfect power and goodness. This approach steps back from the intricately woven web of language, practice, and belief in particular religious traditions. There are some advantages to this way of framing the discussion; as we see in the sciences, abstraction and idealization promote clarity of analysis and contribute to theory formation, and a great deal has been accomplished along these lines in recent Anglo-American philosophy of religion. There are also, of course, disadvantages to this approach. Perplexity about God's relation to evil first arises within traditions of religious practice and reflection that are more complex and richly textured than is conveyed by the typical formal philosophical expressions of the problem of evil. Religious believers usually are not simply generic theists, but rather participate in one of the historical traditions that speak about God in ways not adequately captured by a stripped-down set of propositions regarding a few of the divine attributes.

There can and should be a beneficial cross-fertilization of the philosophical and theological discussions, each gaining clarity and depth from the other. But it is important to pay attention to a significant difference in interest and emphasis between the two. Philosophical discussions of God and evil focus on the question of whether we can generate a morally justifying explanation for God's creation of a world that includes the quantity and distribution of evils that we see around us. This is a matter of concern to many theologians as well, but it is not the starting point or principal preoccupation of religious reflection. Rather, each of the Abrahamic traditions is centrally concerned to articulate a vision of what God is *doing* to address the presence and power of evil in our world. First and foremost, they are religions of salvation, not explanation.

This can clearly be seen in Christianity. The good news proclaimed in the New Testament is that God has acted to liberate and redeem, not that God has offered us a satisfactory accounting of why things are as they are. These two concerns, i.e., with redemption and with explanation, are not unrelated to each other; we may hope that a more complete understanding of the "why" of things will be part of the eschatological fulfillment, the eternal blessedness, of rational creatures. For the time being, however, we and the rest of creation await the consummation of the creative and redemptive work of God. We long for both liberation and comprehension, though neither is within our own power, and it is no surprise that the promise of God's unfailing love is a matter of more urgent concern than the prospect of a fuller explanation.

A focus on redemptive action rather than comprehensive explanation is evident in the New Testament stories of Jesus, which lie at the heart of any distinctively Christian response to the problem of evil. These stories are profoundly paradoxical. Jesus appears on the scene possessing extraordinary power—over demons, over disease, over the forces of the natural world. Where he is present, the sick and the tormented are restored to well-being, the hungry are fed, the despised are returned to community, the dead live. He appears to be the very presence of life in its wholeness. Yet as his story unfolds in the passion narratives, he becomes a powerless victim, destroyed with a vicious carelessness that we know all too well as the familiar way of our world. His life is treated as something of no importance and is extinguished as a matter of course by the indiscriminate machinery of human power. Then comes the "surprise ending"; death is transformed into life. But here we find not simply a negation and reversal of his victimization, an erasure of his suffering. Rather, the stories make a point of insisting that his transformed life incorporates the wounds of his crucifixion. His suffering as a victim of evil is an integral element in the new creation that God brings out of his death.

There is much to puzzle over here, as the history of Christian theology demonstrates. Not least is the question of why we encounter in Jesus not just liberating power and life, but also powerlessness, suffering, and death. Why has God made a world so much in the grip of powers that ensnare, overwhelm, and oppress? We immediately recognize the cast of characters we meet in these stories: the sick, the impoverished, the grieving, the despised, the arrogant, the hateful, the cruel. We are told an enigmatic story of what God does about these evils, and we are called to participate in the new creation that God initiates in the midst of them. But we are not given an explanation of why these evils occur in the first place, why they persist, and why they have such prominence and power that, as the New Testament stories unfold, even the Son of God must suffer them.

1.2 Should We Refuse to Explain?

It may be tempting to refuse any attempt at explanation. It is certainly possible, and perhaps inevitable, for explanation to go wrong, as the story of Job and his disappointing comforters makes clear. Moved by compassion, they sit in silence with him for seven days. Ironically, they are never more articulate than when they say nothing and simply keep him company, for when the conversation begins, it quickly takes a nasty turn, and the friends who came to give comfort end up accusing and condemning him. This just adds insult ("you must be a very great sinner") to the injury (e.g., deaths of his children and the loss of his health) that he has already suffered as a victim of the politics of heaven, a pawn in a cruel game of one-upmanship between God and "the Accuser." The comforters are so eager to vindicate God that they endorse Job's suffering, and their theodicy of rewards and punishments only deepens Job's misery and further isolates him. He has become the bearer of an unwelcome testimony: namely, that the innocent do in fact suffer terribly in the world God has made. Job warns his friends against misrepresenting the way things are in order to make God look good, and in one of the most striking passages in the text, God vindicates Job's warning, telling the comforters, "You have not spoken the truth about me as my servant Job has."[408]

We have here a powerful presentation of the pitfalls of trying to explain the world's evils. This enterprise runs the risk of obscuring or denying the awful depths of suffering and loss, of fitting all suffering into a system of moral instruction, so that we find ourselves approving of it (since after all, it is good for you) and "taking sides" with God against the victims of evil. Given these hazards, it may seem wisest to borrow a strategy from Buddhism for dealing with troublesome theoretical matters. When pressed persistently on unanswerable questions (e.g., "What becomes of an enlightened one after death?"), the Buddha replied that seeking an explanation of these things "profits not."[409] The monk who is inclined toward excessive speculation is admonished not to waste energy wrestling with unresolvable theoretical puzzles, but rather to focus on aligning his life with the path that leads to salvation. The *practice* of liberation from suffering takes precedence over working out the *theory* that underlies it. This same point could be pressed in the Christian context, urging us to participate more fully in God's work of repairing the broken world, rather than expending our best intellectual efforts in trying to figure out how the damage was done in the first place.

As a matter of setting priorities this is surely right. But practice and theory are intimately related, and we need to attend to both. As I just noted, Christian narrative patterns quite naturally lead to a series of questions about the place of evil in the world. If we say that God acts to redeem us, then we need to ask how it is that God's good creation needs to be redeemed. Theologians from Paul's time to the present have in fact offered answers to this question, and if there are hazards in doing so, then there also are dangers in failing to take up these questions. Negatively, there is the dubiousness of a belief system that arbitrarily refuses to grapple with substantial problems of coherence and plausibility. Challenges to the coherence of Christian claims about God and evil arise not simply from skeptical "outsiders" but also from reflective "insiders" who are concerned to harmonize (to the extent one reasonably can be expected to do so) the vari-

---

[408] Mitchell, *Book of Job,* 91.

[409] "Questions Which Tend Not to Edification," in Warren, *Buddhism in Translation.*

ous elements in their web of beliefs. Positively, the failure to address these questions is a missed opportunity to extend and deepen an understanding of the internal logic of the scheme of Christian beliefs.

2 The Explanatory and Justificatory Project

2.1 Filling in the Story

So how might we go about offering such an explanation in a way that is attentive to the hazards and limits of this project? We can think of this as a matter of filling in the background story (the "prequel") that is needed in order to prepare for the action in the central Christian narratives. Put more philosophically, we are attempting to think through the conditions for the possibility that God should create such a world and act in this way within it. What does a Christian story about God and the world presuppose about the place of evil within that world? There will, of course, be a number of different ways of constructing this background story; theologians have developed a range of interpretations of the central ideas in Christian talk about God's saving action (i.e., various Christologies, soteriologies, and eschatologies), and these doctrines will have various entailment relations to claims about sin and evil and to views about God's action in the creation and governance of the world. A fully developed theological account of evil would need to work out views on all these doctrinal topics, i.e., it would be an integral part of a wider theological structure of ideas. In focusing principally on God's relation to evil we trace one thread in this fabric of ideas. We take a step back from the larger theological context, while also assuming some elements of it for the sake of argument. The God whose relation to evil I wish to explore is the God of a Trinitarian Christianity, who creates all things *ex nihilo,* and who acts redemptively within the world to move creation toward the full realization of the good that God intends. Given this conception of God, what can be said about why God creates a world that includes the sorts of evils we see around us?

2.2 Problems of Evil, and Defenses Both Thin and Thick

It has become standard practice to distinguish two forms of the problem of evil. First, the logical version of the problem asserts that one or more of the believer's claims about God (claims which can vary significantly, as in classical and neo-classical theisms and in different versions of Christian theology) are *logically incompatible* with acknowledging the facts about evil in our world. Second, the evidential version of the problem claims that the facts about evil constitute *sufficient evidence to reject* the believer's claims (i.e., reason to hold that it is more probable than not that these claims are false).

The logical objection presents the most radical challenge; if it succeeds, it would show that the beliefs in question are not merely poorly supported, but are necessarily false. It also is the most difficult to sustain, since this challenge can be decisively rebutted by showing that there is at least one logically possible set of circumstances (however bizarre or implausible) under which God and evil are logically compatible. Constructing such an account is the task of a "defense," in the sense in which Alvin Plantinga uses that term.[410] A defense has very limited purposes. It does not assert the truth of the explanation it provides for God's creation of a world containing evil, only its logical possibility. So a defense may include claims that the defender himself holds to be highly dubious or false. The religious believer who struggles with the problem of evil typically seeks more than this. She wants to know not simply that her claims about God's goodness and power might somehow (under some purely theoretical and perhaps quite peculiar set of circumstances) be consistent with what we see of evil in the world. Rather, she seeks some understanding of the way these beliefs cohere within an explanatory story that is at least a viable candidate, a "live option," for winning her assent. This story will not only need to be internally consistent, but also consistent with the network of beliefs that she actually holds. Her concern, after all, is with God's relation to evil in the actual world, and not simply with what obtains in at least one possible world.

In contrast, then, to Plantinga's logically sufficient but theologically "thin" defense, we need to wrestle with the more demanding project of a theologically "thick" defense, i.e., an explanation of evil formulated with an eye to its Christian assertability. Some authors apply the term "theodicy" to any attempt, however tentative, to say what God's reasons actually are for permitting evil.[411] It is useful to distinguish, however, between (1) theodicy as a thick defense that attempts to construct a story about God and evil within the context of a Christian (or other) theology and (2) theodicy as a part of a natural theology (such as a design argument) that seeks to show on nontheological grounds that theism is well supported, given the rest of what we believe about the world.[412] The former is an expression of faith seeking to articulate its own content *(fides quaerens intellectum)*. The latter is a matter of a religious hypothesis seeking epistemic vindication. My interest here is in the former, rather than the latter.

2.3 The Tasks of a Thick Defense

Any attempt to give a justifying account of God's permission of evils will need to address three key questions:

1. What is the good for the sake of which God permits or produces evils?
2. What is the relation of evils to this good?
3. It is consistent with God's perfect goodness to pursue this end in this way?

The internal structure of defenses/theodicies can be mapped by noting how they answer each of these questions. The second question dominates most analyses of the problem of evil, but the first is crucial in setting the frame of reference for the discussion. The third question, though often overlooked, recognizes that the moral challenge is not adequately addressed simply by showing that evils are necessary for the realization of a good. There is the further moral question (raised so forcefully, for example, by Dostoevski

---

[410] For example, see Plantinga, *God, Freedom, and Evil,* 26–29.

[411] See ibid. and Adams and Adams, *Problem of Evil,* "Introduction," 3.

[412] See Terrance Tilley's critique of theodicy in this second sense in *The Evils of Theodicy.*

through the character of Ivan Karamazov[413]) about the justifiability of attaining this good if doing so involves appalling suffering and the destruction of the innocent. The differences between philosophical defenses/theodicies offered on behalf of a generic theism and a thicker theological defense emerge most clearly in answering the first and third questions, where the theological tradition has additional resources available to it in its central affirmations about God's redemptive action.

3 What Is the Good that God Seeks in Creation?

3.1 Moral Freedom in a Thin Defense

What can we say about the good that God seeks in creation? Philosophical discussions of the problem of evil often appeal to moral freedom. Alvin Plantinga, for example, says that, "A world containing creatures who are significantly free (and freely perform more good than evil actions) is more valuable, all else being equal, than a world containing no free creatures at all."[414] The parenthetical phrase plays an important role here. By adding this qualification, Plantinga avoids saying that moral freedom alone is the justifying good. It is easy to see why he does this, since it would be difficult to defend the moral wisdom of knowingly causing the existence of free creatures who predominantly use their powers of action to torment each other (though this may sound altogether too much like a description of the actual world). Moral freedom may be an intrinsic good, but it apparently is not, in itself, a great enough good to bear the full weight of justifying God's permission of evils. Plantinga's additional condition, viz., that there must be a balance of moral good over moral evil, does not improve the situation very much, however. It is not clear whether this condition applies to each individual over the course of that agent's life or to the sum total of free actions across the world's history as a whole. In either case, the appeal simply to a net balance of moral good is vulnerable to the objection that it ignores the wrongness of sacrificing innocent lives simply so that others can exercise (often spectacularly faulty) free moral judgment. This approach is likely to fail, that is, in grappling with our third question.[415]

3.2 Communion with God in a Thick Defense

This is one of the points at which discussion of the problem of evil can be deepened by attending more fully to the central concerns of the theological traditions. As classically understood, the good that God seeks in creation is a great deal richer than simply that there be free creatures who make morally correct decisions. Christianity typically affirms that we are created for the sake of loving communion with God. This has a number of important implications for efforts to wrestle with the challenge of evil. First, in its fullest expression, the good consists in the inexhaustible enjoyment of an Infinite Good (understood, in classical Christian theology, as the Trinitarian life of God made available in Christ). The good of this relationship

---

[413] See the chapter titled "Rebellion" from *The Brothers Karamazov*.

[414] *God, Freedom, and Evil*, 30.

[415] I have discussed this problem in "Victimization and the Problem of Evil."

vastly exceeds in value all other goods and outweighs all evils. Second, though we experience some measure of this good here and now, its realization need not, if God so chooses, be limited to the span of our present life, but rather can be brought to completion in a life with God beyond death. The goods and evils relevant to considering the problem of evil, therefore, should not be limited to those realized in the course of the world's history as we presently experience it. Third, we cannot fully conceive of what this eschatological relationship will be, and that means that our efforts to explain evil must be constructed with reference to a good that exceeds the reach of our current understanding.

On this account, moral freedom will have a place in our explanation of evil only if it plays some necessary role in preparing us to participate in the relational good that God intends for us. The idea of moral freedom has a very complex and controversial history in Christian theology. Questions about the relation of God's gracious action and free human self-determination have been a critical area of theological debate since very early in Christian theological history. Part of what drives this discussion is a subtle recognition, in the theology of sin, of the ways in which our moral freedom has been compromised and distorted. One classical expression of this can be found in theological discussions of the "bound will."[416] We can be harmed (and can do ourselves harm) in our very capacity for moral choice, so that our passions and appetites, our attachments and values become deeply misdirected, with the result that our ability to imagine alternative courses of action and our readiness to recognize and respond to the other's good are tragically truncated. In this understanding of moral life, we are both victims and agents of evil. On the one hand we are subjected to "powers and principalities" (e.g., a received culture of values and attitudes that is not of our choosing but that shapes and distorts our choices). On the other hand, we make responsible choices about how to act in this morally dangerous world.

This is a much more complex and qualified account of human freedom than is usually found in generic philosophical free-will defenses. Those defenses focus on the abstract (incompatibilist) definition of freedom as an undetermined power of choice between alternatives. If we make the bold claim that we possess such freedom, we should also acknowledge that we exercise it under "real world" conditions that profoundly shape and constrain it. This both complicates and enriches responses to the problem of evil. It complicates them because one of the evils that the theologian must now try to explain is God's creation of a world in which moral freedom comes to be profoundly disrupted in these ways. It also enriches theological responses to evil, however, because these ideas return us to the heart of Christianity as a religion of salvation. Theological accounts of the good give a crucial place to God's acts of forgiveness and reconciliation (insofar as we are agents of moral evils) and of liberation and healing (insofar as we are victims of evil, damaged in various ways that we cannot ourselves repair). The stories about Jesus depict this action of God, and the full content and *telos* of God's action is suggested in the stories of Christ's resurrection. Here we have a transformation of the humanity of Jesus which incor-

---

[416] For example, see Rom 7:19; Augustine, *On Free Choice of the Will*; Luther, *Bondage of the Will*.

porates his suffering and makes a new creation from the old. Once again, we find that the good which God is bringing about is, in its fullness, a mystery that we cannot grasp adequately.

3.3 Nonanthropocentric Goods

God's purposes in creation obviously include more than a concern with the destiny of free creatures. So far I have mentioned only goods associated with the lives of finite persons. But a classical theology of creation affirms that the whole created world is good by virtue of its relationship with God. This is a fundamental check on our anthropocentrism. The created world was good before we appeared on the scene, and will be good after we are gone. We are only beginning to understand the history of the universe God has called into being, and we must acknowledge that we cannot comprehend the full array of values that God may be realizing within it. In approaching the problem of evil, we have no alternative but to work with the goods we know, and those will principally be the goods available to us. But we also have no basis whatsoever for supposing that the good of finite persons is the only good God seeks in creation. The good of each thing consists in its being what it is as created and sustained by God. It can be argued that wherever life appears in the universe, new centers of value (entities that have a good of their own) arise. The attempt to explain the presence of evil in God's creation must recognize this vast multitude of goods beyond the good of finite persons, acknowledging that gains and losses occur here that are a matter of moral concern in their own right, and not just as part of the backdrop for the realization of the good of persons. This will also leads us to confront, once again, the profound limits on our understanding of the overall context within which goods and evils arise.

4 How Are Natural Evils Related to These Goods?

The central move in any defense or theodicy is to show that evils of the sorts we find in our world must be permitted or produced in the process of realizing the goods that God intends for creation. Classically, natural evil has been explained in terms of moral evil. There is more than one way to do this, but the most familiar theological account contends that natural misfortune and death entered creation through the catastrophic moral fall of original humanity, perhaps anticipated by the prehistorical fall of rebellious angels. Natural evils, on this account, are a divinely imposed consequence of human disobedience, if not also a result of the malicious actions of fallen angels. If we give up this picture, and suppose that injury, disease, and death are built into the structure of the created world, how can natural evils be reconciled with the creator's perfect goodness?

4.1 Physical versus Logical Necessity

One promising strategy of argument is to contend that the various goods realized in living things can be secured only within a system of nature that includes physical vulnerability and death. It is crucially important to note that this claim can be read in two significantly different ways. First, we might contend that natural evils are a necessary consequence of the structures of nature, and that these structures (or ones very much like them) are necessary for the possibility of life in our universe. This is a claim about the physical conditions under which life has arisen, and it can be elaborated in impressive and growing detail as we learn more in the various sciences, especially cosmology, about how finely tuned these conditions appear to be. Second, we might argue that the goods God intends in creation logically cannot be realized without allowing for hardship and death. This is a matter of conceptual rather than causal necessity; that is, it claims that the permission or production of some evils is among the logically necessary conditions for the realization of these goods.

This familiar distinction between physical and logical necessity has an important role to play in considering the problem of evil. It will not be enough to construct a defense of God's permission of evil by appealing to physical necessities that describe the structure of nature in which we find ourselves. God is not bound by these necessities, but rather is their source. The canvas on which God paints in the act of creation is not defined by the space of all physically possible states of affairs (given the laws of nature) or even the space of all consistent sets of natural laws, but rather the space of all logically possible states of affairs. God could create worlds that are not structured according to any set of laws or worlds that are perfectly deterministic or worlds that include puzzling combinations of determinism and chance (like our own). God could create the world described by some conservative Christian critics of evolution, i.e., a world just like ours that looks very old but in fact sprang into existence 6000 years (or five minutes) ago. Again, God could create the world envisioned by Job's friends, a world in which goods and evils are distributed according to a principle of moral retribution. And God could create the peaceable kingdom, in which the lion lies down with the lamb, presumably after a meal provided in some other way than through predation. The only basis for denying that God could do any of these things is either (1) that they are logically impossible or (2) that they are incompatible with God's nature and purposes. So can we make the case that the good that God intends in creation logically requires that natural evils be permitted or produced?

4.2 Free Process Defenses

One approach contends that God creates an order of natural causes and respects the integrity of that order by allowing it to operate according to its own immanent lawful structure with little or no divine intervention that would disturb its causal history. On this view, creation involves a particular sort of divine *kenosis,* a self-limitation or restraint in the uses of God's power that is the concomitant of empowering creatures to operate as causes in their own right within a structure of lawful relationships. John Polkinghorne calls this the "free-process" defense, suggesting a parallel to the free-will defense.[417] Free-will defenses, as we have seen, claim either that (1) freedom itself is an intrinsic good of sufficient magnitude to outweigh all the evils that accompany it, or that (2) freedom is a necessary condition for some further good. A free-process defense might make claims parallel to

---

[417] E.g., Polkinghorne, *Science and Providence,* chap. 5; *Science and Christian Belief,* chap. 4; *Belief in God in an Age of Science,* chap. 1.

either of these, substituting "a lawful structure of natural causes" for "freedom."

Is the undisturbed operation of a system of natural law a good worth having even at the price of the suffering and death it produces? In summarizing his own views, Polkinghorne says that "A world allowed to make itself through the evolutionary exploration of its potentiality is a better world than one produced ready-made by divine fiat."[418] It is not clear how we should understand the distinction between a world that makes itself and one that is made by divine fiat. Polkinghorne affirms the doctrine of *creatio ex niliho,* and this entails that the world God calls into being will depend at every moment upon a direct act of God that sustains its existence, i.e., upon an act of divine fiat. What Polkinghorne seems to have in mind is a distinction between a world whose history unfolds through the operation of created things according to the natures God has given them, and one in which God makes more or less frequent adjustments in the course of events that override or exceed the causal powers of creatures. Is the former intrinsically more valuable than the latter? It is hard to know how to approach such a question, but clearly a world of the second sort might include a great deal less suffering and death than the first. Let us grant that it is an intrinsic good to preserve lawful relationships within the system of nature and to let the created order unfold its inherent potentialities across time. This may strike us as more elegant strategy of creation than one in which the natural order must be supplemented by extensive divine modifications in order to achieve God's purposes. It is far from clear, however, that the autonomy and integrity of the natural order is, in itself, a *great enough good* to justify all the misery and loss that results from events running their course untouched. We may have good evidential grounds to deny that God in fact typically acts by miraculous intervention. But the question here concerns *why* (in principle) God does not do so, if this course of action would diminish the world's evils.

4.3 Natural Law and Moral Personhood

If we cannot plausibly contend that the existence and operation of a lawful natural order is an intrinsic good great enough to justify all the harm that occurs within it, then can we regard this structure as a logically necessary condition for the realization of some further good? One way of making this case appeals to the good of finite persons as rational free agents. This folds the idea of a free-process defense back into a free-will defense, and it returns to the ancient idea that natural evils are linked to moral goods and evils, though it does so without invoking the idea of an historical "fall." The argument depends upon an analysis of the conditions for the possibility of finite persons coming to share in the good that God intends for them. Free will defenses, of course, insist that moral freedom is among these necessary conditions. I noted above the complex analysis of moral freedom in Christian theology, one important strand of which asserts that our freedom itself has become bound and needs to be set free. Since we cannot expect to untangle here the venerable theological knot about divine grace and human freedom, let us suppose, for the sake of argument, that moral freedom is a

---

[418] *Science and Theology,* 94.

constitutive element in personhood, and leave open the question of what role free-will plays in bringing about the realization of communion with God. If we affirm that persons possess moral freedom, we need not also assert that they freely choose or reject their own highest good; an agent's moral history profoundly shapes the person she becomes, but the completion and fulfillment of her personhood in relation to God may be a gift that exceeds her powers of action (or even imagination). Borrowing and modifying John Hick's language, we might say that we contribute in our moral freedom to the making (and mis-making or unmaking) of our souls, but the final good of the soul need not be understood simply as the outcome of free action. [419]

4.3.1 Compatibilist and Incompatibilist Freedom

Freedom, of course, can be understood in more than one way. Ordinarily, free will defenses take free action to be incompatible with causal determination by antecedent events. A free act will have necessary but not sufficient conditions in the prior causal history of the universe; given precisely the same causal conditions, the free agent could choose an alternative course of action. This failure of causal determination is a necessary but not sufficient condition for human freedom; a free action is not merely a chance event, but rather an exercise of the agent's capacity to direct the course of her own life. One of the principal challenges facing "incompatibilist" theories of freedom is to give an account of agent self-determination that does not reduce to a functional pattern in a closed series of efficient causes. The leading alternative account of freedom offers just this kind of analysis. On this view, free action is compatible with causal determination; for example, it can be argued that a free action just is one undertaken in accordance with the agent's (causally determining) beliefs and desires without compulsion or constraint. [420] "Compatibilist" accounts of this type appear to offer no help to someone trying to construct a defense of God's goodness. In such a universe, God could fully specify the course of every agent's life, determining each free choice. No finite agent would act in any way other than as God decrees, and God would be the agent of all moral evils.

4.3.2 The Nomic Condition

The point of appealing to incompatibilist moral freedom is that it makes it possible to regard moral evil as a product of responsible human choices. For my purposes, however, we need to see that freedom also brings with it the inevitability of natural evils. Rational free agents must be able to make sense of the world in which they act, anticipate the consequences of action, and consider alternative possibilities. This requires that their world be or-

---

[419] On "soul-making" see Hick, *Evil and the God of Love,* pt. 4. Hick's theodicy is premised on the claim that we freely enter into or reject loving relationship with God. I think it is possible to retain a significantly stronger conception of divine grace while still affirming the necessity of freedom and soul-making. My initial thoughts on how this might be done can be found in Tracy, "Evil, Human Freedom, and Divine Grace," 165–90.

[420] For an important and creative effort to move beyond these alternatives, see Murphy and Brown, *Did My Neurons Make Me Do It?* chap. 8.

ganized in a way that is consistent and predictable; rational action would not be possible if events succeeded one another randomly, failing to form any reliable patterns. The finite agent's world, then, must necessarily have an intelligible structure.

It is debatable, however, whether this alone entails that natural evils will occur; that will depend on the particular natural structure that God creates. Perhaps God could establish a natural order organized in accordance with a principle of retributive justice, so that only those who deserve to suffer from natural misfortune do so. Events in such a world might or might not display the sorts of patterns familiar to us as natural law, but they would all conform to some putative principle of moral law—the *lex talionis,* for example. A little reflection, however, shows that such a world would not be well suited to the moral formation of rational agents. In this world of exacting rewards and punishments, no one would suffer unless they deserved precisely the misery that descends upon them. Compassion would be possible, perhaps, but it would be morally suspect to the extent that it expressed sympathy with the malefactor in his wrongdoing. Further, action that seeks to prevent or ameliorate the physical sources of suffering would be bound to fail, since a new source of suffering would just replace the old one until the underlying moral defect was addressed. It would not even be clear that we should avoid undertaking actions that cause harm to others, since these actions would succeed in doing damage only when the other person deserved to suffer in this way. Finally, it would be difficult in such a world ever to do what is right simply for its own sake, since truly virtuous actions would bring a prompt and proportionate reward.[421]

It appears, therefore, that any world in which rational moral agency is possible must have a lawful, impersonal, and amoral structure. I will refer to this as the *nomic condition* for the existence of finite persons. In a world that satisfies this condition, it will be possible to get hurt simply by virtue of the lawful operation of the natural order. Suffering will not always be deserved; if it rains at all, it will rain on both the just and the unjust. The mistake of Job's comforters, whether ancient or modern, is that they feel compelled to provide a morally justifying reason for *each instance* of natural misfortune. Ironically, this gives natural evils too much meaning. In a system of impersonal natural law, physical hardship and suffering will sometimes occur for no reason other than that the causal structures of nature generate them. These evils will be a concomitant, or by-product, of meeting the nomic condition for the existence of finite persons. Evils of this type must be permitted if the good is to be realized. But individual instances of such evils need not always function as the means to a specific good end for the individual sufferer or for any other individual. This is not to deny that natural evils *may* become the means by which a particular good is realized, e.g., suffering may engender courage or insight in the sufferer and compassion in the witness (though it is also clear that suffering can be destructive, inducing despair in the sufferer and prurience and cruelty in the witness). It is to deny, however, that the explanation of natural evils must be constructed in such a way that they always lead to a specific good outcome in a means/ends relationship. If the universe is to include creatures suited for personal communion with God, natural evils must be

[421] Hick, *Evil and the God of Love* (rev ed.), 334–35.

distributed according to natural law, and not in conformity to a moral principle. There will be an unavoidable arbitrariness about the hazards of life in the system of nature, and this will be one of the sources of suffering for rational creatures like us who long for meaning and purpose in the events that cause us so much sorrow.

4.3.3 The Anthropic Condition

An important challenge immediately arises here, however. Couldn't God devise a lawful natural order that made possible the good of finite persons without so high a cost in misery? The specific sorts of creatures that exist and the forms of natural evil they suffer will depend on the particular system of lawful natural relationships that God creates. We cannot conclude from the nomic condition alone that the existence of rational moral agents requires the system of natural law that we observe in our world (or one substantially like it), with the full burden of miseries it brings about. Can we say anything further about the necessary conditions for finite moral agency that would help explain why the universe takes the shape it does? It is clear that not just any system of nature will support the existence of finite persons; among possible worlds that meet the nomic condition there will be a subset that also satisfies additional conditions necessary for the presence of rational moral agents. Perhaps, by exploring the content of this *"anthropic condition,"* we can reinforce and extend the argument that a world containing finite persons must also include natural evils of the sorts we see around us.

Given recent developments in fundamental physics and cosmology, this holds out some promise as an explanatory strategy. As we learn more about the deep structure of our world, it appears that the conditions under which life can appear in the cosmos are extraordinarily limited. Vanishingly small changes in any of a large number of variables (e.g., the gravitational constant, the mass density and expansion rate of the early universe) would have consequences in the unfolding history of the cosmos that rule out forming the structures that make life possible. Yet there is nothing in our (current) physical theories that explains why these parameters have the values they do; they appear to be contingent matters of fact rather than necessary consequences of underlying laws. This is the phenomenon of "fine-tuning" and "anthropic coincidences," and it suggests that if God wants to make a world significantly like ours in its basic structures, then the anthropic condition imposes startlingly precise limits on how it can be done.[422] In order for the world to engender intelligent life, it must be put together in very much the way we find it, and it will, therefore, contain nearly the current range and volume of natural evils.

Arguments of this kind are illuminating; they indicate that life and death, flourishing and perishing are closely linked at a deep level in the structures of nature. But in grappling with the problem of evil, this argument is of limited value. If it succeeds, it will show that even God cannot get the desired result (intelligent life) from this structure (the system of relationships described by physics) without these concomitants (the magni-

---

[422] For example, see Barrow and Tipler, *Anthropic Cosmological Principle*; Barrow, *Constants of Nature*.

tude of natural suffering we observe). But this is a conditional chain of reasoning, premised upon the physics we find in our universe. The objector concerned with evil wants to know why God creates a universe structured by this physics. The creator's options, we noted above, are not limited by the physical necessities of the actual world. It is certainly not obvious that God, in creating an orderly system of nature in which finite personal life can exist, is limited to variants of the system we see around us, and therefore to a world in which roughly the present volume and intensity of natural evils occur. In order to reach this conclusion, we would need to argue that any possible world that satisfies both the nomic and anthropic conditions (and that God can actualize) must be constructed in such a way that natural evil of this magnitude will occur. This would settle, in the theists favor, the question of whether God could improve on the system of nature, but I do not currently see a way to make this argument.

In sum, the considerations we have so far introduced support the conclusion that a universe which makes possible the existence of free moral agents will also make possible the occurrence moral and natural evils. But it remains an open question whether this good could be attained in a way that involves *less* evil of either sort. The discussion of our second programmatic question (viz., What is the relation of evils to the good?) is not yet complete; we must go on to grapple with the challenge presented by apparently excessive evil in the world God has made.

5 Are There Excessive Evils?

A recent vigorous philosophical debate has focused on the claim that we are rationally justified in concluding that the world does in fact include many evils which God could just as well have prevented. This is an intuitively appealing claim; most of us can readily generate a list of improvements we would make in the design or operation of the world if only we had the power to do so. This intuition can be turned into an argument, and it is one of the most challenging forms of the problem of evil.

5.1 Rowe's Evidential Argument

William Rowe has put forward several versions of an evidential argument, one of which makes use of the following premises:
1. There exist instances of intense suffering which an omnipotent, omniscient being could have prevented without thereby losing some greater good or permitting some evil equally bad or worse.
2. An omniscient, wholly good being would prevent the occurrence of any intense suffering it could, unless it could not do so without thereby losing some greater good or permitting some evil equally bad or worse.
3. There does not exist an omnipotent, omniscient, and wholly good being.[423]

The second premise is presented as a necessary truth entailed by the theist's concept of God as omniscient and perfectly good. The first premise is

---

[423]"The Problem of Evil and Some Varieties of Atheism," in Adams and Adams, *Problem of Evil*, 127.

offered as an empirical claim that can be supported from our experience of the world's goods and evils and the relations between them. It is not possible to prove the truth of *1*. Rowe acknowledges that we would need a comprehensive understanding of the entire world system in order to establish definitively that an instance of intense suffering in fact serves no greater good. But he argues that it is reasonable to affirm premise *1* based on consideration of familiar (indeed, ubiquitous) cases of apparently purposeless suffering. As an example, he describes an instance of natural evil; lightning starts a forest fire, terribly burning a fawn, which dies slowly in agony over a period of days. It appears that this natural evil could readily be prevented by God, and as far as we can see, doing so would not result in losing any greater good or incurring any comparable evil. This of course is just one of innumerable such cases, multiplied beyond imagining throughout the long evolutionary history of life on Earth.

> In the light of our experience and knowledge of the variety and scale of human and animal suffering in our world, the idea that none of this suffering could have been prevented by an omnipotent being without thereby losing a greater good or permitting an evil at least as bad seems an extraordinarily absurd idea, quite beyond our belief.[424]

5.2 *Judging Appearances*

How might we reply? Rowe's argument relies on moving from "It appears that God could have prevented this intense suffering without losing a greater good" to "It is reasonable to believe that God could have prevented this intense suffering without losing a greater good." This move from appearing that p (or not-p) to justified belief that p (or not-p) is a familiar feature of our lives as knowers, but several authors have challenged Rowe's use of this evidential connection in this context. One of the most effective of these challenges is offered by Stephen Wykstra who argues that we are entitled to claim "it appears to me that p" only if it is reasonable to believe that we could tell the difference between p-situations and not-p-situations. If I am asked whether the milk is sour, but have a bad head cold that renders me largely unable to smell or taste, I am not entitled to say, "It appears that the milk is just fine." Wykstra calls this the condition of reasonable epistemic access (or CORNEA), and he holds we have reason to think

---

[424] Ibid., 131. Wesley Wildman's "argument from neglect," in this volume, has this same structure. He points to apparent improvements that God could make in the structure or governance of the world, and contends that this provides evidence of divine negligence that is incompatible with the claim that God is omniscient, omnipotent, and perfectly good. One of the virtues of Wildman's preliminary conference paper is that he presents in considerable detail the alternative pathways for the world's natural history suggested by various fields of current scientific knowledge. Just as in Rowe's argument, however, this involves moving from "it appears that God could, without loss, prevent various evils" to the conclusion that "God could, without loss, prevent various evils." The strategy of reply will be the same to both arguments, namely, to undercut this evidential move.

that this condition is not satisfied in cases like the one Rowe presents.[425] This is because basic theistic beliefs, on Wykstra's account, entail that the goods God intends are generally beyond our comprehension, so that even if there is a justifying good served by a troubling case of intense suffering, we cannot expect that it will be apparent to us. As a result, we are not in a position to make the claim that, "It appears God could have prevented this suffering without the loss of a greater good," and therefore we have no evidential basis for the first premise of Rowe's argument.[426]

It seems to me that Wykstra is right about the condition of epistemic access. His case for the conclusion that this condition is not met, however, is problematic. It depends on two claims: (1) "*if* theism is true, the outweighing goods by virtue of which God allows suffering would generally be beyond our ken,"[427] and (2) these unknown goods are the ones relevant to instances of intense suffering. It must be granted that God's nature and purposes in creation vastly exceed the comprehension of finite minds. But this acknowledgment of human finitude and divine mystery need not lead to the conclusion that God's reasons for permitting suffering are generally incomprehensible to us. The living theistic traditions, after all, understand themselves to be a response to God's self-revealing engagement with us, and they have a good deal to say in practice about God's good purposes in creation. Theological reflection in these traditions has generated a variety of artful strategies for balancing affirmation and negation, comprehension and incomprehension, speech and silence. It would be a mistake to deny

---

[425]"The Humean Obstacle to Evidential Arguments from Suffering: On Avoiding the Evils of 'Appearance'," in Adams and Adams, *Problem of Evil,* 138–60. Wykstra's formal expression of CORNEA is as follows: "On the basis of cognized situation *s,* human *H* is entitled to claim 'It appears that *p*' only if it is reasonable for *H* to believe that, given her cognitive faculties and the use she has made of them, if *p* were not the case, *s* would likely be different than it is in some way discernible by her" (Ibid., 152).

[426] It might be objected that we *are* entitled to make the appearance claim under these and similar circumstances, but that we are not justified in drawing a conclusion on this basis about what is the case. It is important to note that Wykstra is using "appears" in a quasi-technical sense. Sometimes a statement about how something appears is simply a report of my subjective state. An observer is always entitled to this merely *phenomenal* appearance claim. Wykstra, however, is concerned with appearance claims that are presumed to possess "evidential import" (Ibid., 146–47, 152–55). Appearance claims in this *epistemic* sense ordinarily warrant a move from (1) "it appears that *p*" to (2) "it is the case that *p*." The familiar presumption in favor of concluding (2) on the basis of (1) has been expressed in various formulations of the "principle of credulity." Wykstra's strategy of argument is to block the path to (2) by undercutting the assertion of (1), when this is taken as having epistemic import. Another approach is to allow (1) and prevent the further step to (2) by showing that the presumption in favor of this move can be defeated in this case. This would involve arguing that the principle of credulity applies under a set of conditions (e.g., of epistemic access that would enable us to tell the difference between p and not-p situations) that are not met in this instance. Wykstra argues carefully for the first strategy, but for my purposes it does not matter which approach is taken. My discussion below of our epistemic limits could be used to carry out either strategy for blocking the move to (2).

[427] Ibid., 157. It may be that the first claim entails the second; that will depend on how complete an ignorance of God's purposes is asserted in the first claim.

that the suffering we see around us may be necessary for the realization of goods that, at least for the time being, we cannot recognize. But it is not clear that we should base the response to Rowe on this claim. This approach would leave with us with a particularly spare version of what I earlier called a "thin defense"; it defeats the objection Rowe raises, but it provides no insight into why God creates a world that includes intense suffering. If we have little or no conception of the goods God seeks in creation, then the entire enterprise of trying to explain God's permission of evil collapses, and we are left with a bare assertion that God is up to something good, though we can say next to nothing about what it is.[428]

5.3 Another Strategy of Reply

We do not need to adopt this agnosticism about the good, however. We already have identified a network of interconnected goods linked to the occurrence of natural suffering, namely, the existence and operation of a lawful natural order that makes possible the life of finite persons, understood as free moral agents. Might these familiar goods be called upon in replying to Rowe's argument? This will involve a subtle shift in the strategy of argument. We are seeking to block the move from, "There appear to be evils that God could prevent without the loss of an outweighing good," to "There are such evils." The key to doing so is to pay attention to our limits as knowers, and there are at least two such limits that we might consider. Wykstra claims that we have a radically incomplete grasp of the goods God may intend in creation. Alternatively, we can point to our limited understanding of the conditions that must be met to secure even those goods of which we do have some comprehension. That is, rather than appealing to unknown goods, we can cite known goods, but note the incompleteness of our understanding of their conditions of actualization. One advantage of this argument is that it builds on beliefs theists typically hold about God's purposes in creation, and it avoids committing us to a skepticism about the good that undercuts the entire project of seeking to make (even partial) sense of God's permission of evil.

It is crucially important here to underscore a point that we noted in discussing the nomic and anthropic conditions for the existence of finite

---

[428] Some philosophers have argued that this insistence upon ignorance of the good leads to a wider skepticism that undermines other central claims of theism. If, as Wykstra appears to claim, theism entails that the goods we can envision provide little or no guidance in understanding God's purposes in creation, then theists are vulnerable to the argument that God might have some unknown good reason for, say, operating as a Cartesian deceiver who causes us to have globally false beliefs, including false beliefs about God. See, for example, Beaudoin, "Skepticism and the Skeptical Theist," 42–56. It might be argued that in order to block this infectious skepticism, we should affirm that our understanding of the good provides reliable, if limited, guidance to the goods that are possible. But if this is granted, then Wykstra's response to Rowe will fail (because he can no longer appeal to our ignorance of the good to undercut the epistemic force of the claim that "it appears God could prevent this evil"). I contend that if we give up Wykstra's appeal to unknown goods, and argue instead from the limits on our understanding of the necessary conditions for the realization of known goods, it is possible to block Rowe's argument without generating (at least this version of) global skepticism.

persons. Evils are not always related to goods as means to ends. The means/ends relationship is what first springs to mind when we look for a morally sufficient reason for God's permission of the fawn's suffering; we ask whether this evil leads directly to a good that (1) outweighs the evil, and (2) could not be attained by any less costly means. When the relation of goods and evils is understood this way, it is indeed difficult to identify any good that meets these conditions, and it may appear that the best response to this problem is to contend that this evil serves some mysterious good known to God but not to us. As we have seen, however, evils may be related to goods not as means to ends but as the concomitant, or by-product, of meeting necessary conditions for the realization of the good. A system of nature that satisfies the nomic and anthropic conditions makes possible a great variety of goods, but it also generates myriad forms of natural suffering. Some of this suffering may serve as the means by which a good, directly related to a particular evil, is achieved. But it is not necessary, and I think not plausible, to contend that suffering always stands in an instrumental relation to an outweighing good. Rather, natural evils may occur simply as a necessary consequence of establishing a lawful system of nature that provides the required context, or background conditions, for achieving the good that God intends. Further, we noted that this system of nature must be amoral; the forms of moral life available to rational creatures would be significantly truncated if suffering were distributed according to a recognizable principle of justice. But this entails that suffering will not always be directed to any good end that we can see, e.g., as a means of moral correction or teaching. Sometimes suffering will occur, whether as a result of operations of natural law or of the wrong actions of human beings, that is not for the sufferer's own good. God will have a purpose for permitting such suffering, but the suffering itself may have no purpose (i.e., as the means to a specific good outcome).

This has important implications as we consider Rowe's argument. It makes it clear that we should not always expect a one-to-one correspondence between the occurrence of a natural evil and an outweighing good that is produced by means of that evil. If that is what we are looking for when we ask about the good that would be lost if any particular instance of intense suffering were prevented, then it may not be there to be found. Rather, the good in such cases may simply be the continued functioning of a lawful system of natural relationships, along with the goods such a system makes possible.

The challenge to the theist, then, centers on the question of whether particular instances of natural suffering could be prevented by God without losing the goods associated with the system of nature. Surely, the objector will say, it is implausible to claim that if this fawn does not suffer and die in this horrible way then the natural order as whole will come unhinged. That sounds right, but on closer examination, the matter is considerably more complicated. We can bring out this complexity by exploring in a bit more detail the question of *how* God would prevent an individual instance of natural suffering. Rowe does not discuss this, and he seems to suppose that we can, without too much trouble, think of God acting to add or subtract particular evils from the history of the universe. Consider two possibilities.

5.3.1 Divine Action through Natural Law

First, God might act through the lawful structures of nature to prevent particular evils. In a deterministic natural order, this would require that God act in the initial design of the world to adjust the laws of nature or reset the initial conditions. This might entail extensive changes in the unfolding history of the universe both backward and forward in time from the particular event that is to be modified or omitted. If, on the other hand, the natural order is in some respects underdetermined (i.e., it includes events with necessary but not sufficient causal conditions in the prior history of the universe), then God might act by determining some or all of what would otherwise be left to chance. If these chance events are located in the structures of nature in the right way, God could act through them to affect the developing course of the world's history, and yet do so without disrupting the web of natural law.[429] Once again, however, these would not be isolated events. This noninterventionist particular divine action would have causal antecedents in the prior history of the world and it would have causal consequences in the events that follow.

In either a deterministic or indeterministic universe, then, the embeddedness of events in an extensively interconnected network gives us good reason to wonder whether preventing a particular evil may have a bearing on the realization of the goods that God intends in creation. This only becomes more likely if we think of God acting this way in many instances of natural suffering. We cannot demonstrate that any particular divine action of this kind would have deleterious effects on attaining the good that God intends. On the contrary, for all we know some divine actions that prevent particular instances of suffering might have additional consequences that advance the realization of the good. But the fact that we cannot reliably assess these matters undercuts the evidential value of the claim that "It appears that God could just as well prevent this natural evil."

5.3.2 Divine Action Outside Natural Law

Second, God could act outside the ordinary course of nature to eliminate at least some instances of intense suffering. This is probably what Rowe has in mind. Why doesn't God miraculously intervene to interrupt the chain of events that leads to the fawn's injury and lingering death? If we affirm that God is the omnipotent creator of all finite things, then surely God can act in this way, contravening if need be the causal laws that describe the ordinary operations of nature. If God refrains from doing so, it will be out of regard for God's own purposes in creation. We may be convinced that we have evidential grounds for denying that God routinely performs such actions, but as we have already seen (in section 4.2), the problem of evil challenges us to give good reasons *why* God does not do so.

Phillip Clayton and Steven Knapp argue in this volume that if God abrogates natural law to ameliorate suffering even on one occasion, God is morally obligated to do so on every occasion, with the result that the lawful

---

[429] For detailed discussion of the issues in science and theology that are raised by this possible mode of divine action, see the essays in Part 3 of Russell, et.al., *Quantum Mechanics*.

structure of nature would break down.[430] This is because God would have no principled reason for acting in some cases but not others; on pain of moral inconsistency, God must either act in every instance or in none at all. As welcome as this conclusion might be in dealing with the problem of evil, there are difficulties with the argument for it. First, there may be principled grounds for God to act in some cases but not in others.[431] Divine moral consistency is preserved if God intervenes whenever it is possible to act in morally equivalent ways in morally similar circumstances. There is no reason to think that this rules out *all* divine intervention. Second, consider the moral comparison of the following possible worlds: (1) a world that contains a great deal of suffering that God could have intervened to prevent without loss of any goods (except those associated with perfect consistency of divine action) and (2) the closest possible world to the first in which God intervenes to reduce suffering, and in which at least some of these divine acts are not distributed according to a moral principle. In the first of these worlds, God will display perfect fairness by fastidiously treating like cases alike, but God will allow creatures to suffer for no reason other than a commitment to consistency of (in)action. The second world will include less suffering, but at least some acts of divine mercy will be gratuitous, i.e., they will not be distributed according to a principle of justice. It is at least not obvious that the former world is morally better than the latter, particularly not if God's love triumphs in the end over *all* suffering. There is no clear moral advantage gained by resisting the claim that God could help but does so only on some occasions, and contending instead that God adopts a uniform policy of never helping at all.

It appears, therefore, that we cannot readily avoid the question about why God does not act outside the laws of nature to diminish suffering in the world. The first move in reply, of course, is to note that there are limits on how extensively God can intervene miraculously and still maintain the lawful character of the created order. It is almost certainly too simple to imagine that there is a specific balance point at which, if God performs one more miracle, the good of a lawful context for rational moral life will be lost. But suppose for the sake of argument that there is such a point. A world of this sort need not be the best of all possible worlds (there may be no such thing), but it will be one of a class of worlds that achieves the best possible balance of goods attained and evils permitted. I want to argue that we are not in an epistemic position to judge whether or not our world is a member of this class. Indeed, even if our world is poised at this balance point, it will appear to us that it includes an excess of evils.

In order to see why this is the case, we need to note a distinction between evils that can be eliminated only by substitution and evils that can be eliminated *simpliciter*. If we suppose that the world contains an optimal balance of goods and evils, then the elimination of one evil will require that some comparable evil be permitted elsewhere (e.g., if God intervenes to spare the fawn, then God must let nature run its course when a spring flood drowns fox kits in their den). In such a world there may be many individual instances of suffering that are preventable by God, but only by al-

---

[430]"Divine Action and the 'Argument from Neglect.'"

[431] For an elegant reflection on this possibility see Robert Merrihew Adams, "Theodicy and Divine Intervention," in Tracy, ed., *God Who Acts,* 31–40.

lowing something to happen that is equally bad. Some number of evils belonging to a *class* of substitutable evils must be permitted, but many particular *members* of the class (considered individually) need not be. Rowe's premise 2 allows God to permit an evil under these circumstances. But this has an important consequence. The question about whether an evil is preventable now has two distinct forms: (1) we can ask whether an evil is a member of a substitutable class of comparably bad evils, and (2) we can ask whether God has permitted a larger number of instances of this class of evils than is necessary in order to make possible the goods that God intends. It is only the latter question, the question of whether there is an excess of such evils, that poses a problem for God's goodness or power.

The issue, then, is whether we might be entitled to conclude that an instance of evil is preventable, in the sense that it is excessive and should have been eliminated by God, based on the fact that this appears to us to be so. I think it clear that we are not entitled to draw this conclusion. For first, substitutable instances of suffering will always appear to us to be preventable by God; indeed, they will always *be* preventable—but only by substitution. We are in no position to tell the difference between evils that are merely substitutable and evils that are excessive. Even if the world includes no excess evils, it will appear to us that it does, and so the condition of reasonable epistemic access is not met. Second, the judgment that an individual evil could be eliminated as excessive is inextricably linked to a global judgment about the total volume of evil that God must permit for the sake of the good. We cannot decide the question about excessive evils on a case-by-case basis, but rather must assess the overall balance of goods and evils in the universe across it history. This global judgment obviously exceeds our epistemic reach; we would need something approaching omniscience in order to make it. Perhaps it might be objected that the global judgment can be made in at least an approximate form; it may seem obvious that the world around us is in no danger of reaching the tipping point between lawful structure and anomic disruption, and that God could intervene quite a bit more often without the natural order falling apart. It helps in overcoming this intuition, however, to note that miracles will have consequences in the ongoing course of nature (just as we saw in considering the first form of divine action). These consequences will ramify through succeeding events for good or ill, and may require further miraculous adjustments to the course of events. This is a curious effect; miracles will tend to "infect" the natural order and undercut its immanent structures in ways that are not apparent to us. We are not, of course, in an epistemic position to map out these relationships in any detail, but that is precisely the point; we should not suppose that God can simply insert from above a single miraculous change that leaves the rest of the world's history untouched.

6 Conclusion: Some Tentative Thoughts on the Crucial Third Question

Where does this leave us? If my response to Rowe is right, then the objector from evil cannot claim rationally sufficient evidential support for the assertion that there are instances of intense suffering that God could have prevented without loss of a greater good. But neither is the theist in a position to show that there are no excessive evils or to tell a detailed story about why each evil must be permitted. The most that can be done is to explain that the world God has made includes goods that cannot be realized without the creation of a lawful natural order that generates natural evils. Just as the objector cannot make an evidential case that there are evils that God could have prevented without loss of the good, so the theist cannot make the parallel case that there are no such evils. We have an epistemic stalemate that reflects intrinsic limits on human understanding. This is fatal to theodicy understood as a form of natural theology. But it is something that a faith seeking understanding can live with. Indeed, this limit on our ability to carry out the project of theodicy is entailed by the concept of God that generates the problem of evil in the first place. Given theistic claims about God's knowledge and power as creator (and evident facts about human finitude), it should come as no surprise that our efforts to account for evil must remain incomplete. No matter how fine-grained we make our theoretical explanation of evils, we will be unable to show that all the conditions of this theory have been met in the world God has made. By paying attention to these epistemic limits, we defeat the evidential objection to theism, but we also make clear the inevitable collapse of efforts to vindicate God's goodness by showing that the world conforms to the terms of our justificatory explanation. Confidence in God's goodness cannot be grounded in an ability to show that God must permit each evil that occurs.[432] Evidential objections and evidential theodicies both must fail.

The project of explaining evil, therefore, must remain incomplete and hypothetical. It need not be restricted simply to a thin defense, however, but rather can draw upon substantive claims of the theological traditions, as I have tried to do. The line of argument I have sketched allows for a limited explanation of why God permits the kinds of natural evils we find in the world. It does so by reference to the goods that we are especially suited to recognize and participate in, namely, the good of personal communion with God. These considerations do not put us in a position to explain why the world includes *so much* natural suffering. And it does not allow us to assign a moral meaning to *each instance* of natural suffering, either as a form of retributive justice (as Job's friends insisted) or as a way of teaching us a lesson (in a universal pedagogy of pain). On the contrary, the account I have given blocks the way to these explanations, if they are offered as applying to every instance of suffering. It is a mistake to offer blanket explanations of this kind, and when we do so it is no surprise that the result is morally implausible, particularly to the sufferer. Persons sometimes suffer through no fault of their own simply by virtue of being the kind of animals we are in the natural structures that generated us. This suffering is arbitrary and unfair; but the theistic faiths affirm that the goods God intends in creation cannot be realized without permitting it to take place.

This raises with great poignancy the third question in my initial agenda of issues. If the goods achieved in creation (including the existence

---

[432] Does this entail that there is no basis at all for affirming the goodness of God? Clearly not. Theism itself, I have argued, entails that we will not be able to complete the project of explanatory theodicy. But theists claim other grounds, both evidential and nonevidential, for affirming the goodness of God. For an excellent discussion of epistemological issues in religion and science, see Murphy, *Anglo-American Postmodernity*.

of finite persons) cannot be secured without the permission of so much un-deserved suffering, would it be better not to seek these goods at all? Can such a world be an expression of God's perfect goodness?

6.1 Nonhuman Fellow Sufferers

This question arises not just with regard to the human victims of natural misfortune, but also with regard to the suffering of other sentient life forms. The explanatory justification I have given is constructed by refer-ence to the necessary conditions for the good of finite moral agents. Unless something more is said, this leaves us with an account in which all of the suffering of other living things is treated merely as a background condition for the good of persons. So we need to ask on behalf of *all* victims of natural suffering whether the good is worth having at so high a cost.

In constructing an explanation of natural suffering we have no choice but to appeal to goods with which we are familiar—there is no point in of-fering an explanation that begins with ignorance, and promptly ends there. In appealing to what we know, however, we do not deny that there are val-ues over and above (and perhaps largely independent from) those with which we are familiar. So in response to the problem of the suffering of other living things, it is important to recognize that there will be goods realized in their lives that are distinct from (if analogous to) those we ex-perience. The system of nature unfolding in the universe makes these goods possible, just at it does our own. The question is whether the balance of good and evil in their lives makes it worthwhile, from their point of view, to have had the life they did.[433] We obviously are not in a position to make this judgment globally about the life forms that have arisen on Earth. Anthropomorphic projection is probably inevitable here, and in assessing the life experience of other creatures, we should acknowledge that we cannot comprehend the depths of either their satisfactions or their sorrows. They possess capacities of sensation and action that we can only dimly imagine, and if they do not experience the goods associated with rational moral life, neither do they experience its distinctive evils. We ought not to project uncritically onto them our characteristic anxieties and discontents, much less our hostilities and cruelties. The opportunity to inhabit their life-world in their distinctive way may be as rich a good as can be achieved for them. We know in our own species, however, that lives can be dominated by struggle and misery in ways that overwhelm the opportunity to participate in the goods that are proper to our kind. If this is true of us, there is no reason to deny that it is true of our fellow sentient creatures.

6.2 The End of Evil

At this point we must return to the observation I made at the outset; Chris-tianity is a religion of salvation. The goods in terms of which evils should be assessed are not exclusively those realized in the world as we now find it.

---

[433] By referring to "their point of view" I do not mean to imply that they can make this judgment themselves—if they could they would be persons. I am claiming only that they *have* a good-of-their-own and that this is the relevant frame of refer-ence in which to make this judgment.

Indeed, any response to the problem of evil that is constructed solely by reference to immanent goods secured in the present course of the world's history will be hard pressed to provide a morally plausible reply to the question about whether this good is worth having, given the price at which it comes. At the heart of the difficulty is the challenge expressed so powerfully by Ivan Karamazov.

> Imagine that it is you yourself who are erecting the edifice of human destiny with the aim of making men happy in the end, of giving them peace and contentment at last, but that to do that it is absolutely necessary, and indeed quite inevitable, to torture to death only one tiny creature, . . . would you consent to be the architect on those conditions?[434]

Ivan's brother, Alyosha, quietly answers "No" to this question, and it is not easy to see how this answer can be avoided if the innocent children in Ivan's stories are simply sacrificed as "dung on the fields" for a greater good enjoyed by others. Ivan's challenge is one form of a classical objection to consequentialist ethics; Ivan contends, in effect, that God ought not to be a utilitarian (though it has been remarked that only God, possessing perfect foreknowledge, *could* be one with full success). Some important moral issues arise here that I have discussed elsewhere,[435] but it is enough for the moment to note that the moral context shifts significantly if we suppose that the child's life is not simply sacrificed for the sake of a good in which he or she never participates. It is in God's power to make the good available to this child in spite of her suffering, so that the sad fact of her victimization is not the last word on her life.

Further, and more difficult to comprehend, Christianity has affirmed that God is able to take up the victim's history of sorrows and incorporate it into the good, so that suffering is no longer an utter loss to the sufferer. This is not to say that suffering becomes something good in its own right, but rather that it is redeemed from meaninglessness, from sheer destructiveness, and is embraced within the good that God creates for this individual. As classically understood, this is accomplished through God's unreserved participation in the moral and physical hazards of the created world. God does not remain at a safe distance from the destructiveness unleashed along the way to realizing the good. Rather, God meets us in our loss and grief, suffering as a victim of evil and fashioning in this unexpected way the bonds of a relationship in which our ultimate good is found. This is not simply to say that God is a universally sympathetic fellow sufferer. Over and above this passive compassion, God's entry into the world's alienation and misery is a redemptive action through which the good is brought into being.[436] On this account, redemption is the continuation and completion of creation.

---

[434] Dostoevski, *Brothers Karamazov,* pt. 2, bk. 5, chap. 4, 287.

[435] "Victimization and the Problem of Evil."

[436] There are two aspects to this redemptive action. It is not only a response to the evil we suffer, but also to the evil we do—and to the predicament in which we find ourselves ambiguously entangled as both victims and victimizers. Given my focus on the problem of innocent suffering within the system of nature, I focus here on just one side of this complex and subtle network of ideas.

Though we now participate in the good for the sake of which evils are permitted in creation, the full character of this good is not yet apparent to us, and so its ultimate relation to suffering remains hidden. The Gospel stories make a point of insisting that the wounds of the risen Christ do not vanish; this image affirms that suffering is not simply ignored or negated in the final good that God brings into being. But these marks of suffering and evil no longer have in them the power of death; here suffering is taken up into the life of God, stripped of its destructiveness, and suffused by life. This is a vision not simply of the restoration of a good that has been lost, but rather of the transformation of the creature and the completion of God's creative work. A developed theological response to the problem of natural evil will need to draw upon these eschatological themes and think through what might be said about how other living things participate in this good. The vision of a completed creation must necessarily remain enigmatic; if we cannot fully grasp what God intends for us, we are even less able to spell out just how the rest of creation will be included in this final good. It is consonant with the central themes of the Christian faith, however, to affirm that God's unlimited generosity extends to all that God has made. If we affirm that God creates the world and participates in its history for the sake of love, then we have strong theological grounds for expecting that the whole created order will share in the fulfillment of God's good purposes.

DIVINE ACTION AND THE "ARGUMENT FROM NEGLECT"

Philip Clayton and Steven Knapp

1 The Argument from Neglect

Not everyone thinks the universe is grounded in an ultimate reality of the kind described by theists. Among those who do, not everyone agrees that the ultimate reality is (something like) a person, insofar as personhood is understood as entailing rational agency, that is, the capacity and disposition to perform finite intentional actions that bring about particular states of affairs.

More than any other single factor, we believe, what accounts for the reluctance even of some convinced theists to embrace a personalistic picture of the ultimate reality is the perception that the acts we would expect a divine agent to perform do not occur. Above all, the acts we would expect to occur but that seem not to are acts that would prevent or relieve the egregious cases of innocent suffering that we observe all around us. These are acts, at least, that we would expect a *benevolent* divine agent to perform; their non-occurrence raises the possibility that the ultimate reality is a malevolent person, or perhaps an amoral one. But, for most versions of personalistic theism, an evil or amoral divine reality would presumably be even worse news than an impersonal one.

The inference from the apparent nonoccurrence of expected divine actions to the rejection of personalistic theism has been usefully dubbed "the argument from neglect" by one prominent nonpersonalistic theist, Wesley Wildman. Here is a passage in which he eloquently summarizes that argument:

> Of course, it is not actually the existence of suffering that is the problem for personal ideas of God. That is a shared challenge for all religions and all theologies. It is what a supposedly personal active God doesn't do about it that is the problem. Consider the following analogy. When my children endanger themselves through their ignorance or willfulness, I do not hesitate as one trying to be a good father to intervene, to protect them from themselves, to teach them what they don't know, and thereby to help them become responsible people. I needed to do that a lot more when they were little than I do now but I believe that my love for those children can be measured as much by my interventions as by my allowing them space to experience making their own decisions independently. They do need to experience the effects of their choices, whether good or bad, but I would rightly be held negligible as a parent if I allowed them such freedom that they hurt themselves or others out of ignorance or misplaced curiosity or wickedness.
>
> To the extent that we think of God as a personal active being, we inevitably apply these standards. Frankly, and I say this with the utmost reverence, the personal God does not pass the test of parental moral responsibility. If God really is personal in this way, then we must conclude that God has a morally abysmal record of inaction or ineffective action. This I shall call the *argument from neglect,* and I take it to be the strongest moral argument against

most forms of personal theism. It applies most obviously to versions
of personal theism in which God is omnipotent. But the argument
from neglect also applies to views of personal theism that deny omnipotence, such as process theology, because the argument establishes that God's ability to influence the world is so sorely limited as
to make God virtually irrelevant when it comes to the practical
moral struggles of our deeply unjust world.[437]

To meet this objection, a defender of personalistic theism has to do two
things: first, to show that there may be a good reason why a personal and
active God, if there is one, either cannot or chooses not to perform the acts
we would expect a benevolent God to perform; second, to avoid what is in
effect the reductio ad absurdum of so constraining divine action that it becomes pointless or irrelevant.

The set of possible divine motives is presumably infinite, and the set of
possible constraints on divine action may also be infinite. We are not interested, however, in determining whether it is merely logically possible that
there is a good reason for what looks like divine neglect.[438] Nor, at the
opposite extreme, would we be so bold as to claim to *know* what reason God
actually has, or even to claim that we can establish this with, say, a probability of greater than 0.5. Instead, we would like to know whether there is
an account of divine motives and divine action that would constitute a
plausible explanation for apparent divine neglect—plausible, that is, in the
eyes of the relevant community of inquiry, which in this case means a
community not already closed to the possibility of personalistic theism.
Ideally, such an explanation would be sufficiently plausible that, at least in
the long run, the community would come to regard it as superior to its
competitors. But as long as it gives a plausible and consistent explanation
of why a benevolent God would refrain from intervening in cases where we
would otherwise expect such a being to do so, it will count, in our view, as
successfully defeating Wildman's objection to the claim that ultimate reality is more aptly conceived in personal than in nonpersonal terms.

2 Responses to the Argument from Neglect

With that aim in mind, it seems to us that the best place to begin is with a
hypothesis general enough to embrace the many phenomena that might be
mentioned in support of the argument from neglect but concrete enough to
embody those assumptions about the nature and intentions of God that
have emerged from the theistic traditions. As it turns out, defending this
hypothesis requires one not only to reflect on innocent suffering but also to
appeal to conclusions in a number of other fields, including scientific
knowledge about the natural world, fundamental considerations in the
philosophy of science, and the role of kenotic thinking in recent christologies. Although in the present paper we cannot develop the full formal argument across all these fields, we do indicate the range of topics that must

---

[437] See Wildman, "Review and Critique," MS p. 3, used with permission.

[438] As an example of this approach see Plantinga, "Free Will Defense," in Brody
(1992), 292–304.

be addressed by parties on both sides of the debate. Here, then, is the hypothesis with which we propose to answer the argument from neglect:

Suppose the purpose, or at least one purpose, of God's creating our universe was to bring about the existence of finite rational agents capable of entering into communion with God.[439] Suppose that the way God achieved that purpose was by creating a universe in which events would be consistently governed by regularities of the kind described by the laws of physics or, more broadly, the laws of nature. Because the universe operates according to its own internal regularities, beings who evolve through the operation of those regularities are not simply the direct expressions of the divine will (as would be the case if they were directly created by divine fiat) but partake of the (relative) autonomy with which God has endowed the universe as a whole.[440]

Everything else that we have to say about the argument from neglect follows from this fundamental hypothesis, which of course will require elaboration before its relevance to that argument can fully emerge. But questions about it immediately arise. Perhaps the most important are these: First, how regular do the regularities have to be? Why can't God preserve the regularities just enough to bring about the evolution of rational and autonomous beings but suspend their operation whenever necessary to prevent innocent suffering? Second, if God *can't* suspend the regularities of nature, why suppose that God can perform any actions at all within the created universe?

2.1 First Response

The first of these questions turns out to be the easier one to answer, because the answer follows rather quickly from an analysis of the conditions under which autonomous agents can evolve. The second is more challenging, because it requires us both to identify at least one sphere of activity in which divine acts would not constitute suspensions of natural laws and then to show why God's ability to act in that sphere does not raise, all over again, the specter of divine neglect.

Turning, then, to the first question: why can't God suspend the regularities of nature whenever God has reason to do so? Consider a universe in which God could and did suspend at will the operation of (to begin with) physical regularities, a world (say) in which the murderer's bullet turned into a flower when it left his gun and floated gently to the ground. Could rational and autonomous beings—beings, that is, at least as rational and autonomous as we human beings appear to be—evolve in such a universe? Perhaps they could reach our level of biological complexity, as long as God preserved enough physical regularity to sustain the right mechanisms of

---

[439] At this general level, our hypothesis resembles what John Hick has called the "Iranaean" theodicy, which explains divine nonaction in terms of the conditions required for "soul-making," i.e., for the formation and moral development of finite persons; see Hick, *Evil and the God of Love*.

[440] Although some will be inclined to accept a libertarian account of human agency, the argument in this chapter does not presuppose it; a compatibilist should be able to endorse our thesis as well. What we are most concerned with is the minimal conditions for finite agency, that is, the conditions that must be met for finite agents to carry out actions with relative autonomy.

mutation and selection. But could they function in the universe with the kind of self-determination or autonomy we seem to have?

It is hard to see how the evolution of rational agency, not to speak of moral agency, would be possible in such a world. One wonders, for example, could beings in such a world develop natural science in anything like the form we know it? It seems that they could not, because they would have no basis for developing, or at least developing fully, the appreciation of natural regularities on which science depends. Imagine that scientists became convinced that God regularly changed the outcomes of physical events in a counterfactual and non-lawlike manner, and imagine (for the sake of argument) that they were right. That is, imagine they knew that God regularly set aside natural regularities and intervened in the natural world according to hidden divine purposes and that, were God *not* to act in this manner, the phenomena that they observed would not have occurred. Even if the scientists could never in fact catch God red-handed, as it were (because no observation or experiment could actually detect these exceptions to natural law), it seems clear that they could not pursue the expansion of knowledge in the natural sciences in the way we now do. For one would always be unsure whether some natural pattern was really lawlike or merely an artifact of God's happening to intervene in similar ways over a particular span of time. Moreover, one would never know when the actual laws of nature might suddenly be set aside for the sake of some overriding divine purpose. The self-understanding of science would be radically changed, and indeed its motivation would be radically diminished.

Why should it matter, however, whether science as we know it is possible? After all, the institution of science is, in one sense, the contingent product of a certain strand of human social and cultural history. In another sense, however, science as we know it is a systematic mechanism for managing a far broader and arguably inescapable process: namely, the fixation of human belief through the shared investigation of a universe that is sufficiently stable to permit such investigation.[441] In that sense, science is merely one institutional expression of the more general human project of individual and collective self-definition and self-determination, which proceeds by our interacting with a reality that we can come to understand, in no small measure because it is *not* subject to arbitrary alteration by human—or more than human—fiat. (Intentional alterations by human agents who have evolved within the world and remain dependent on its physical conditions are not, of course, excluded by this principle.)

In short, we assume that the rationality and autonomy of finite individuals both depend upon and entail participation in what C. S. Peirce would have called a "community of inquiry."[442] A community of inquiry can foster and express the development of rationality only if it proceeds on the assumption that whatever it investigates is, at least in principle, capable of

---

[441] Cf. Peirce, "Fixation of Belief," 109–23.

[442] Peirce thought that the feedback from the relevant communities of inquiry was equally crucial to, and appropriate for, matters of religious concern: "if religious life is to ameliorate the world, it must . . . hold an abiding respect for truth. Such respect involves an openness to growth, to development. Thus, as ideas develop through the community of inquirers, they will have a gradual effect on religious belief and subsequently on religious practices." See Peirce, *Collected Papers of Charles Sanders Peirce,* 1:184.

being investigated, that is, progressively examined and re-examined by successive members of the community in question. In this case, what is true of the community is also true of individuals: we can only develop a sense of (relatively) autonomous selfhood if the objects around us have reasonably stable properties.

Suppose we stipulate, then, that both individuals and communities of inquiry can only develop if they assume that the world around them is immune from arbitrary alteration on a regular basis by supernatural forces. Let's call this the regularity argument. This argument still leaves unclear why a sufficiently powerful divine agent could not at least *occasionally* intervene to override physical regularities. One can perhaps see that if God intervened whenever necessary to prevent innocent suffering of any magnitude whatsoever, such frequent abrogations of natural law would sooner or later deprive us of the ability to perceive ourselves as persons separate from others, or indeed as beings separate from God. But why should God refrain from intervening on at least some of those occasions where a brief suspension of natural regularities would prevent a tremendous evil? For instance, God might not act to prevent my every injury and yet still might act to prevent the kind of brief and local seismic event that triggered the Indian Ocean tsunami in December 2004, which in a matter of a few hours took more than a quarter million lives and inflicted unimaginable suffering on the even larger number of their survivors.

The question is a reasonable one, and in fact it goes right to the heart of the argument from neglect. The answer, we suggest, is that creating and sustaining a universe in which free rational agents can evolve and act turns out, upon reflection, to be an either/or affair. A benevolent God could not intervene *even once* without incurring the responsibility to intervene in every case where doing so would prevent an instance of innocent suffering. Call it the "not even once" principle.

Why, however, should that be the case? What kind of necessity is it that would compel a benevolent God to act with such consistency? Is the necessity a *forensic* one (because God would have no way of explaining to others why God had not also intervened in other cases)? Would it in fact be *unethical,* because unfair, for God to intervene only in certain cases but not all? Or is the God-universe relation *metaphysically* such that the universe can only remain autonomous if God never acts upon it coercively, perhaps because a single act of divine coercion would somehow (through some kind of necessary chain reaction?) automatically subject the universe as a whole to the direct and therefore absolute control of the divine will?

The forensic option seems too anthropocentric; why should God worry about being criticized by the likes of us? The ethical explanation is better but, as initially stated, perhaps not fully convincing; why couldn't God remove any unfairness by adjusting the degree of divine intervention according to such criteria as the severity of the suffering, or the amount of disruption needed to prevent or stop it—much as we humans do when establishing our own policies for intervention in cases ranging from child protection to disaster relief to full-scale regime-change? The fact that we do this so badly does not permit us to eschew intervention altogether; still less, one might argue, should God be let off the moral hook. (We will return to this issue of what might be called "proportional intervention" in a moment.)

The third option is perhaps the most intriguing to those of a metaphysical bent, but it poses the conceptual as well as theological risk of positing a quasi-Gnostic gulf between God and God's creation. Why, for instance, would a benevolent God create a universe with which that God had no intention of interacting? And this option has the further disadvantage of invoking a rather fanciful causal mechanism—the metaphysical "chain reaction"—of which we have no actual concept or evidence.

It's not obvious that the forensic, ethical, and metaphysical responses are fatally flawed; each one may offer some support for the "not even once" principle. But given the sorts of reservations that one might raise against the three explanations just canvassed, it behooves the theist to look for a more rigorous response. We believe that a combination of the ethical and metaphysical responses provides a more compelling defense of the "not even once" principle than either by itself. Suppose we begin by looking more closely at the reason we mentioned for resisting the ethical defense of the principle. The critic wonders why God can't do what human agents at least *try* to do. We humans proportion the scale and frequency of our interventions to the magnitude of the harms they are intended to stop or prevent and to the risk of greater harm they pose. Does it make sense to impose on God a similar responsibility to intervene when the conditions are right for doing so?

Here is a reason to think that this does *not* make sense. From a human point of view, it is relatively easy to identify what we regard as cases of actual or potential suffering that are so great that we consider ourselves bound to intervene to stop or prevent them. But, as is often pointed out in a case like the international response to the Asian tsunami, our ability to make such discriminations is powerfully dependent on the limitations both of our attention and our resources. We quickly suffer from "passion fatigue," which is all the more reason why we not only can but must be discriminating in choosing which emergencies warrant a response. For all these reasons, a policy of proportionate intervention is both necessary and possible for finite agents like us. It is therefore permissible for us to intervene in some cases without thereby automatically acquiring an obligation to intervene in all.

God, on the other hand, has (as far as we have reason to suppose) no such limitations of attention, resources, or compassion. On most versions of personalistic theism, God is compassionately aware of absolutely every case of suffering of every sentient creature, and indeed of each creature's degree and precise qualitative experience of suffering. Where we humans see a clear difference between, on one hand, an emergency that cries out for a dramatic and immediate reaction and, on the other hand, a general state of misery that warrants a longer-term, more subtle, and more gentle response, God sees no such dichotomy but a vast continuum of suffering far more pervasive, intense, and immediate in its need for relief than we could ever allow ourselves to appreciate.[443]

---

[443] Cf. George Eliot's famous comment in *Middlemarch*, chapter 20: "we do not expect people to be deeply moved by what is not unusual. That element of tragedy which lies in the very fact of frequency, has not yet wrought itself into the coarse emotion of mankind; and perhaps our frames could hardly bear much of it. If we had a keen vision and feeling of all ordinary human life, it would be like hearing the grass grow and the squirrel's heart beat, and we should die of that roar which lies on

It makes sense, then, to say that human beings can intervene sometimes without being obligated always to intervene, because the occasions and the capacities of human intervention are so drastically limited by our finitude (as well as our sins!). Yet God lacks the luxury of our finitude. God therefore has no reason to intervene in the case of what we regard as unusual suffering while tolerating the less visible suffering that God perceives across the spectrum of sentient life.

But what about "hidden" interventions? Why couldn't God intervene in ways that are not "humanly distinguishable,"[444] so that the world still has the *appearance* of regularity required by the regularity argument above? If this response works, God would not be limited by the "not even once" principle and could act in subtle and diverse ways to reduce suffering—as long as the natural regularities still appeared to hold. This suggestion faces a few difficulties, however. For one, those who advance this position admit that God acts to reduce the suffering of some but not others, which means that they attribute to God precisely the inconsistency that Wildman has identified. Further, it's not clear that they avoid the specter of arbitrariness that, we have argued, threatens to undercut science and rational agency. Scientists who endorse this response would have to say that, although they act *as if* the world were regular and do not expect ever to actually catch God working a miracle, they in fact believe that God is constantly setting aside natural regularities just below the surface of human detection. To believe this is to believe that the natural order is in fact laden with irregularities, however lawlike it might appear to us in practice.

2.2 Second Response

So far we have been addressing the first of the two questions provoked by our main hypothesis: why can't God at least *sometimes* override the regularities of nature when doing so is needed to prevent innocent suffering? We have answered this question in two ways: first, by showing why the development of rational and autonomous agents requires a greater degree of regularity than might initially be obvious; and second, by arguing that occasional divine abrogations of natural law, even if metaphysically possible, turn out to be morally inconsistent with the capacities of a divine agent— not because breaking natural laws is inherently immoral, but because by doing so God would incur a responsibility to intervene in most or all cases of suffering. But this would make it impossible for God to limit the frequency of such interventions and therefore to preserve a universe in which beings like us could evolve.

But have we now, as the saying goes, "proven too much"? In showing why God cannot intervene in such a way as to suspend the operation of natural regularities, have we also shown that God is unable to act at all within the created universe, and therefore that, as Wildman argues, the personal agency of God is simply irrelevant to human experience and action? This is the second question raised by our hypothesis. In contrast to

---

the other side of silence. As it is, the quickest of us walk about well wadded with stupidity."

[444] Nancey Murphy has helpfully formulated the objection in this way in correspondence.

the first, answering this one will compel us to go beyond the logical implications of the hypothesis itself by introducing a substantive claim about the nature of the universe in which we happen to find ourselves.

If God is to be able to perform finite actions that bring about events in the created universe—events that would not have occurred merely as a result of the universe's own internal processes—then, given the restrictions on divine agency we have already established, there must be at least one area or sphere of existence within the created universe where events are not determined by the operation of natural regularities. And indeed there is such a sphere of existence: the sphere of thought, or mental activity, or let us simply say, of mind.

2.2.1 The Anomalous Nature of the Mental

Consider, to be begin with, Donald Davidson's famous defense of "anomalous monism."[445] The term "monism" signals the importance of avoiding dualist theories of mind, that is, those that advocate the existence of two separate yet interacting substances (say, mind and body). Davidson expresses this insistence in the form of (token-token) identity theory: an individual mental event is not a different event from the physical state of affairs on which it supervenes. He nonetheless recognizes the distinction in *type* of event: mental events qua mental are different in kind from the events properly studied by physics. Thus Davidson's position has been described as a form of property dualism in the philosophy of mind.[446]

"Anomalous" signals that mental events are not nomological, not law-governed. Although one expects there to be significant patterns and detectable regularities among mental phenomena, those phenomena are not merely instances of overarching laws that govern all mental events. Intentional explanations have irreducibly holistic features; one cannot reduce them to their neurophysiological components without losing the very thing one hopes to explain: the agent's intentions. For example, there are no independent means for specifying what factors are or are not relevant to an intentional act; hence there is no way to link such acts to a specific part of the physical world in a supervenient relationship.

As Davidson notes, "any effort at increasing the accuracy and power of a theory of behavior forces us to bring more and more of the whole system of the agent's beliefs into account."[447] Of course, generalizations can be drawn across many instances of human behavior or across many behaviors by a given individual. Thus we speak of character dispositions, patterns of behavior, and tendencies manifested by particular groups, societies, and cultures. But there are no grounds for concluding that human behaviors (and perhaps behaviors of other life forms as well) are instantiations of some underlying set of mental laws. By contrast, we have every reason to accept the existence of laws in physics, and at least some biologists support the existence of "general laws" of biology.[448]

---

[445] See e.g., Davidson, "Mental Events," 207–24.

[446] See e.g., Silcox, "Mind and Anomalous Monism."

[447] Davidson, "Irreducibility of Psychological and Physiological Description," 321.

[448] See Kauffman, *Investigations*.

We do not here take a position on exactly how constrained or how "free" human actions are. To accept the holism of Davidson's account of mind is to maintain that, despite the dependence of the mental on the physical, human actions are not determined by the operation of natural laws or regularities. Yet it is not to deny that such actions manifest significant regularities. Patterns of human action may be law-*like,* and rigorous forms of social science may well be possible. Yet they will not be equivalent to, much less reducible to, the nomological natural sciences.[449]

Davidson's materialism compelled him to accept epiphenomenalism, the view that the mental is without causal efficacy in the world. It is not clear to us, however, that either scientific or philosophical arguments require one to accept this conclusion. Jaegwon Kim has argued for the dilemma, "either epiphenomenalism or the rejection of the causal closure of the physical world."[450] At a 2002 conference, David Chalmers responded that the problem of consciousness requires one to select one of three major options: epiphenomenalism, dualism, or panpsychism.[451] Since by "dualism" Chalmers presumably meant, minimally, denying the causal closure of the physical world (CCP), his trilemma naturally supplements Kim's dilemma. What then of CCP?

If one accepts the framework of reductionism in the philosophy of science, the costs of *not* accepting CCP are enormous. After all, if all valid scientific knowledge must be expressed in terms of idealized, purely physical systems, then to challenge CCP would be to challenge the possibility of carrying out the computations on which such knowledge depends. But, as it turned out, reductionist philosophies of science were not able to tell the whole story of scientific knowledge. Distinct explanatory principles come into play at different levels in the hierarchy of complexity (or: at different stages in the emergent process of evolution). Although discovering the precise interconnections and interdependencies between these levels remains one of the important tasks for science, formulating the distinct organizing principles *at each particular level* (e.g., structures and morphologies, functions, purposes) is also a primary task. Only at the "lowest" levels of the hierarchy does one need to rely on models of ideal physical systems that manifest CCP. By the time one begins to study interacting systems that are far from thermodynamic equilibrium—and all living systems fall into this

---

[449] We have argued that the properties of the mental are different enough from physical properties that the two cannot be identified as instances of a single type of property. Among the qualities of mental events, it appears, are their "first personal" or subjective character, their intentional and hence teleological structure, and their non-nomological nature. This difference is more accurately expressed by the term "property emergentism" than "property dualism," however. The latter suggests that there is one and only one fundamental ontological divide in the natural world, that between mental and physical phenomena. (This seems to be the view, for example, of Hasker in *The Emergent Self.*) By contrast, recent studies of emergent phenomena in science and the philosophy of science show that multiple levels of emergence are manifested across natural history. If in addition to the mental many other levels of distinct properties exist in the natural world, "dualism" is not an accurate label for the position.

[450] See Kim, "Making Sense of Emergence," 3–36.

[451] "Theories of Emergence," a conference held at Granada, Spain, in August 2002; see Clayton and Davies, *Re-emergence of Emergence.*

category—the assumption of CCP plays no significant role in the day-to-day quest for scientific explanations. Hence the allegedly disastrous consequences of denying CCP do not materialize.

The post-reductionist or "emergentist" framework that we believe represents the best contemporary account of the scientific process thus supports the rejection of CCP.[452] Dispensing with CCP also allows one to formulate a notion of downward causation that supplements Davidson's anomalous monism in the philosophy of mind. By downward causation we do not mean that a substantial entity such as soul exercises agential causation on the level below it (say, the brain). Instead, we mean that agents with certain *types* of properties—in this case mental properties—carry out actions in the world, and these actions are only adequately explained by an ontology rich enough to include the particular types of causal influences that these agents exercise.

Nancey Murphy has helpfully argued for including environmental factors in accounting for the downward causal effects of the mental in a way that supplements Davidson's holism.[453] These factors are obscured in the standard diagrams, which show a specific idea or mental state M influencing a given bodily state B. M-predicates are a highly complex type of property; they cannot be defined only with reference to a single body and the neurophysiological processes that produce a particular instance or "token" mental state. The ideas of marriage, success, or justice require reference to broad social networks, and such networks play a causal role in bringing about the occurrence of such ideas in persons.[454]

Understanding these "horizontal" influences is indispensable for a viable theory of downward mental causation. The overall psycho-physical state of a person—including her social location, her culturally identifiable range of possible actions, and her beliefs about herself—are causal factors affecting the physical state of her body. The chemistry of her neural synapses will be different from what it would have been if her overall psychological, mental, and social state had been different.[455] Referring to a person's mental states and the various cultural and intellectual influences to which they respond is part of explaining the physical state of her body. Such factors can be included only if explanations are given also in terms of type-type influences rather than in the (to us conceptually problematic) language of "this idea" influencing "this brain state."

What relevance does this account have to debates about divine action? The emergentist philosophy of mind just sketched makes it possible to con-

---

[452] Distinctive properties emerge at many different levels in the process of explaining the natural world. It is not possible to explain the behavior of intentional human agents, nor presumably that of other complex organisms, as manifestations of underlying "covering laws"; Hempel's deductive-nomological or "DN" model for explanation is simply not relevant to scientific practice at these levels. But CCP was alleged to be necessary at all levels of explanation precisely in order that DN explanations would be possible. Drop the DN requirement on explanations, and the main argument for CCP disappears.

[453] See Murphy, "Emergence and Mental Causation."

[454] The roots of this theory of "embedded mind" go back to the early work of Humberto Maturana and Francisco Varela; see their *The Tree of Knowledge.*

[455] Presumably this conclusion is true not only for persons, but for any entities that manifest holistic properties as complex as mental properties.

ceive how there could be a divine influence on human thought that was non-interventionist, that broke no physical laws, and that was consistent with the ongoing pursuit of science. Suppose that, above the level of the mental, there is a yet higher type of property; call it the spiritual. (Theists believe that this property is identified with a specific spiritual being. The situation is more complex for panentheists; if the world is located within God, spiritual properties are also in some sense properties of the one universe.) If the emergentist account of mental causation is correct, then it is possible to apply the same logic to this new level. Just as no natural laws are broken when one explains the behavior of human beings in terms of their thoughts, will, and intentions, so also no laws are broken when one explains their behavior in terms that include the causal influence or "lure" of certain higher spiritual values on their thinking and consequent actions. If a type of property that is irreducibly spiritual does in fact permeate this universe and have a causal influence upon it, then the level of the mental is not the highest type of causality relevant to explaining human thought and action. It may well be difficult to prove that a God exists who lures humanity (and perhaps other life forms) in this way. But the possibility of this additional type of causal influence need pose no more difficulties for scientists and philosophers of science than does the level of the mental.

An emergentist theory of mind thus opens up the possibility of a divine influence at the mental or spiritual level that does not require making an exception to any natural laws. Could a critic then use this conclusion to undercut the "not even once" principle? If human mental life can have some effect on the underlying neurophysiological structures on which it is dependent, why couldn't God—whom we also assume to have mental properties—*also* directly cause changes in the physical world? But attempts to extend divine causal influence in this way clash with a fundamental assumption of the emergence position, an assumption that enables emergence to explain the evolution of new causal properties while preserving the natural regularities on which science depends. Emergence arguments assume fundamental continuities across the various contiguous levels of the natural world. Chemical properties emerge from physical systems, attributes that we associate with life are holistic properties of natural structures (cells and groups of cells), and mental properties are predicated of brains and the persons who have them rather than of a human soul or spirit. Emergence is a scientific hypothesis because the causal relations in each case can be reconstructed and studied. Where bottom-up explanation and prediction are not possible, empirical reasons can be given (e.g., increases in complexity, phase transitions, system effects) for introducing a new explanatory level. In every case, the result is a specified field of empirical study and the seamlessness of natural explanation is preserved.

Thus a theory of emergent mental causation does not lend support to claims for direct divine influence within the domains of (say) physics or chemistry. As one ascends the ladder of emergence, however, the patterns gradually change from the strict exceptionless laws of physics, through the law-based but less precise adaptationist arguments of biology, to the general principles used to explain cultural, social, and psychological phenomena. Once that point has been reached, seamlessness of explanation—and equally downward dependence!—no longer has a clear definition. At that point it becomes possible for divine agency, which is certainly not down-

wardly dependent on, or even contiguous with, any lower level of reality, to
have an influence on events within the finite universe, and to do so without
disrupting the seamless explicability of natural phenomena.

We have argued in this section that the sphere of mental activity is one
area of existence in which God can act without incurring an obligation to
prevent events like the Indian Ocean tsunami. We have also suggested
that, if there are any emergent levels above the mental, they would be open
to divine influence in a similar way. But is such influence ever possible *below* that level? Might it be, for example, that there are phenomena in the
biological sphere that have "anomalous" features, features sufficiently
similar to the anomalous nature of human mental functioning that divine
influence might be operative there as well? Answering these questions
means sifting through complex empirical and conceptual material in the
biological sciences and would require a much longer presentation. One
should note, however, that the position we have defended would allow for
divine influence on organisms prior to *Homo sapiens* as long as the same
conditions were met. Nonetheless, our sole aim here has been to show that
the realm of the mental represents at least one natural sphere in which
divine influence can occur, without overriding the regularities whose preservation is a necessary condition for the emergence of finite rational agents.

2.2.2 Does the Problem of Evil Return in New Form?

Suppose all this is correct. Suppose it is indeed the case that God can perform intentional acts of communication with the minds of those creatures
that are capable of receiving what God conveys. Are we not faced, once
again, with Wildman's "argument from neglect"? It isn't as if a divine agent
could only prevent innocent suffering by performing physical miracles, such
as stopping or reconfiguring the seismic movement that caused the December 2004 tsunamis. Why would a benevolent God have refrained from
somehow warning those in its path that the wave was on its way?

The tsunami example, tragic as it is, is perhaps worth dwelling on.
After all, our whole account entails the existence of a God who, on one
hand, refrains from altering the regularity of natural processes and, on the
other hand, performs purposeful actions in and through the thoughts of
human beings. Unless we suppose that God's refusal to interfere with
physical processes entails that God is simply unaware of those processes,
we have to assume that God knew instantly that the Sumatra-Andaman
earthquake had occurred, even if this was (on some accounts) something
God could not have known ahead of time. But even if we were to suppose
that God was unaware of physical events like undersea earthquakes, the
very hypothesis that God performs acts of communication with human
minds entails that God is aware of what human beings think, say, and experience; and that entails, for instance, that God must have known the
tsunami was headed for Sri Lanka within moments of its striking Sumatra.

Assuming, then, that God knew about the impending disaster in Sri
Lanka some two hours before it occurred, how do we explain why so many
thousands of Sri Lankans (and their visitors) continued to work and play
along the beaches without, apparently, the slightest hint that death was on
its way? Should we say, for instance, that God is aware of our thoughts but
can't perform the inferences we can and therefore could not extrapolate from

the wave's hitting Sumatra that it must be on its way to Sri Lanka as well? Or should we suggest instead that God is aware of our thoughts but, lacking our referential framework, is simply unable to *understand* our thoughts? But in either of those cases, what would it mean to claim that God was capable of performing purposeful acts of communication with human minds?

We seem to face a dilemma: either God is capable of conveying information to human minds, in which case God's failure to do so in a situation like that of the tsunami leaves the theist defenseless against the argument from neglect; or divine and human thoughts are so incommensurable, and hence so mutually untranslatable, that we are compelled to abandon any meaningful notion of communication between God and human minds. One can escape this dilemma, it would seem, only by identifying a *kind* of information that God might convey to human minds whose content would be such that it could not serve as a means of warning humans of impending disaster or some other particular cause of innocent suffering.

Formulating this dilemma helps to clarify both the logic and the limitations of the argument we have been making. It points first to the intentionally formal nature of the account of divine action we have been providing. The truth is that one cannot infer from general constraints of the sort we have explored what *specific* constraints might apply to divine communication in any given instance. The objector to personal theism argues that if all divine action is impossible, God becomes "virtually irrelevant" to human action. And yet if God were to intervene in such a way as to impart information that *directly and inescapably* affects an individual's psychological state—for instance, by conveying an unambiguous warning of an impending disaster—then God would be obliged to intervene in every case in which an intervention of this kind would yield a better outcome. The dilemma can be resolved as long as it is in principle possible for God to communicate with finite agents in such a way that God does not incur an obligation to prevent or relieve all cases of innocent suffering.

How, then, might God communicate with human minds without incurring such an obligation? Perhaps divine communication takes an *axiological* form, where God presents to a person's consciousness a value that she is free to embrace, pursue, reject, or ignore. Or it might take the form of God's bringing about the kind of religious experience in which the subject becomes aware of God's presence. Or those two kinds of communicative action might be combined, in the form of an experience in which a sense of the divine presence leads to an apprehension of axiological truths.[456]

Perhaps an example will help to make this possibility clear. Consider the case of a medical technician who unwittingly prepares a fatal overdose of what would otherwise have been a life-saving medicine, because he failed to pay close enough attention to the placement of a decimal point in the physician's hastily composed instructions. God might have prevented this accident in countless ways, for instance by placing in the technician's mind a command to double-check the prescription; or perhaps by causing him to feel intensely anxious at some appropriate juncture in the act of measuring out the dosage. Had the technician ignored either of these divine signals, God could have responded by, as it were, turning up the volume, until the technician was left with no choice but to comply!

---

[456] James, *Varieties of Religious Experience.*

In these and no doubt innumerable other ways, once again, God could intentionally act in the mind of a finite agent to prevent an error from having its fatal effect. In so doing, however, God would incur an obligation to intervene in similar ways whenever a human failing was about to have disastrous consequences. The result would be a negation of finite rational agency at least as complete as the negation that would result from continual divine intervention in the sphere of physical events. By contrast, God could present to an agent the general value of being careful as an expression of love for one's neighbor, and the agent would be free to embrace, or not to embrace, that value as a component of her motivational set.[457] In that case, no negation of finite agency would occur.

As this example shows, what matters is not so much the mode of divine communication as whether the effects of that communication are mediated through the agency of the individual with whom God communicates. It turns out not to matter, apparently, whether the divine influence comes in propositional or affective form. What God must refrain from doing, if the problem of evil is to be answerable, is giving us thoughts or feelings that compel an automatic or reflexive response, because otherwise God would incur an obligation to prevent or correct our mistakes whenever they might occur. And a world in which mistakes were impossible would be a world in which finite rational agency was also impossible.

By restricting divine agency in this way, have we once again impaled ourselves on the second horn of Wildman's dilemma? Have we reduced the scope and efficacy of divine action to the point where it indeed becomes irrelevant to human moral struggles? We contend that we have not. For on our account, not only has God purposely created a universe in which there could evolve beings capable of making moral choices and entering into communion with God. God also purposely and graciously responds to, and interacts with, those beings, accompanying them on their journeys, inspiring their joys, and luring them, gently, into harmony with the divine will.

Is all of that "virtually irrelevant" to our moral struggles, as Wildman argues? Does moral relevance consist only in the power to perform dramatic acts of intervention? If so, not only is God morally irrelevant; so are all the great moral exemplars of the past who, being dead, are even more powerless to act on behalf of justice!

God, on the hypothesis we have been exploring, is not only the creator of those natural regularities that enable finite moral agents to exist in the first place; God is also engaged with us, once again, through modes of gentle guidance, growing illumination, and persistent attraction. Such a God may not be able to stop a fatal mudslide, or warn the villagers of its impending arrival. But it by no means follows that such a God is morally irrelevant.

3 The Eschatological Dimension

We have another response, however, to the moral prong of the argument from neglect. If, despite all that we have argued, Wildman's argument should turn out to be correct after all, and if one were forced to the conclu-

---

[457] Of course, if the agent takes on this general value, it will become part of his pattern of agency and will presumably affect other of his actions as well.

sion that God was not a personal agent, what would one have to say about the moral relevance of theism in what would then be its non-personal form?

At the very least, it seems to us that the moral relevance of that version of theism would tend to be confined to those who already, as the Gospel saying goes, "have their reward." It might in various ways inspire the small minority of those with leisure to contemplate the faceless mystery it posits. But if God has not acted to create a universe containing finite persons about whose future God cares—say, because God is incapable of acting or caring at all—then suffering in this life, and indeed the fate of the vast majority of all human beings who have ever lived, is unredeemed and unredeemable, and their hope is not only false but cruel. There can be no hope of any future consummation. It seems a not unreasonable *tu quoque* to ask the arguer from neglect about the moral contribution of such despair to "the practical moral struggles of our deeply unjust world."

Evoking the hope of an eschatological fulfillment, however, raises one final possible objection to the somewhat minimalistic picture of divine agency that we have developed, partly although not exclusively in response to the argument from neglect. If innocent suffering is a necessary part of a universe in which rational and autonomous agents can evolve, and if God for good reason refrains from preventing it, why should we suppose that God has the power or disposition to bring about a different "cosmological epoch" in which innocent suffering would be absent? In correspondence with the authors, Wesley Wildman (source of the phrase just quoted) has posed this challenge in the following lucid terms:

> . . . God's nature[,] including morally relevant aspects of God's nature[,] must be consistent across all cosmological epochs because of your view of divine creation. Pictures of that nature derived from one epoch must be relevant somehow to all epochs, accordingly. If God can and does act in a way you approve in a cosmological epoch you want to imagine but don't actually live in, then God should be able to do that in this cosmological epoch, also, on pain of moral inconsistency.

This is a powerful objection, among other reasons because the need to preserve God's moral consistency has played so crucial a role in our response to the argument from neglect. We appear to face one final dilemma: we can either give up the notion that God is morally consistent, thereby jettisoning our explanation of why God cannot at least *occasionally* intervene to prevent innocent suffering; or give up the notion of different cosmological epochs, at the cost of undermining the plausibility of our claim that God seeks communion with finite persons.

Is it clear, however, that God's altering the conditions of finite existence in a later cosmological epoch would entail the kind of moral inconsistency Wildman supposes? What if God has a good reason to alter the conditions of finite existence between this and a subsequent epoch, precisely because so doing is the best way of realizing God's overarching moral purpose? On the hypothesis we have been developing and defending throughout this paper, God's ultimate purpose is to bring about the existence of finite persons with the capacity to enter freely into a loving fellowship with God. Again on our hypothesis, persons fitting that description can only come into existence if they are not simply projections of the divine will but have emerged from a process designed to endow them with (relative) autonomy.[458]

According to John Hick, who employed a similar approach in his early book *Evil and the God of Love,* it is reasonable to suppose that such a process would have to continue in worlds beyond this one (because so many persons in this world lack the opportunity to develop morally and religiously at all, or to develop as fully as God would presumably desire that they should), and this future epoch (or these epochs) would indeed exhibit conditions identical or at least functionally similar to the ones that cause suffering in ours. Perhaps the process of development is never-ending, and God's actions in all future epochs are restricted by the very same considerations that restrict them in this one, although the persons who move through those successive worlds continue to grow in their love of God and neighbor. Or it could be that, in the next epoch (or perhaps at the end of a very long succession), each person reaches a stage where further "soul-making" (in Hick's Keatsian phrase) is no longer needed, at which point she may enter an epoch with an entirely different set of cosmological conditions.[459]

Again, the underlying assumption here is that God has "in mind" the same desideratum across however many epochs it takes to develop persons to the point to which God intends that they develop. Across all these epochs, divine action is limited only by the cosmological constraints required to produce and sustain autonomous persons. It is not obvious that the conditions necessary for a world in which autonomous persons can develop (i.e., the conditions necessary for *this* world) are also necessary for a future world in which fully developed persons can flourish in the communion for which they were intended. Hence the evils inherent in this world might not be entailed by the next.

In short, it is perfectly possible for God to create other and better worlds without contradicting what, on our hypothesis, was God's purpose in creating this one. And that hypothesis, we submit, provides a sufficiently plausible answer to the argument from neglect.

---

[458] Note that the emergentist theory of mind developed in section 2.2.1 coheres with and at least indirectly supports this processual understanding of the divine creative intent.

[459] Hick develops his "soul-making" theodicy in *Evil and the God of Love* (see n. 3 above).

TOWARDS A CREATIVITY DEFENSE
OF BELIEF IN GOD IN THE FACE OF EVIL

Terrence W. Tilley

> Catholic theology may plausibly suppose that the random, experimental character of evolution is consistent with a divine love that longs for the world to make something of itself instead of being constantly tinkered with or pulled like a puppet.   —John F. Haught[460]

1 Introduction

The present chapter seeks to bring insights about God and evils, especially "natural" evils, by critically introducing some strategies from recent work in analytical philosophy of religion into the discussion of God and evil in the context of science. I begin with preliminary comments setting out the basic orientation from which I work. I then introduce the dominant strand of work on God and evil in the analytical tradition, the Free Will Defense (FWD), in the form developed by Alvin Plantinga over the last forty years. After acknowledging that FWD solves the logical problem of evil, I discuss philosophical and theological problems with FWD. I then sketch a different form of defense, the Creativity Defense (CD). I explore how CD can defend faith in God in the face of natural evils in a way that avoids some of the problems of FWD. I show how CD can account for a robust understanding of social evils, absent from accounts of FWD. The conclusions suggest how this approach might be congruent with others who take natural science seriously as a dialogue partner for Christian theology.

2 Preliminary Comments

In a preliminary chapter, I sketched a number of claims that I (and others) have argued for elsewhere. I will not discuss them further here, but simply summarize them, with references to supporting published arguments. My basic contention is that the problems of evil generally need to be dissolved, not resolved. The main thesis of my major work in this field, *The Evils of Theodicy,* is that attempting to explain evils, that is, doing theodicy, is itself an evil, as it contributes to the problems, not the solution, of the problems of evil.

2.1 Theodicies and Defenses

It is crucial to note the difference between *defenses* of belief in God against the accusation that believers are incoherent if they believe in God in the face of evil, and *theodicies* that attempt to explain the reasons God has for there to be evil in the world. I have argued elsewhere that the former is a legitimate academic enterprise and the latter a practice that creates more evils than it explains.[461]

---

[460] Haught, "Darwin and the Cardinal," 39.

[461] Thomas Tracy, "Lawfulness of Nature and the Problem of Evil," 2.2 (this vol.) seeks to form a "theologically 'thick' defense, i.e., an explanation of evil formulated

It is most important to note that these are *academic* practices. Few, if any, will be consoled or strengthened in moments of grief by either theodicies or defenses. The appropriate response to the grief, regret, anger, and remorse people feel at times when God seems utterly absent to their suffering, loss, or victimization is not to explain or defend God. Rather, appropriate responses are more in terms of "constant in companionship," a practice necessary, but never sufficient, to sustain a fellow-sufferer in faith.[462]

2.2 The Problem of Evil Emerges in the Enlightenment

Second, I have argued that the contemporary form of problem of evil and the response of theodicy is a discourse practice distinctive to the Enlightenment.[463] Just as Aristotelian biology is incomprehensible, clearly wrong, or simply nonsense if understood as commensurable with contemporary biology, so is Augustinian and Thomistic "theodicy" incomprehensible, wrong, or nonsense if these theologians are thought to be replying to the sorts of questions raised by David Hume or J. L. Mackie. The "demystification" of the world by scientific and humanistic discourses, and the unwarranted extension of the practices and concepts of science to be used as the proper pattern for all ways of understanding what there is, give modernity a quite different problematic from the ancient and medieval worlds. This profound shift is all too frequently ignored. Making empirical, verifiable science a monolithic paradigm for knowledge resulted in a "flatness" of method and description and a "monochromatic" vision of being and beings not found in earlier times. Such modern flatness resulted, in turn, in conceiving God's agency as essentially no different from creaturely agency and forcing God into the increasingly minute "gaps" in human

---

with an eye to its Christian assertability." Tracy has suggested in discussion that the creativity defense discussed below is such a "thick" defense. However, because his "thick defenses" are explanations, they are de facto theodicies. The creativity defense mounted below is *not* an explanation, but rather a defense that is not as much at odds with the rest of Christian doctrine as the Free Will Defense is. Hence, I find the tasks Tracy outlines for a "thick defense" indistinguishable from those for a theodicy, subject to the strictures I have discussed in *Evils of Theodicy*.

[462] D. Z. Phillips has recently argued that the separation of logical from existential problems of evil is "spurious," and that the former is rooted in the latter; see his *Problem of Evil and the Problem of God*, xi–xii. While I agree with him about the rooting of the logical problem in the "existential" (*not* "evidential") problems, that does not imply that the *academic* solution to the logical problem should be thrown out. Rather, logical problems require logical solutions, i.e., the learned accusation of believers' incoherence in believing in God in the face of evil deserves a logical defense from believing philosophers and theologians. The errors emerge when someone tries to develop a theodicy, to explain to what good purpose God has put evils into the world—thereby ignoring the real problems.

[463] The burden of showing that ancient, medieval, and modern discourses on God and evil are incommensurable is borne by Kenneth Surin, *Theology and the Problem of Evil*; Tilley, *Evils of Theodicy*, develops this point by showing how some modern readings of ancient texts generate confusions, not clarification. Arguments developing the points made in these preliminary remarks are developed in the texts of Surin and Tilley; Phillips goes too far in his accounts, although we are all fellow-travelers in rejecting theodicy.

knowledge.[464] Thus, the supernatural (the uncreated Divine Agent) and the preternatural (created, but extraordinary, often spiritual agents like angels, demons and "gifted" or "possessed" humans) were collapsed into one category practically irrelevant to serious investigation in the face of the rise of science (and scientific history).[465] Few, if any, today advocate returning to a medieval cosmology or anthropology. Yet the loss in modernity of a robust sense of the supernatural perfecting the natural ("*gratia perficit natura*") characteristic of the medieval worldview makes it at least difficult to hold a reasonable faith. Nothing can be the same after the loss of any relevant concept of the supernatural for both scientific and humanistic disciplines—and that means that the problems of evil are not the same before and after the Enlightenment's demystification of the world.

2.3 Theodicy as a Discourse Has Bad Results

Third, I have argued that the modern discourse practice of theodicy as explanation of why God allows or causes evils in the world (which is *not* a "defense") tends to participate in the scientific paradigm of construing all reasonable belief, including faith in God, as based in hypotheses that require proof. This treatment of rational belief in God as hypothetical has two profound consequences for construing the world: It leads both to the effacing of social evils and to the construing of all evils (at least as a class, if not each of the individual evils in the class) as instrumental to an ultimate good and, thus, not genuinely evil.

The modern discourse of theodicy takes the classic understanding of evils as *poena* and *culpa* (rather roughly "penalty" or "bad result," and "sin" or "fault," respectively) as if these are equivalent of "suffering" and "sin." But this modern pair of concepts is quite different from the ancient ones. Classic Christian discourses on the problem of evil construed all suffering as deserved *poenae*; modern discourses presume that some suffering is undeserved—and that's the problem central to modern discourses, the problem of "gratuitous" evil, and the problem they try to solve! Their concepts and their problematic are essentially different from ancient discourses on evil. In so doing, modern discourses on evil simply efface the notion of social evil or social sin, in large part because of the inability of modern individualistic thought to ascribe responsibility for bad acts (sin) to groups, while ancient thought can ascribe *poena* and *culpa* to groups. Hence, modern theodicies—and some defenses, as well—efface social evils and have difficulty dealing with "natural evils." I will develop the significance of these points below.

Modern philosophy of religion in general and the discourse of theodicy in particular construes the reality of evils as evidence against "the God-hypothesis." Treating belief in God as if it were a hypothesis subject to verifi-

---

[464] Elizabeth A. Johnson persuasively argues that inserting God into the "gaps" confuses something "missing" in nature with the "openness of natural systems"; see "Does God Play Dice? Divine Providence and Chance," 9. The result of theologians' accepting of the flattening of being noted above is the loss of the distinction between primary and secondary agency, thus collapsing divine agency into secondary, creaturely agency, a point also explored briefly by Johnson.

[465] Lorraine Datson traces this development in "Marvelous Facts and Miraculous Evidence in Early Modern Europe."

cation or falsification and as acceptable only upon sufficient evidence is itself a falsification of the commitment of religious faith and a misconstrual
in the basic issues involved in religious commitment.[466]

The modern discourse of theodicy tends to claim that all evils, or types
of evil, are instrumental. They implicitly deny that real evils are genuine
evil. An evil is a genuine evil just in case it is an event, act, person, or state
of affairs without which the world would have been better. Genuine evils
are gratuitous evils. By explaining evil, theodicists construe evil as an instrumental good. While they recognize some particular evils as incomprehensible, they find that each class of evils is construed as necessary for the
occurrence of a greater good, thus rendering them as merely apparent evils.
Modern theodicists in effect declare what is evil to be good. Rather than
recognizing the genuine evils in the world as a challenge to the constancy of
faith, modern theodicy construes evils as evidence against the God-
hypothesis.

In an attempt to respond to the accusation that religious believers are
incoherent in believing both in God and that there are genuine evils in the
actual world, Christian philosophers have developed a defense of the coherence of belief in God in the face of evils, the FWD.

3 The Free Will Defense

I have supported the aims of the FWD in the past.[467] However, the lack of
connection between its conception of free will with other historic and contemporary concepts central to a lively Christian faith is worrisome, to say
the least.

In brief, FWD, as developed by its most effective contemporary propo-
nent, Alvin Plantinga, argues against the coherence of the claims of the
objectors and for the possibility of theism. Crucial to his position is his argument that it is logically possible that even an omnipotent God with middle knowledge could not actualize possible worlds that contain only free
beings that only do what is right.[468] If God makes creatures free, then God
cannot make them choose only the good. If God makes creatures choose
only the good, then they do not choose freely. If God creates them free, God
is responsible for creating them, but not for their free choices; if God creates them good, then they are not responsible for choosing between good
and evil.

---

[466] For a sketch of an alternative religious epistemology, see T. W. Tilley, *The
Wisdom of Religious Commitment.* For my recent work on the relation of scholarship
to the commitment of faith, see my *History, Theology and Faith: Dissolving the Modern Problematic,* chapters 9–11.

[467] See Tilley, *Evils of Theodicy,* especially 130–33, 252, and "Use and Abuse of
Theodicy."

[468] For an early formulation of FWD, see Alvin Plantinga, "Free Will Defense." A
more developed formulation is in his *God, Freedom and Evil.* A further development,
a defense against the evidentialist challenge, can be found in his "Probabilistic
Argument from Evil." More recent literature abounds and is often exceedingly technical; thanks to Thomas Tracy for his help in formulating this brief account accurately.

3.1 The Philosophically Problematic Status of the Free Will Defense

FWD requires that entities be free in a libertarian sense; that is, for a choice to be free it must be uncaused (even though the conditions in which the choice is made in some way may affect the choice). The libertarian sense is opposed to the "compatibilist" sense of freedom, which finds free will "compatible" with the will being determined by factors such as a person's character. It is not clear to me that this duality exhausts the possibilities if the universe is characterized by quantum indeterminacy and chaotic complexity, but that issue is beyond the scope of this chapter. FWD requires libertarian freedom, no matter what other accounts there may be.

God knows what is good for each entity, human or otherwise. Why does God not just determine each entity to attain that good? The traditional answer, whether convincing or not, is that some goods are not attainable without libertarian freedom. The problem then becomes specifying that good or those goods, for it would have to be a good or goods attainable for beings free in a libertarian understanding of free will, but not attainable for entities that are free in a compatibilist understanding of the relationship of determinism and freedom.

I am not convinced that this problem "behind" FWD has been resolved. Is libertarian freedom a coherent concept? What are the goods that require that God create beings with libertarian freedom to achieve them? While it is possible that libertarian freedom is coherent (the issue is argued by philosophers) and it is possible that some goods are such that libertarian freedom is necessary to achieve them (and, by implication, that all the evils in the actual world are "worth it" to a God who might have created free beings fettered instead), FWD does rely on a pyramid of possibilities.

These issues do *not* show the inadequacy of FWD. Rather, the question is the logical status of concepts used in FWD. Logically, FWD is undefeated. Yet if there are successful counterarguments to the cogency of its presuppositions, that is, if they could be shown to be rationally untenable, such counterarguments would undermine FWD.

While FWD does not require that one take faith in God as a hypothesis, it does not explicitly challenge or deepen the "flattening" of the world brought about by modern, scientifically oriented discourses, a flattening incompatible with a robust Christian faith.[469] As such it tends to see God as

---

[469] Philip Clayton and Steven Knapp have raised the following question: Was "God metaphysically constrained or limited" in creating free (or creative, as I argue below) creatures, "or are the limitations a result of a free divine choice? Since this is perhaps the major faultline in the discussion, it will be crucial to take (and to defend) a position on this question." My position is that the question is irrelevant to the task at hand because of the nature of a defense. I am not offering a metaphysical explanation of what God can or cannot do. I am offering a logical and rhetorical analysis of what we can or cannot say about God, evil, and creativity. I am making a number of specific claims I believe constituent in a robust Christian faith. I am *not* attempting to explain why the propositions articulating these claims function in a system, but only to show that they are not incompatible with each other as convictions expressing Christian faith. I also need at least minimally to suggest and ideally to show that these claims are not undermined by plausible claims legitimately extended from other practices in which we engage, that is, that Christian faith is not found in a watertight epistemic compartment with an impermeable membrane between "it" and all the other practices in which we engage (such as scientific inquiry and research)

if God were another (secondary) cause alongside other causes in the world, an infinite entity with attributes somehow comparable with finite entities. This move renders God as in competition with creatures. While FWD does not support a god of the gaps, neither does it suggest an alternative.

3.2 Theological Problems of FWD: Grace and Freedom

The FWD has problems with grace and with the relationship of grace and freedom. The first problem is with the Christian tradition. Augustine's and Aquinas's understandings do not fit very well with contemporary ways of putting the problem of human freedom to choose. To say that they are "compatibilists" is anachronistic, though probably the best "fit" of Augustinian and Thomistic positions in contemporary categories.[470] Moreover, Augustine's understanding of the will varied considerably over the course of his writings.[471] Augustine evidently thought the "uncaused" choice for evil inexplicable; if he was a compatibilist, then it would follow that he would find FWD to rely on an inexplicable concept and to aim at providing an answer that, by its nature, cannot be given. Aquinas's distinctions between primary and secondary causality mean that his views do not fit easily with contemporary discussions in this area.

Personal evils, usually identified as moral evils, that is, evil deeds and sin, are usually thought to be well-handled by FWD. The literature on this issue over the last forty years is immense. My judgment is that, given the waning of debate over the logical problem of evil over the last decade or so, the logical problem of evil can be taken as resolved because it is undefeated despite many challenges from skeptics. The solution, however, is not very satisfactory from a theological perspective. As William Wainwright once concluded, "The free will defense is . . . incompatible with some traditional understandings of God's power and sovereignty—that of orthodox Islam, of some types of theistic Hinduism (Dvaita Vedanta) and of Christian theologians like Augustine, Aquinas, Luther and Calvin."[472] Wainwright clearly portrays the cost—at least for orthodox Muslims, some Vedantists, and traditional Christians—of accepting FWD, specifically its requirement of libertarian freedom. The cost to Christian theology is especially acute. The key theological cost seems to be that "grace" simply drops out of the picture, and that is a major problem, for a classic Christian claim is that freedom-

---

and the various beliefs that they generate in us as a result of that engagement. For an exploration of my practices and the generation of beliefs in a particular context, see my *Inventing Catholic Tradition*, 75–87; for the relationship between history as an academic practice, theology as a practice, and faith as a practice, see my *History, Theology and Faith*, 106–91; I would assume similar status for scientific practices as I do for the practices of historians in that text.

[470] Wainwright, *Philosophy of Religion*, 88.

[471] For instance, T. D. J. Chappell finds that Augustine has three different views in his writings: that *voluntas* is uncaused, caused by itself, and caused by causes other than the *voluntas* itself; he finds the last to be Augustine's "most convincingly argued view about normal voluntary actions," although the original will to evil (alone) may be best understood as *ex nihilo* as in the view that *voluntas* is uncaused; see *Aristotle and Augustine on Freedom*, quotation at 149; also see 176–207.

[472] Wainwright, *Philosophy of Religion*, 88.

without-grace is finally not freedom of the will at all, but a "bondage of the will," a determination of the will that actually brings about the destruction of the possibility of human free acts, a bondage not intrinsic to human nature, but brought about by "original sin" (a concept open to construal in many ways).

Interestingly, "grace" seems to drop out of Plantinga's work. In *Warranted Christian Belief,* Plantinga writes extensively about the noetic and affective effects of sin.[473] They are pretty dreadful. Original sin "carries with it a sort of *blindness,* a sort of imperceptiveness, dullness, stupidity" (207). It also skews our affections, giving us "a sort of madness of the will," (208). Although we may know in some way what is right, we "find ourselves drawn to what is wrong" (208). What Plantinga does not do at this point is to show how the key propositions in FWD about free acts/choices are compatible with the blindness and madness caused by sin. If our intellects are tremendously darkened and confused and our affections substantially disordered so we are drawn to what is wrong, how can we be said to be sufficiently free to choose between good and evil?

What is the assertion of "free will" worth if our epistemic and affective status is so disordered by the universal effects of sin? Indeed, on this account, sin seems so pervasive and human life so disordered that libertarian freedom seems an unactualizable possibility. It may be logically possible that all evils are due to the choices of libertarianly free creatures, but whether this is more than a mere logical possibility—perhaps unactualizable in the actual world described in *Warranted Christian Belief*—it is not clear how Plantinga's concepts of sin in his understanding of epistemology (what he calls the extended Aquinas/Calvin [A/C] model) is consistent with the fully libertarian concept of human freedom necessary for FWD.

Moreover, although Plantinga frequently speaks of the testimony of the Holy Spirit and the gift of faith, what is remarkably absent from his discussion is the term "grace." He writes of regeneration "curing the will" (304). He utilizes Edwards's *Religious Affections*—a treatise in large part devoted to distinguish indeterminate from truly *gracious* affections. Yet he writes without deploying the concept *grace* in any significant way as far as I can see.[474]

---

[473] Plantinga, *Warranted Christian Belief,* 206–28. Further citations to this work in this section are given parenthetically in the text.

[474] If freedom is construed contracausally (or libertarianly), as in Plantinga's FWD, a problem with the causal effectiveness of grace also arises. One could, of course, assert a universal gift of grace, so that all are enabled to act freely, even though all do not actually choose what is good, of course, when they make free choices. Such universal prevenient grace would be construed as just what makes contracausal freedom possible. But then "grace" could have no other "causal" effect because "grace" would then interfere with the exercise of libertarian freedom, a position that would not fit well with Plantinga's reformed (in the Calvinian tradition) theological stance. If grace were said to "strengthen" the will of some, but not others, questions of divine preference would arise. So would questions of whether such wills were free enough to choose evil. Once issues of grace are connected to issues of freedom, as in classic Christian theologies, it is not clear how supporters of FWD could have a robust concept of grace consistent with their concept of freedom and divine sovereignty.

One might say that Plantinga is not doing Christian theology in his work, but a fundamental Christian philosophy. A philosopher need not deploy all the Christian theological concepts. Indeed, to do so might undermine this philosopher's most important goals: to show belief in God not rendered irrational by the problem of evil and to show belief, even faith, in God rendered rational by the extended A/C model. And if the philosopher's work is separate from the practice of religion and the systematic theologian's work, this view makes sense.

Yet Plantinga's omission of grace from what is de facto a theological anthropology is a significant problem for those who find grace a central Christian concept. Perhaps one could say that the concept is inchoately present. But it is unambiguously absent from FWD—and evidently has to be for FWD to work, unless revised as suggested above with a minimalist notion of prevenient grace—and it is unmentioned when Plantinga is writing of the work of the Holy Spirit and the production of faith in *Warranted Christian Belief.*

It is not an argument against FWD itself that it severs connections with so much of the Christian theological tradition. All that it needs to be to be effective is to be internally consistent, and thus possibly true, to defend belief in God as traditionally conceived being compatible with actual evils in the real world. Libertarian freedom is necessary for the defense, but logically it needs only be possible, not actual, for the defense to work. That severance evidently also affects Plantinga's formulation of his extended A/C model in religious epistemology. The consistency of FWD with that work is unclear (at best). However, it does mean that believers cannot really use FWD to solve the logical problem of evil unless they part with Augustine, Aquinas, Luther, Calvin, and others and accept a libertarian notion of freedom and, perhaps, avoid a robust doctrine of grace. Is developing a logical defense worth the price? That is not a properly philosophical question, but it strikes me as a relevant one for believers who wish to defend their faith philosophically.

Hence, I propose an alternative, a Creativity Defense. The basic reason for formulating CD is that FWD simply does not handle either natural evil or social evil plausibly. CD also has a place for grace.

4 A Creativity Defense

A "creativity defense" of the compatibility of believing in God with recognizing the reality of evil and evils in the world God created does not reduce God to a "secondary cause" within the world, an agent whose freedom and creativity is somehow commensurable with and in competition with the free (and creative) agency of creatures. It also does not attempt to say what God's reasons are for creating a world with evils in it, even horrendous evils.[475] Finally, it suggests that God's grace is in the world as the gift of

---

[475] Throughout this chapter, I utilize some concepts evolved in process philosophy and theology. I do not, however, accept process thought as a system. First, defenses do not entail commitment to specific systems. It could be argued that some philosophical systems are incompatible with some defenses, but the necessary claim that all systems save one are incompatible with a defense seems implausible. Second, process philosophers tend to work in, rather than question, the modern episteme in a way that I find incompatible with any robust Christian faith. Whitehead

creativity, a gift visible in a universe that evolves by chance in the natural realm and by constructions in the social realm. Chance has yielded increasing complexity and consciousness in the universe that makes the social possible.

The Creativity Defense (CD) sketched here can, from a certain perspective, be seen as a version or an extension of FWD. Certainly, CD owes much to the formulations and arguments of FWD. But CD is compatible, I believe, with most accounts of human freedom, only presuming that the accounts proffered have room for the reality of creativity.

FWD and CD formulate the logical problem of evil in the same way, using the following sorts of propositions:

P: God is omniscient, omnipotent, and wholly good.[476]

Q: The actual world contains genuine evils, natural, personal and social.[477]

The logical problem of evil claims that P and Q are incompatible propositions. The logical problem can be resolved if one can find another *possibly* true proposition r, (note that this does not have to be an *actually* true proposition or even one that the proposer believes; but it helps rhetorically if the r proposition is at least plausible), such that p + r entail q. One such proposition is supplied by FWD. Another is R:

---

wrote, "In the first place, God is not to be treated as an exception to metaphysical principles, invoked to save their collapse. He is their chief exemplification" (*Process and Reality* [1929], 521). Taken as an expression of the medieval *via eminentiae,* this claim is unexceptionable. However, process thought tends to treat God as another actual entity, only a big one, and our talk about God as univocal with talk of other actual entities rather than analogical. Whitehead is surely correct to say that God is not to be invoked to rescue metaphysical claims; but neither is God's freedom or creative power to be understood univocally with that of created actual entities without reducing God to the ontological level of creatures—the very move which generates the modern version of the problem of evil. Whitehead's claims that it is "as true to say that God creates the world, as that the world creates God" and that both God and the world "are in the grip of the ultimate metaphysical ground, the creative advance into novelty" (*Process and Reality* [1929], 528, 529) indicate that he does participate in the modern monochrome ontology. Hence, I do utilize the insights of process theism, but do not accept process philosophy or theology as a system.

[476] Numerous caveats could be made about each of these concepts, and the body of literature on their terms is simply immense. For present purposes, I will take it that "omnipotence" means that we can say that God can do whatever is logically possible for God to do, that "omniscience" means that we can say that God knows all there is to know, and that "omnibenevolence" means that we can say that God wishes each and every entity in the world to thrive. Assumed in this proposition is another: that we can say that God is the creator of the world, but the mode of creation can remain unspecified as the present work is not to present a theory or philosophy of evil, but to defend religious views.

[477] For present purposes, I treat the logical problem of evil; thus, the question of the quantification of evils and the problems the amount of evil in the world creates for belief in God is beyond the present chapter. Moreover, the quantification issue really is at home in discussions of the "evidential problem of evil," the logic of which assumes that belief in God is a hypothesis defeasible by contrary evidence, a position that seems incompatible with a robust religious faith.

> R: It is not possible for God create a world in which
>   entities could exercise creativity and which did not
>   contain genuine evils.

Properly understood, this r combined with p entails q. This is a "creativity
defense," or CD for short.[478] Like FWD, we have options. We can deny p or
one of its constituents; that renders the problem irrelevant. We can deny q,
which is the tactic finally taken by theodicists. Or we can explore r to see if
it is possibly true and, with p, entails q.

CD envisions that God creates a universe that can evolve entities that
reflect God's own image and likeness, the image of God as creator. University of Georgia philosopher Will Power once wrote, "Given any conceivable
or describable socio-temporal order, God does all that is possible to do to
eliminate or minimize evil and maximize good within the optimal metaphysical conditions which God has primordially established wherein the
possibilities for good outweigh the risks of evil."[479] If the possibility of
evolving beings with authentic creativity is a primordial condition of the
universe, then it is possible that God created such a universe, that God
loved the universe enough to take the chance that creative beings would
evolve in it.

What I am suggesting here verges on hubris. In debt to process theists'
insights, I am suggesting that all of creation has been graced by its Creator
with creativity. God creates, redeems, and sustains not a universe that is
static, nor one that is inanimate, nor one that is dependent on God for interminable meddling and guidance, but one that has the capacity to evolve
creatively. This universe has the gift of a certain autonomy.

This understanding of creation is not incompatible with neo-Darwinian
theories of evolution or quantum indeterminacy. As Elizabeth Johnson put
it:

> Without chance, the potentialities of this universe would go unactu-
> alized. The movement of particles at the subatomic level, the initial
> conditions of nonlinear dynamic systems, the mutations of genes in
> evolutionary history, all are necessary for the universe's becoming,
> though none can be predicted or controlled. It seems that the full
> gamut of the potentialities of matter can be explored only through

---

[478] Nancey Murphy, in one version of her portrayal of God's "decision tree"
shows the structure of this sort of defense. However, by writing it as if it were about
what God was thinking once upon a time—or before or beyond all time, whatever
that would mean—(material mode), rather than about what we can say about God
(formal or grammatical mode), she tends to anthropomorphize God and may reduce
divine agency to the level of creaturely agency. Both FWD and CD are not as much
about what God did or does, but what we can say even God cannot do (who can
plumb God's ways?); that is, what we can or cannot assert without incoherence about
the God we worship. While I think Murphy has the structure of argument about
right, I am not persuaded by her mode of argument or the content of the argument
she makes. Her argument needs to show that we cannot say that God could have
(logically) chosen to create a world that takes the "other option" at each "stage."

[479] Power, "Splendor of Divine Perfection," 22. I am also indebted to the preface
to the second edition of David Ray Griffin, *God Power and Evil*.

the agency of rapid and frequent randomization. This role of chance is what one would expect if the universe were so constituted as to be able to explore all the potential forms of the organization of matter, both living and nonliving, which it contains.[480]

The maximization of good that Power mentioned, then, is not the maximization of some good things or good acts, but the creating of a universe that itself is creative, exploring *all the potential forms* of nature, including those forms which are not merely replicative, or subject to change by chance alone, but which become creatures who participate in the creative goodness of God. Some creatures are not only able to reproduce, but to create novelties. Matter has the "wondrous ability . . . so to organize itself as to bring forth the truly new from within itself."[481] This is creativity. This is grace.

The production of novelty ranges from the inanimate through the social. Such creativity is not limited to the blind whims of biological evolution or the gentle direction of a divine Guide in which minimally free actual entities create new realities very slowly, but to the intentional creation of social practices and structures by humans as well. God did not create governments. They are social structures, created by humans, and are nonetheless good insofar as they are structures that promote human thriving. The evil that they do is parasitic on their goodness. So far, the height of the divine gift of God's own creativity makes possible the human creation of social practices and structures—but also the creation of social practices and structures that are destructive of the possibility of agents' individually or collectively exercising creativity. The possibility of constructing sinful structures is parasitic on the possibility of constructing social structures. As personal sin is a *culpable* wrong act of God's good creatures, so social sin is a destructive use of social structures created by God's good, creative creatures.

Powerful creativity may not be only an attribute of single entities, but also of groups. No individual created monarchy or democracy or communism alone. And the practices and structures that resulted solely from groups of individuals' acts are both novel and powerfully effective. Some of those are practices and structures for good, others may be heavily perverted for evil, but none alone the result of individuals' acts—sinful or graceful—alone. The creativity of the universe means that even our creations literally come to have a life of their own. Our creativity is a form of creaturely participation in the gracious agency of God available for complex entities.[482]

---

[480] Johnson, "Does God Play Dice?" 7–8.

[481] Ibid., 6.

[482] One might continue developing this vision by suggesting that every entity is redeemed and that every entity has been endowed, proportionate to its complexity, with reconcilability. The redemption of the cosmos and humanity is not a divine task alone, but a task in which redeemed human and other agents are empowered to participate. We are essentially social beings, called to be reconcilers who can use the our abilities to reconcile authentically or to thwart the powers that make for reconciliation in an estranged world.

One might further suggest that all entities sustained by the Holy Comforter are empowered to "keep hope alive."[483] Sometimes this is to comfort the afflicted, and at other times to afflict the comfortable. That we are not responsible for comforting all the afflicted—that's God's task—but are able to heal, to feed, to soothe, to stir up and to challenge when appropriate those to whom and to which we are related.

If this sort of vision of the way in which humanity is created in God's image is possible, then R: "It was not within God's power to create a world in which free entities could exercise creativity and which did not contain genuine evils" is possibly true and could account for natural and social evils. God could, perhaps, thwart human creativity. But then God would not merely be thwarting the ability of humans to choose, as in FWD. God would be withdrawing that innate ability of matter that reflects the divine creativity.

R does not require a rejection of divine omnipotence. We cannot coherently say that God can do what is logically impossible. If creativity is an essential attribute of the universe, then God could not make the universe happen in a certain way. Either God could make the universe evolve in a certain way or God could let the universe exercise its own creativity. But, like FWD, the idea that God could make a creative universe always create only good is at least prima facie incoherent. But an omnipotent God still can do what is logically possible for God to do. This has a chilling implication. As Charles Journet stated in a text whose terror I have only just begun to understand, "if ever evil, at any time in history, should threaten to surpass the good, God would annihilate the world and all its workings."[484] As primary creative agent, God could remove his sustenance of each and every secondary creative agent—each and every actual entity—in the universe. This dictum plumbs the extent of the universe's relative autonomy.

Scientific materialism offers an alternative to r. Perhaps the most difficult problem for the scientific materialist is to offer an explanation of how at least one planet in the universe developed beings of sentience, intelligence, and creativity from inanimate physical elements. Of course, chance, random drift, and natural selection play their roles. But these are *descriptions* of the process—descriptions that get better and better as science develops—but not *explanations* of why the process happens as it does. That it happens by "chance" is certainly a possible explanation as well as a description. But given that the anthropic principle, perhaps in one of the "soft" versions, has some plausibility, there is good reason to suggest that "chance" is not the only possible explanation, however well it may work as a description. Hence, that the universe is created such that God endowed it with genuine creativity is not ruled out as possibly true; scientific materialism is not a better explanation, and both materialism and theism are compatible with the description of natural processes proffered by science.

For the creativity defense to work, it has to be possibly true that God created a universe that could and did develop creativity and that in so do-

---

[483] I take this phrase from Dermot A. Lane, *Keeping Hope Alive: Stirrings in Christian Theology.* In the United States, the phrase is associated with the Rev. Jesse Jackson.

[484] Journet, *Meaning of Evil,* 289.

ing the entities that universe could, proportional to the development of creativity, develop the potential for destruction. If God created a universe with "open natural systems," this seems obvious.

As a way of showing the advantage of a CD, so briefly sketched here, we can show how it can carry traditional concepts of natural evil and account for social evils.[485]

5 Natural Evil

Plantinga's version of FWD accounts for natural evil by showing that it is possible that all natural evil or patterns of evil are attributable to nonnatural agents (the "fall of the angels"). This is, as many have noted, an ad hoc hypothesis. Logically speaking, however, it is sufficient; it is possibly true that what we call "natural evil" is due to the actions of malevolent beings. But this is rather far-fetched. These devils would have to be preternatural, secondary causes, a concept that certainly does not fit with science. We have here a case of the devil-of-the-gaps, invoked as a possibility to account for the inexplicable, i.e., natural evil. With regard to natural disasters (see section 5.1 below), this claim is especially incredible.

Of course, it is possibly true that unicorns exist and we haven't found them yet; such possible truths may be quite implausible, even if they are logically sufficient to do the logical work that the defense needs.[486] But like much of pre-Enlightenment thought, FWD presumes the possibility of preternatural "spiritual" entities whose choices before or beyond the creation of the world have determined that the natural world must be afflicted with evil, effectively making them agents in the world. Given modern cosmology, it is difficult to understand what sense the "time before" or "place beyond" the creation of the world could mean. Yet it is possibly true that the course of the world has been distorted by demonic acts—or that God has not forestalled the results of such acts. But, if so, then we have not only another case of "hidden causes" in the world, but also another case of "gappiness," with the devils-of-the-gap being increasingly forced into that part of our world of which we do not or cannot have knowledge.

The decrease in the belief in and acceptance of the preternatural in modernity described by Lorraine Datson (see n. 6) was a gradual process. Just as modern science squeezed God into the gaps of explanation, so it squeezed the devil into those same gaps. No longer do we explain some

---

[485] I do not discuss sin here because FWD deals with sin well enough. It is also possible that in a fully developed CD, analogous moves to dealing with natural and social evil would cover individual evils as well.

[486] Wallace A. Murphree, in "Can Theism Survive Without the Devil?" argued that Plantinga's FWD is not committed to the *possibility* of the devil, but to the *actuality* of the devil, as a cause of natural evils. Murphree is not alone in thinking that theism requires the devil. John Wesley evidently thought so and Richard Swinburne has also noted a diabolical origin for natural evil is an ancient part of the Christian tradition and suggested that it "may indeed be indispensable if the theist is to reconcile with the existence of certain animal pain"; see Swinburne, "Problem of Evil," 93. Another contribution of various forms of process theodicy and CD is that they can present other possibly true propositions to take the place of r with regard to natural evil in FWD which shows that a belief that it is actually true that the devil is the source of natural evil is not necessary, as Murphree argued, to support FWD.

mental illness or witchcraft practice as brought on by demonic possession. The devil has become a scapegoat ("The devil made me do it!") for irresponsibility. While there are forces for evil in the world, it is not clear that they are preternatural. They may be labeled by some as "demonic" as a trope rather than as a description. But in fact they are human creations, social structures, some of which are or generate actual evils (see section 6 below).

5.1 Understanding Natural Evil in the Christian Tradition

Natural evil is typically denominated in the discourses of theodicy to be "pain and suffering."[487] But this is not the only understanding in the Christian tradition. Indeed, it is not the dominant one.

The classic understanding of natural evil is rooted in Augustine's work. It is much more subtle.[488] First, Augustine affirms that whatever is, is good, because God made it. Never, after his conversion to Christianity, did Augustine deny the fundamental equation of created *esse* with created goodness, uncreated *esse* with uncreated goodness. The goodness of creation derives from the Goodness of God who made it.

Second, this affirmation leads to an obvious problem: "Why do we call some things 'evil'?" The first question—all too often taken for granted in modernity—is a descriptive epistemological or linguistic question. Augustine uncovers two different sets of circumstances that properly lead to such a description. In *Enchiridion,* Augustine writes:

> In this world, what is called evil, well ordered and properly placed, more highly commends good things so that they may be more pleasing and laudable when compared with evil things. For the omnipotent God, who . . . is the greatest good, would not allow anything evil in his works, unless he were so omnipotent and good as to make good out of evil. But what is that which is called evil but the privation of good?[489]

Here Augustine denies that evil has a "nature"; rather it is a deprivation of what is. He then follows this discussion with illustrations of evils as defects, corruptions, or deprivations of naturally good things. But his description needs to be unpacked.

In the first sentence, "evil" applies to one of two or more entities *compared to each other.* Augustine gives no examples here. Perhaps one might think of the dull plants in a garden setting off the brilliant flowers, or of the plain framing which sets off the brilliance of a complex mosaic, or of the "blue collar" athletes on a team without whom the "star" could not be brilliant. We might call *malum* some things that are not very interesting or complex if they are out of place, or some supporting athletes who are not skilled if they are expected to star. However, in their proper places, these are not in themselves *mala,* but have their proper roles as setting off or enhancing the greater. Each is called *malum* when it is misplaced or compared with the greater. This first comparative use of "evil" undergirds what

---

[487] See Hick, *Evil and the God of Love*; Swinburne, "Problem of Evil," *passim.*

[488] The following paragraphs reconsider material used in Tilley, *Evils of Theodicy,* 118–20.

[489] Augustine, *Enchiridion* 11.

Hick has called the "aesthetic" theme in Augustine, especially as one finds such comparative "goods" and "evils" in the rhythms of nature.[490] However, it is not an important motif in *Enchiridion*.

In the last sentence of the quotation, evil applies to an actual entity *in comparison to what it should be/is by nature*. This second comparative use of the term, as Augustine's illustrations show, is much different. Here the evil is seen not by comparing two things (legitimately or illegitimately), but by comparing a thing to its nature. It is less than it should be, deprived of its full measure of goodness. "Therefore, every entity [*natura*], even if it is defective, is good in so far as it is an entity and evil in so far as it is defective."[491]

It is important to distinguish these two uses of *malum*. David Ray Griffin's distinction between *prima facie* and *genuine* evils comes close.[492] The first use of "*malum*" is ambiguous. It may indicate that an entity *is* corrupted or it may indicate that an entity *appears* evil from certain perspectives or in certain roles or places. This is prima facie evil. For example, the sufferings which animals undergo as they are attacked for food by other animals as part of the natural food chain are prima facie evil. However, these animals are not certainly *corrupted*. It may be part of their proper nature in the natural environment to eat and to be eaten by others. If it be part of an animal's (or plant's) nature to be eaten by other animals, then the prima facie evils of their sufferings and deaths are not genuine evils, for these prima facie deprivations are not losses of goods which are natural to them.

The second use of *malum* is not ambiguous. It indicates that an entity *is* corrupted, no matter how it may appear. This is genuine evil. If animals suffer *because* their natures have been corrupted, then such *prima facie* evil is *genuine* evil, because these prima facie deprivations *are* losses of goods which are natural to them. Unfortunately, commentators do not always seem to keep these separate;[493] but then, it seems, neither did Augustine in other places.[494]

Moreover, Augustine implied that an ability to disambiguate prima facie evils in the natural realm is irrelevant to theology. "Therefore, when it is asked what is to be believed with regard to religion, the nature of things need not be investigated as it is by those the Greeks call *physikos* [natural scientists]."[495] This includes not being concerned with the "origins and natures of animals." Yet if knowing the *natures* of animals is irrelevant to religion, then knowing if animals' natures have been or have not been corrupted must be irrelevant to theology. Thus, knowing whether evil in the natural realm is genuine or merely apparent is irrelevant to religious belief and its theological explication. And, indeed, Augustine makes this plain. After commenting on the multiple opinions of the *physikoi*, he concludes:

---

[490] Hick, *Evil and the God of Love* (rev. ed.), 82–85.

[491] Augustine, *Enchiridion*, 13.

[492] Griffin, *God, Power, and Evil* (2d ed.), 22.

[493] See Hick, *Evil and the God of Love* (rev. ed.), 87–89.

[494] Augustine, *City of God* 11.18, 23; trans. Henry Bettenson.

[495] Augustine, *Enchiridion* 9; also see 16.

It is sufficient for a Christian to believe that the cause of created things, whether celestial or terrestrial, whether visible or invisible, is nothing but the goodness of the creator, who is the one true God; that there is no entity [*natura*] which is neither him nor from him; and that he is the Trinity, namely the Father and the Son generated from that same Father, and the Holy Spirit proceeding from the Father, but the very one and same Spirit of Father and Son.[496]

In sum, Augustine teaches that evil is the privation of a good natural to an entity; but whether what appears to be evil in the rhythm of nature is genuinely evil is irrelevant to religious belief.

5.2 *Understanding Natural Evil Today*

Augustine lived in a static, premodern universe, with stable *naturae*. Can this traditional understanding be revamped for a defense in a universe characterized by quantum indeterminacy and chaos on a planet whose evolution is characterized by natural selection and random drift? I think it can.

The first issue is to clarify the concepts we use in talking of "natural" evils. Some of these, of course, may be genuine evils, others not so. Roughly speaking, we need to recognize at least three interconnected patterns of "natural" evil. Entities can experience evil in proportion to the level of sensation, consciousness, and memory that they have developed. The potential for pain is correlate with the potential for pleasure. The potential for suffering is correlate with the potential for consciousness. The potential for extended suffering that goes beyond the suffering induced by pain is correlate with the potential for memory. Rocks can feel no pain; organisms with some form of nervous system can feel pain. Organisms without some level of consciousness cannot suffer; the more conscious an organism is, the more the organism can suffer. Organisms without some level of memory cannot have ongoing suffering, but only occasional sufferings; without memory, higher levels of suffering, such as fear, grief, self-disgust, etc., are not possible.

Can we say that an omnipotent God could and should have created a universe in which entities could have pleasure but not pain, consciousness but not suffering, memory without grief? Such a universe would be so different from the one we know that we could not say anything about it using the concepts we have developed in order to understand this universe. We can certainly imagine an "anaesthetic" universe in which conscious entities could feel no pain. Perhaps they would be like angels and not have bodies. But they could also feel no pleasure in the sense that physical pleasure correlates with physical pain. Could they suffer? What would cause them to suffer? Could they grieve? What would cause them to grieve? Could they die? What could that mean without a body? These *poenae* are integral to a physical universe. Analogous questions could be developed to show the difficulty, and perhaps impossibility, of our conceiving such a universe.

All of these questions are certainly not decisive arguments against the possibility that God could have created such a universe; they do not show

---

[496] Ibid., 9.

that it is logically impossible for God to have made such a universe. But they do strongly suggest that it is conceptually impossible for us to think that he might have done so. And I would thus say that the burden of proof is on someone who could demonstrate the logical possibility that a universe with sentience and consciousness but without the potential for pain, suffering and grieving were possible.

In Griffin's terms, actual occurrences of pain, suffering and grieving are prima facie evils. But the potential for entities to have such occurrences is not an evil, unless one thinks that the development of sentience, consciousness and memory which carry the potential for such evils, are themselves evils. The question is whether actual occurrences of pain, suffering, and grieving are *genuine* evils. Of course, some are not. The pains that warn one or remind one that a movement is potentially injurious are useful, even good. Perhaps some sufferings and griefs are also instrumental to higher goods. But the issue that has emerged is that of "gratuitous" evils, evils that serve no purpose and that God could have or should have prevented.

But how could God have created a universe that eliminated gratuitous evil? As Johnson put it, "The movement of particles at the subatomic level, the initial conditions of nonlinear dynamic systems, the mutations of genes in evolutionary history, are all necessary for the universe's becoming, though none can be predicted or controlled."[497] If the universe is truly random, and if it can be creative, then even an omniscient and omnipotent being could not predict or control all that will happen. And here we can rely on Augustine's insight: this is not a scientific problem, but a problem for steadfastness of faith. We finally do not have the ability to distinguish scientifically between gratuitous, genuine evils and other evils that are *poenae* but not gratuitous.

That some entities fail to be what they could be is an evil. Some of those failures are simply *poenae*. Others are *culpae,* but those need not concern us here. Some actual evils are due to chance, some to malfeasance. All are, in some sense, failures. Failures are at least prima facie evils in any universe. That this universe has brought forth entities that can sense, be conscious and remember is a necessity if there are to be creatures that can be creative as well. And if there this universe is characterized by creativity and chance, God cannot (logically cannot) exclude the reality of failures. At a high level of creative development, creative creatures with memory are crucial to the evolution of the universe at a social level, and also bring the possibility of profound failure.

6 Social Evil

Social evils, those institutionalized patterns of practices that harm people directly or indirectly, are treated by FWD as an agglomeration of individual sinful choices and their results. Again, that social evils may be reducible without remainder to individual choices seems possibly true, but this proposal is so far out of harmony both with the intuitions of many theologians and philosophers and the evident reality of various patterns ("racism," "sexism," "homophobia," "economic exploitation," etc.) that seem to skew

---

[497] Johnson, "Does God Play Dice?" 8.

people's judgment and even diminish radically their responsibility for their acts that a different solution may be needed. [498]

A main argument of *The Evils of Theodicy* was that theodicists' conceptual constructions obscured those structures of evil variously labeled social sin or social evil. Social evils are complex structures of prejudice, oppression, and injustice which are part of the context in which we live. These structures can defeat our best-intentioned attempts to do what is right. They can redirect our aim for the good beyond our wills. They are not reducible to individual sinful acts.

It is not clearly a sin for me to buy Nike sport shoes. Yet the conditions under which the shoes are produced for our purchase are exploitative and even oppressive. Some overseas factory workers spend nearly twice the time of U.S. workers and receive perhaps a tenth of the U.S. wages. That they make more money than their compatriots in agriculture does not exonerate the Nike system, but even more deeply indicts international agribusiness. Each time each of us buys a pair of Nikes, even if we do not realize it or have no other viable option, we support the system of oppression. It was not designed as an oppressive system, but as one which minimized production costs and maximized profits and shareholder value. Its results, despite any and all good intentions of management, shareholders, and consumers, are the creation of a system of economic exploitation in which we are implicated each time we purchase these sport shoes. The system is the problem, not my individual acts, whether my acts are knowingly and willfully sinful acts of cooperation with evil or morally innocent acts because of ignorance or lack of real freedom to choose other options.

Moreover, the system cannot be forgiven; nor will individuals' refusal to buy one specific brand of sport shoes atone for or stop the social evil of the exploitation. To stop the evil wreaked by the system, a radical overhaul is necessary—and perhaps will come with the internationalization of labor, presuming that governments abroad do not suppress unionization (an unlikely hope, given the pattern of union-busting and oppression in the First World a century ago—on which business in the developing world is today modeled).

If you cannot see this fraction of the global economic system as a powerful social evil, I invite you to consider others: war, prostitution, the creation of a permanent underclass, etc. Indeed, I would aver that the current pattern of runaway reproductive technologies with designer babies, breeding embryos for their stem cells and then destroying them, and soon, uncontrolled human cloning for those who have the resources to buy clones, all while one-third of the population of the United States is denied access to much basic health care is another form of social evil not reducible solely to individual human sinfulness. Perhaps there are better examples than mine: no one would argue that war is a good thing to do and few would argue that prostitution is a fair-wage entrepreneurial system. War is, at best, a justifiable evil in certain circumstances—and its destruction is hardly re-

---

[498] To attempt to argue that FWD cannot account for social evils would require that it be shown that the notion of social evil attributable to individual choices is incoherent. That seems a task that could never be accomplished as excluding all the theories of individual choice as logically unable to account for social evils seems an impossible task. Nonetheless, the required theories seem implausible to many within and without the Christian tradition.

ducible to individual choices and even deadly contagion of sin; the politically correct label "sex-worker" whitewashes a system of exploitation that goes far beyond the sexual sins committed by the typical "johns" who pay for sex.

Like the theodicists criticized in my earlier book, Plantinga's account of social sin reduces it to individual sins committed by people in a diseased state. He has a "contagion model" of social evil, a model that understates the power of "the social side of sin." He writes:

> We human beings are deeply communal; we learn from parents, teachers, peers, and others, both by imitation and by precept. We acquire beliefs in this way, but just as important (and perhaps less self-consciously), we acquire attitudes and affections, loves and hates. Because of our social nature, sin and its effects can be like a contagion that spreads from one to another, eventually corrupting an entire society or segment of it.[499]

Here we have the "social disease" or "contagion" image of sin. Social evil is reduced to an agglomeration of individual sins which spread contagiously from individual to individual. But the social evil of worldwide exploitation is more than the result of a "contagion."

The strategy of most responses to the problem of evil is to construe evil practices as parasitic on what is basically good. Human beings are basically good, but terribly warped by sin, including social sin. Sin is construed as an affliction primarily of the individual. Only individuals commit sin. In this view "sin" is equated with "*culpa*." What is missing here is a view that our social structures have "a life of their own" independent of any number of human actions. They can create *poenae*; they too can have *culpae* not reducible to the individual sins that people may perform in creating them. Social evil needs to be construed as parasitic on social structures and social groups, not on individuals' actions. Social evils are an affliction of and the fault of our communities and societies, not of individuals.

Contagion models of social sin simply fail to do justice to great social evils from genocide, including the Shoah, to contemporary patterns of oppression. Contagion models obscure the huge number of economic, social, religious, industrial, and other structures and practices that combine to make an engine of social evil so powerful. To reduce social evils merely to individuals' sins, their effects, and their infection of others is, in my judgment, to distort our understanding of the power of such sinful structures, a power parasitic on the good our social, political, and economic structures are designed to create.

The gravest *culpae* of all are the thwarting or distorting of human thriving in general and creativity in particular. On this view, not all *culpae* are individuals' sins, but any act, entity, state of affairs, or structure that actively thwarts or distorts the process of creating what is good are *culpae*. Whether nonconscious entities can do so is, as Augustine might say, beyond our ken. Certain conscious entities and communities of conscious entities can. Perhaps sentient entities can as well. And structures that make our creative impulses destructive are clearly *culpae* as they prohibit each *natura* from becoming what it could be, defects or failures that are *poenae*

---

[499] Plantinga, Warranted Christian Belief, 207.

resulting from *culpae* rather than chance. In an evolutionary universe, *naturae* are not static as Augustine presumed; those that do not creatively preserve or bring about the good—which may include their own end or death—may be characterized as *mala*. They would not be out of their static place, but "sports" in the evolutionary process or evolutionary failures. In a universe characterized by chance and creativity, tragedy must be possible even within the realm of divine providence. They would be less than they might have been. In this way there seems to be a linkage unnoticed by much modern discussion between "natural," "personal," and "social" evils.

7 Natural Disasters: Both "Natural" and "Social" Evil

One implication of the discussion of natural and social evil above is that natural disasters are not necessarily simply "natural evils." Natural disasters are evils insofar as they affect sentient creatures. However, at least some of the devastation due to catastrophic natural events is due to social failures. The potential for the devastation and loss of life caused by the December 26, 2004 tsunami which resulted from an extremely powerful earthquake under the Indian Ocean near Sumatra was undetected because no governments had placed warning buoys of any tsunami warning system in the Indian Ocean; the only buoys were in the Pacific Basin—where a tsunami warning system was begun over fifty years ago.[500] With regard to another natural disaster, Hurricane Katrina, a devastating hurricane that struck the Louisiana coastline on August 29, 2005, Lawrence M. Hinman has noted, "When we look more closely, however, we see that much of the loss of life and devastation from Katrina was as much the result of human decisions as it was of raging winds."[501] Many of these decisions were political decisions—not to fund improved protection for New Orleans when scientific evidence clearly demonstrated that a major hurricane would flood much of the city. Some of the inadequacies in the dike system are attributable to faulty construction, perhaps even criminal activity by sleazy construction companies, not to mention latent and patent racism.[502]

Natural evils in the classic sense—*poenae*—or in the modern sense—individuals' pain and suffering—are quite different from natural disasters. Much of the devastation from such events needs to be understood as an effect of social and personal evils. Hence, the notion that there is one all-inclusive class of evils that can be characterized as "natural"—typically in opposition to "sin"—simply can't be sustained in view of the facts about most disasters. While it is beyond the limits of this chapter to explore these issues, any attempts to reduce natural disaster to a natural evil of "pain and suffering" without a socio-political component must fail in the face of

---

[500] See "Tsunami Warning System Calls Grow" http://news.bbc.co.uk/1/hi/world/asia-pacific/4127927.stm (accessed November 12, 2005).

[501] Lawrence N. Hinman, "A Natural Disaster," http://www.signonsandiego.com/uniontrib/20050908/news_lz1e7natural.html (accessed November 12, 2005).

[502] See Tom Brewton, "Media Fingers in the Dike," http://www.americandaily.com/article/10044 (accessed November 12, 2005). Brewton notes that different media blame different political causes, but no one denies that there is a social and political contribution to this natural disaster.

the facts about what makes such events into disasters, rather than simply transformative events.

8 Conclusion

CD is based on a vision that offers a different account of both natural and social evils from that of FWD. It is more compatible with contemporary science and the Christian tradition than is FWD. CD can account for the power of social evils and the fact that they have "a life of their own" in a way that FWD does not. CD can also salvage the insights of those who tell the universe story for defenses in face of the problems of evil.

The CD does not entail process theism, as it need not accept the co-dependence of God and the world or the process theists' account of divine power. It is compatible with this view, but also with a "dual agency" view characteristic of some contemporary forms of Thomism.[503] Nonetheless, CD learns from process theism about the importance of creativity and social nexus of actual entities. Pure natural evils result from a universe suffused with creativity (as process theism suggests). Social evils are the result of the corruption of human creativity in the creation of social structures analogous to the way individual sin is the corruption of human freedom; as such social evils are a constituent of the "natural evils" that are "natural" disasters. Thus, a CD has the potential to provide a defense of God's goodness in the face of manifold forms of evil beyond the hackneyed "sin and suffering" typically found in theodicies and defenses and accepting a traditional understanding of *poena et culpa*.

---

[503] Stoeger's "Epistemological and Ontological Issues Arising from Quantum Theory," in *QM*, 95–98 (and the essays cited in his note 32) and Johnson's "Does God Play Dice?" are interventions in the ongoing discussions of divine agency that approach the problem in ways that utilize the resources of a theology of dual agency. They develop insights along the lines explicated more fully by authors such as Austin Farrer, Robert Sokolowski, and Norris Clarke. I find the positions on divine agency taken by Nancey Murphy and Robert John Russell in this volume and in other essays to make the problems of evil impossible to resolve or dissolve. While Murphy and Russell deny that God "intervenes" in the world, they find that God can act at the quantum level. Their position also seems like an updated "God-of-the-gaps" approach, as indeterminate quantum variations seem the last gap cosmology has left open if God is to exercise divine power in the world. Their description of divine action at that level (which seems like an intervention to me, but not to them because they find it adds no matter nor energy to the universe, but merely changes the information in the universe) makes God in some sense not merely omnipotent, but omniresponsible (and perhaps even omnificent), for an omnipotent God could have affected sufficient quantum events to avoid any, and perhaps every, evil in the universe. Rather than make the problems of evil infinitely worse, as their views do, it seems wiser to follow a different theological and philosophical account of divine agency.

III. RESPONSES: THE GOD-WORLD RELATION

Kirk Wegter-McNelly
Denis Edwards
Wesley J. Wildman

NATURAL EVIL IN A DIVINELY ENTANGLED WORLD

Kirk Wegter-McNelly

1 Introduction

Discussions of the problem of natural evil often focus on the issue of whether or not God can rightly be said to allow the occurrence of natural evil for the sake of some greater good. If natural evil is not to be made the direct result of God's creative act, on the one hand, or left completely unexplained, on the other, then something like this strategy appears to be an unavoidable if perilous endeavor.[1] Peril notwithstanding, though, I think there is a good case to be made that this strategy, precisely as a theological strategy, does not go far enough. Any theological account of natural evil will need to consider not only the relation between natural evil and God's creative will but also the relation between natural evil and God's creative activity, that is, between natural evil and the mode(s) of God's presence in creation. Theological explanations of *why* God allows natural evil will need to be supplemented by corresponding theological accounts of *how* God allows natural evil to occur. Without this additional explanatory dimension, one is left to puzzle over the relation between what God desires for creation and what God actually does in and with the particular world God has created. More importantly, one is left to wonder whether or not affirmations of God's active presence in creation can be maintained in the face of such evils.

In this essay I attempt to provide this additional explanatory dimension by offering a description of God's relation to physical processes through an extended analogy drawn from quantum physics. I suggest that God can be helpfully envisioned as "entangled" with the world after the manner in which quantum physicists speak of physical objects being entangled with one another. While the remarkable phenomenon of quantum entanglement has been interpreted in different ways, it suggests to many physicists and philosophers that the relatively autonomous behavior of the parts of complex quantum systems rides on top of a deeper unity. Entanglement suggests a mode of relationality in which the presence of the relation is hidden within the apparently independent existence and behavior of the *relata*. In this essay I wish to explore the logic of this concept for the sake of positing a theological analogy, namely, that the relation between God and the physical world is effective in such a way that God's sustaining relation to the world is hidden by and within the physical world's relative autonomy.

The significance of this analogy for a discussion of natural evil is that it provides an account of God's relation to physical processes in which the efficacy of the divine presence in those processes can be understood as coalescing entirely on the relation between creation and God. In other words, the analogy creates the possibility of locating the effect of God's ac-

---

[1] It is possible, of course, to refuse to be led out of the dilemma and instead to grasp one of its horns. Wesley Wildman grasps the first (see his chapter in part III of this volume) by making good and evil coprimal aspects of the divine nature, whereas Karl Barth grasped the second by insisting that the problem of evil is ultimately not ours to solve but God's; Barth, *Church Dogmatics* 3/1:381.

tion vis-à-vis physical processes in the relation between God and creation, rather than in the physical processes themselves. In this analogy, the divine act of establishing and maintaining a relation with the processes of the physical world can be understood, counterintuitively, as the very basis for the relative causal independence of those processes. And it is their independence which, in turn, manifests itself in the real possibility of the kind of physical events that typically earn the title "natural evil."

In the following section I locate my own thought within some of the contemporary currents of trinitarian thought and discussions of the problem of evil. I also argue for the importance of drawing theological images and concepts analogically from the natural sciences. The first two subsections of section 3 provide brief overviews of quantum entanglement and spin-½ systems, respectively, for a nonscientific audience (requiring only a basic grasp of algebra and trigonometry). Those already familiar with the history and the formalism of quantum theory can safely proceed to subsection 3.3—the constructive center of the overall argument—where I present a quantum thought experiment that brings to light some of the possibilities for developing a theological analogy of the God–world relation. The final section develops the analogy itself and evaluates it for making theological sense out of how violent and destructive natural processes could possibly exist within a world understood to be both the product of God's creative act and the locus of God's sustaining presence.

2 Some Background on God and Natural Evil

2.1 Divine and Creaturely Relationality

A good deal of recent theological writing on the nature of creation has emphasized the concepts of relationality, interconnectedness, and wholeness.[2] This emphasis emerges from the convergence of a number of politico-theological issues—concern for the environment, for social justice, and for gender equality, to name only a few—but chief among the factors leading in this direction has been the revival of trinitarian language in twentieth-century theology, which has played a surprisingly common role among otherwise diverse agendas.[3] I do not think it is too much to say that many contemporary Christian theologians have come to perceive the relational character of divinity as the lasting value of trinitarian God-language. As one theologian recently put it, "far from being irrelevant, [the Trinity] is central to Christian faith: it expresses the entire God–world dynamic."[4]

---

[2] See, for example, Griffin, ed., *Sacred Interconnections*; Jungerman, *World in Process*; O'Murchu, *Quantum Theology*; Vaught, *Quest for Wholeness*.

[3] See, for example, Boff, *Trinity and Society*; Cunningham, *These Three Are One*; Johnson, *She Who Is*; LaCugna, *God for Us*; Moltmann, *Trinity and the Kingdom*; Peters, *God as Trinity*.

[4] McFague, *Life Abundant,* 143–44, esp. n. 17. According to McFague, lying beneath theology's central aim of expanding models conceptually to provide an ever more coherent and comprehensive account of belief is what she calls a "root metaphor," which for the Christian tradition, she argues, is the "kingdom of God." This notion, referred to in Jesus' parables and pointed to by Jesus himself, suggests to McFague that relationality is the defining feature of the Christian view of God and the world; McFague, *Metaphorical Theology,* 108–11.

God's decision to relate to a world other than Godself is a faithful expression of who God is.

Among those who have written extensively on the topics of divine and created relationality, Wolfhart Pannenberg has developed a powerful synthesis of these themes in his trinitarian doctrine of God and his explicitly trinitarian doctrine of creation. With respect to the divine nature, he argues that the basic theological task is to envision the divine unity as it manifests itself in the concrete and differentiated life of the Trinity, which "requires a concept of essence that is not external to the category of relations."[5] It is this trinitarian God, Pannenberg urges, that we must think of as Creator. Thus he construes the second person of the Trinity as the principle of distinction and the third person of the Trinity as the principle of unity, both within the divine life and in the externalizing divine act of creation.[6] Seeing this act of creation as an outward expression of the inwardly relational life of God, he argues, provides a stronger connection between God and the world than is possible from the perspective of an undifferentiated (nontrinitarian) monotheism.

At the same time, Pannenberg also thinks a trinitarian account of creation can provide a robust view of the world's relative independence (i.e., causal integrity), since it is understood to be the product of a God who is active and complete in God's own being and thus who has no need of creation.[7] Pannenberg locates the unity and completeness of the triune God in divine love, which manifests itself internally in terms of mutual self-distinction and interdependence and externally in the creation of the world. He characterizes the external unity in terms of "the divine love eternally affirm[ing] the creature in its distinctiveness and thus set[ting] aside its separation from God but not its difference from him."[8] In this unity, created plurality does not disappear but is affirmed and sustained by its dynamically and eternally relational divine source.

In what follows, I assume a broadly trinitarian stance consonant with Pannenberg's notion of divine relationality as the source and pattern of created relationality. One benefit of such a perspective is that it can serve as a starting point for mutual affirmation of the world's radical dependence on and relative independence from God. The analogy I develop here for the God–world relation on the basis of quantum entanglement is intended to image this mutual affirmation in such a way that it does not collapse into

---

[5] Pannenberg, *Systematic Theology,* 1:335.

[6] Ibid., 2.20–35. It is interesting to note that McFague has suggested replacing the traditional language of "Father, Son, and Holy Spirit" with "mystery, physicality, and mediation." It is this kind of coalescence that I have in mind when I say that trinitarian language has recently played a surprisingly common role among otherwise diverse theological agendas; McFague, *Body of God,* 193.

[7] As Pannenberg points out, to say that the world is the outward product of a divine act is to assert that it is not God—that it is not an emanation from God which is in any way necessary—and thus that its being is contingent. Pannenberg argues that Christian theology found in the doctrine of the Trinity a way of asserting that God is essentially active, and thus that God does not need the world to be active. The trinitarian relations are themselves rightly understood as inward acts of God. Pannenberg, *Systematic Theology,* 2:1.

[8] Ibid., 1:446. For his early thoughts on the unified nature of the Trinity, see Pannenberg, *Jesus—God and Man,* 179–83.

contradiction. It is worth noting, in this regard, that the traditional characterization of the trinitarian relations in terms of *perichoresis,* or mutual indwelling, suggests close conceptual proximity of the theological notion of trinitarian relationality to the scientific notion of entanglement. On the perichoretic model, the three divine persons exist as distinct persons because of (on the basis of) their relationships to one another, not unlike the way two entangled photons can be said to exist as individual photons only insofar as they exist as a single entangled reality. If Pannenberg is right to insist that we think of creation in explicitly trinitarian terms, then constructing a view of the God–world relation on the basis of an analogy with quantum entanglement may open up interesting new paths for reflecting further on what it means for an essentially relational God to be in relation with this world, which contains such a wide variety of richly relational features.

2.2 The Problem of Natural Evil

The presence of natural evil in the world raises the following important question for theologians engaged in the difficult task of reframing God's relationship to humanity by the light of contemporary science and in terms of God's broader relation to all creation: Is there continued theological wisdom in thinking of death-dealing natural events such as earthquakes, tornadoes, and hurricanes as somehow "covered" or explained generally by the presence of life in the universe, if not more specifically—as with older theologies—by the fact of human sin? Those who wish to bring new clarity and force to this argument in conversation with contemporary science will need to specify not just the *why* but also the *how* of this connection; that is what this essay aims to do.

Providing a cogent theological account of natural evil is an important part of the larger contemporary theology-and-science agenda, but it must be said at the outset that I am not interested in showing natural evil to be ultimately something less than evil—the aim, I take it, of traditional theodicy. The theodicist who wants to make evil something less than its appearance misses the deep theological point that natural evil, like moral evil, is evil precisely in the sense of being that which mysteriously opposes the will of the Creator. And perhaps equally problematic, lessening the evilness of evil risks portraying the need for redemption as less than urgent—since one can infer that God is presently doing the best that can be done. Such are the perils to which I alluded in the introduction, and about which I will have more to say below.

Terrence Tilley has argued forcefully in his *Evils of Theodicy* that theodicies are themselves to be judged morally evil for the ways in which they have misportrayed and effaced the depths of evil.[9] With Tilley, I do not think that God is to be released from responsibility for evil through a demonstration of its inevitability or ultimate nonexistence. Freeing God from this responsibility, however accomplished, takes too lightly the real horror and devastation buried beneath abstract references to "evil." Consider the massive loss of life and culture that occurred as a result of the 2004 tsunami in Southeast Asia. Entire communities were destroyed by

---

[9] Tilley, *Evils of Theodicy,* esp. chap. 9. See also his essay in this volume.

this cataclysm. The overall human death toll climbed to nearly 200,000 (with approximately 1.5 million more displaced), making it the most destructive tsunami in recorded history.[10] An event such as this cannot be tamed theologically by pointing out, for example, the crucial role of shifting tectonic plates in a dynamic world capable of producing life. Explaining the relation between natural evil and God's will and mode of activity in creation cannot be allowed to lessen the evilness of evil, which I take to be tantamount to siding with God against the victims of suffering, to paraphrase Thomas Tracy (in this volume).

But where does this leave us? It does not follow from my (or Tilley's) critique of theodicy that our only remaining option is to sit silently in the face of natural evil. On the contrary, we can and must echo the psalmist's frequent complaints against God for failing to uphold justice or bring an end to suffering, complaints echoed by Jesus himself when he cried out from the cross, "My God, my God, why have you forsaken me?" The real challenge for theologians is to be ever attentive to this cry while engaging in the difficult work of countering the claim that the existence of natural evil undermines the notion of a loving, powerful, and active God. We need to be about the business of bolstering the hope of those who have been harmed by natural processes that God will put an end to their suffering and heal their wounds. What I take to be within the scope of theological acceptability is something less than a full-blown theodicy, something akin to what Tilley labels a "defense," that aims only to defend the compatibility of natural evil and God's active presence in the world.

The particular approach to defense I adopt here is to show that the reality of natural evil is not incompatible with God's active presence in physical processes but, in fact, can be seen as following from the specific character of this presence. Clearly, this kind of approach can provide no leverage against the larger question of why God would allow natural evils to exist. It does not attempt to explain why God acts as God does. But it can do what a defense aims to do, namely, to disarm the charge of incompatibility. It should be noted that demonstrating compatibility does not rise to the level of providing "reasons for." It is just this limitation which makes it a potentially fruitful means of defense. In fact, the compatibility argument, despite being unsatisfactory from the traditional theodicist's point of view, is really the only habitable middle ground—the "theologian's doom," as J. R. R. Tolkien might have called it—between the theodicist's overreaching attempt to take the evilness out of natural evil and the surrender of Job (all of whose afflictions, it could be pointed out, can aptly be described as instances of natural evil).

A number of authors in this volume critically explore an approach to the problem of natural evil that attempts to establish, either physically or metaphysically, that the conditions leading to the possibility of natural evil are necessary for the emergence of life as we know it.[11] According to this "by-product" argument, natural evil is the unwanted but unavoidable concomitant of life: because some of the natural conditions or processes giving

---

[10] "2004 Indian Ocean Earthquake," *Wikipedia: The Free Encyclopedia,* http://en.wikipedia.org/wiki/2004_Indian_Ocean_earthquake (accessed October 19, 2005).

[11] See the essays in this volume by George Ellis, Nancey Murphy, Robert Russell, and Thomas Tracy.

rise to specific natural evils are demonstrably necessary for life's emergence, a world that includes life cannot but be a place in which natural processes lead to suffering. God does not desire natural evil, so the argument goes, but God's desire to create a world that includes life leads inevitably, and thus through no fault of God's own, to its presence.

How strong is the ground upon which the by-product argument rests? For it to succeed, the natural sciences must give us a sufficiently broad understanding of the interrelations among physical processes for distinguishing between those instances of evil which are inevitable, given the existence of some particular good, and those which are not. And if the argument is to be advanced by framing this inevitability in terms of the laws of physics or biology, then one must presuppose that God can rightly be thought of as working entirely within nature's regularities to create life. Alternatively, if one frames this inevitability in terms of metaphysical constraint, then one must presuppose instead that God is bound by what is logically possible. In either case, one must also presuppose that humans have at their disposal a sufficiently robust moral calculus for distinguishing genuine from counterfeit evils. Each of these presuppositions is, to say the least, questionable.

One may also ask whether the by-product argument is a theodicy or a defense. The answer to this question depends upon how one construes the force of the argument. If it is meant to demonstrate the (physical or metaphysical) inevitability of natural evil, given some set of divinely desired goods, then the argument leans toward theodicy. Nancey Murphy's contribution to this volume moves in this direction through its attempt to refute all possible objections to the by-products argument. However, if one accepts that humans lack either a sufficiently robust moral calculus or a sufficiently broad understanding of the interrelatedness of physical processes to see the argument through, then one can deploy these insufficiencies to argue that the atheist is in no better position than the theist to judge the inevitability of nature of natural evils, in which case the argument leans toward defense. Thomas Tracy construes the argument more along this line in his contribution to this volume. In my judgment, Murphy's maximalist use of the by-products argument does indeed let God off the hook with respect to natural evil, even if it does not take the evilness out of evil, whereas Tracy's more minimalist use of the argument allows him to claim that God may have God's own reasons for allowing evil while resisting the claim that we would ever be in a position to understand these reasons. In Tracy's defense-friendly account there is still room for the psalmist's complaint against God, but in Murphy's theodicy-friendly account this room seems to have all but disappeared.

In light of this discussion, and in what I take to be the spirit of Tracy's defense against the problem of natural evil, I wish to supplement a minimalist *why* response to the problem of natural evil (i.e., the theist cannot say why with any certainty, but neither can the atheist appeal to natural evil as evidence against belief in God) with an equally minimalist (i.e., analogical) *how* account of the relation between the God's presence in the world and natural evil.

2.3 Reasoning Analogically from Physics

As a theologian working within the Reformed tradition of *"reformata et semper reformanda,"* I see my task as one of drawing from contemporary and historical theological reflections on God, humanity, and the world while reformulating them in light of a critical assessment of the implications of the sciences for the sake of a more robust theological understanding of the world as God's creation. This perspective finds its *raison d'être* in articulating the coherence of the human experience of God's presence in history through Christ, rather than in the apologetic task of proving the reasonableness of belief in God according to some allegedly general or neutral standard of rationality.

Augustine once asked whether any vestiges of God's triune character were to be found within creation itself.[12] Theologians throughout the ages have wanted to see traces of the Trinity in creation, whether in the external world of nature or the internal world of the human psyche. But these efforts were famously subjected to withering critique in the writings of Karl Barth, who in the face of dire political circumstances at the beginning of the twentieth century resolutely rejected knowledge of God apart from revelation in Jesus Christ. Barth refused to accept the possibility of *vestigia trinitatis* in the world that pointed us to the existence and nature of God.[13] The biblical concept of revelation, he argued, is the only root of the doctrine of the Trinity.

Although there is much of value to be gleaned from Barth's thought on this matter, there can still be a way of making legitimate connections between our knowledge of the world and our understanding of God. As Colin Gunton put it, Barth's rejection of natural theology still leaves room for the idea that "the world in some way speaks of the being of the one who made it."[14] Gunton rightly judged that this can only take place effectively on the basis of prior belief, and not solely on the basis of science. Accordingly, my general mode of engagement with the sciences is not one of moving directly from scientific description to theological affirmation, but rather one of illuminating and reshaping the Christian tradition on the basis of interaction with the sciences. This, I believe, is in line with Barth's more positive assessment of the early meaning of the *vestigium* argument.[15] James Loder and James Neidhardt have eloquently expressed this positive dynamic between theological and scientific knowledge:

> the inner nature of God cannot be pursued extrinsically, abstractly and deductively, as if we could deduce from an analysis of the physical universe the nature of God. However, this does not mean that theology must reject the sciences. Rather, it means that scientific investigation of the natural or the human order must be brought within the body of positive theology, and pursued in indissoluble unity with it. The sciences must become *natural* to the fundamental

---

[12] See, for example Augustine, *City of God* 11.24ff.

[13] Barth, *Church Dogmatics* 1/1:333–47.

[14] Gunton, *Triune Creator,* 144.

[15] Barth, *Church Dogmatics* 1/1:338ff.

subject matter of theology; they will provide the inner material logic that arises in our inquiry and understanding of God.[16]

One way of achieving this kind of dynamic unity, which I pursue in this chapter, is to allow physics to function as a storehouse of concepts and images that can be put to service in the construction of theological analogies. While this is certainly not the only way in which one might bring the two into conversation—and certainly not the most daring—it should not be overlooked as a potentially fruitful mode of cross-fertilization.

My attempt at a defense of the compatibility of natural evil and God's active presence in the natural world takes the form of an extended analogy. It is often noted that analogies prove nothing. To suggest, for example, that God's wrath can be understood as the anger of a loving mother toward a child who is engaged in self-destructive behavior does not prove that God's wrath has this particular quality, or even that God actually "gets angry." Analogies like this one, however, can help bridge the divide between biblical accounts of divine wrath and a theological account of a God who is understood to be neither spiteful nor abusive but whose anger at sin reflects God's loving desire for human flourishing. Analogies also open up personal intuitions and viewpoints for public exploration and debate. They place our hunches out on the table for inspection, so to speak. Their power lies not in their ability to establish truth but in their capacity to reorient a community's efforts at making meaning out of its texts and traditions. "It's like this," we say, as we seek to convey our sense of what is being obscured or misunderstood in the present moment.

Historically speaking, personal analogies for the God–world relation have clearly been central to the Christian tradition. However, there are several distinct advantages to expanding the traditional collection with nonpersonal analogies: they can situate human existence within the wider context of the inorganic world and universe; they can help to reestablish the connection between life and nonlife by supporting a renewed vision of all reality, that is, the entire universe, as God's creation; they can lend plausibility to the view that the physical world's existence is, theologically considered, more than a mere prerequisite for our own; and finally, there is no compelling reason to think that the most common starting point for analogical accounts of divine agency, namely, human agency or intentionality, would be a reliable or even relevant source when it comes to characterizing God's relation to natural processes or evil. Even on the presumption that God is most appropriately described as personal, it is a failure of imagination to suppose that God cannot relate to the physical world in a manner consistent with its own mode of existence. The remarkable judgments of quantum physicists regarding the agency and relationality of the created order make the fruit of their labor genuinely interesting and relevant to the ongoing quest to find faithful ways of articulating God's relation to natural processes and the evils they produce.

---

[16] Loder and Neidhardt, "Barth, Bohr, and Dialectic," 283, emphasis original.

3 God–World Entanglement

I begin this section by reviewing the concept of entanglement as it first appeared in early discussions of the meaning of quantum theory by its founders, then as it resurfaced in the 1960s with the work of John Bell, and finally as it has established itself more recently as a central theoretical and experimental topic among physicists and philosophers of science. I then provide a brief overview of those aspects of the mathematical formalism of quantum mechanics (using a simplified version of the so-called Dirac notation) necessary for understanding the subsequent thought experiment. I conclude this section by presenting the thought experiment, as well as the theological analogy that follows from it. As I mentioned previously, readers already familiar with the basic history and mathematical formalism of quantum theory can safely proceed to section 3.3.

3.1 Brief History of Quantum Entanglement

In 1935, Albert Einstein and two of his colleagues at Princeton's Institute for Advanced Studies, Boris Podolsky and Nathan Rosen, (EPR) published a now-classic paper arguing that quantum theory implies a world in which causal influences can be transmitted "nonlocally."[17] Locality (one of the hallmarks of classical physics) is the principle that physical influences can get from point A to point B only by traveling at some finite speed through every point in between. Einstein posited a version of this principle, the so-called light limit, in his special theory of relativity (SR), according to which no causal influence (i.e., signal) can propagate through spacetime faster than the speed of light. This limit implies, perhaps counterintuitively, that there can be no causal relation between certain events (quantum or otherwise) in the spacetime continuum, namely, those that are "space-like separated" from one another.[18]

However, according to EPR, the generally accepted view at the time (and today as well) that quantum theory provides a complete description of physical reality at the atomic and subatomic levels entails the possibility of a connection between two events no matter how widely separated they might be (which would include, though EPR did not say so explicitly in their paper, space-like separated events). This appeared to violate the locality principle, although it has since been shown that this influence does not strictly violate the light-limit imposed by SR (more on this below). The EPR argument was a *reductio ad absurdum* of quantum theory.[19] No one

---

[17] Einstein, Podolsky, and Rosen, "Quantum-Mechanical Description," 777–80.

[18] Consider the following oversimplified example: The Sun "right now" and the Earth "right now," as two events in the spacetime continuum, are "space-like separated" events, whereas the Sun "right now" and the Earth "ten minutes later" are "time-like separated" events. Time-like separation implies the possibility of a causal connection (signaling) between events, whereas space-like separation implies mutual causal isolation. According to SR, the speed of light, or light-limit, marks the "boundary" between time-like and space-like configurations of events. Hence, a photon emitted by the Sun "right now" influences what happens on the Earth as soon as any physical thing can, which is to say, roughly "eight minutes later."

[19] Specifically: nonlocality is entailed by the claim "quantum theory gives as complete a description of the world as possible, even though it fails to ascribe definite

could be expected to believe in a theory that implied the possibility of such "spooky action at a distance," as Einstein later referred to it.[20] In a subsequently published response to the EPR paper Erwin Schrödinger agreed that the formalism of quantum theory pointed to this strange and troubling reality, which he dubbed "*Verschränkung.*"[21] One possible translation of this word into English, "entanglement," has become common parlance among physicists, although the more technical term "nonlocality" still appears in the literature.

Unfortunately, Einstein's well-known debates with Niels Bohr over the (in)adequacies of quantum theory—the EPR paper was the last of a number of different arguments Einstein produced against the theory—managed to muddy the waters of entanglement more than it cleared them. Their exchanges led many young physicists to dismiss entanglement as a merely "philosophical" and therefore irresolvable issue or, to those more amenable to Bohr's philosophizing, as one having no interesting (i.e., testable) consequences. In 1964, however, John Bell published a remarkable thought experiment extending the original EPR argument (as modified by David Bohm[22]) that uncovered a discrepancy between the empirical predictions of any theory based on the principle of locality and those generated by quantum theory.[23] The first convincing laboratory demonstrations of Bell's thought experiment yielding results in favor of the existence of entanglement in the natural world were reported in 1982.[24] The subsequent history of this topic can best be summarized as an explosion of theoretical and experimental interest among quantum physicists, much of it now focused on the significance of this phenomenon for the emerging field of quantum computation.[25]

Philosophers of physics have also taken a keen interest in entanglement and are currently attempting to sort out its significance for our understanding of the nature of causation and identity at the quantum level. Early debates centered on whether or not entanglement between space-like

---

properties to objects at all times"; nonlocality is an absurd picture of the physical world; therefore, quantum theory must be incomplete.

[20] In German, *spukhafte Fernwirkungen.* Einstein, Born, and Born, *Born-Einstein Letters,* 158.

[21] Schrödinger, "Discussion of Probability Relations," 555–63.

[22] Bohm, *Quantum Theory,* 611–23.

[23] Bell, "Einstein Podolsky Rosen Paradox," 403–8. In subsequent work, Bell strengthened his initial argument, which had considered only deterministic local theories, by showing that even explicitly indeterministic local theories could not match the predictions of quantum theory; see Bell, "Hidden-Variable Question," 29–39.

[24] Aspect, Dalibard, and Roger, "Experimental Test of Bell's Inequalities," 1804–7. For a more recent and rigorous demonstration of Bell's argument, see Weihs et al., "Violation of Bell's Inequality," 5039–43.

[25] See, for example, Cabello, "Ladder Proof of Nonlocality," 1687–93; Greenstein and Zajonc, *Quantum Challenge*; Tittel et al., "Experimental Demonstration," 3229–32; Torgerson et al., "Experimental Demonstration," 323–28; Zbinden et al., "Experimental Test," article no. 022111, pp. 1–10. For a discussion of quantum computation, see Brown, *Quantum Computer*. I have summarized the history of theoretical and experimental work on entanglement through 2002 in Wegter-McNelly, "World, Entanglement, and God."

separated objects did in fact violate SR's light-limit. According to Jon Jarrett's widely discussed analysis, the "locality" of Bell's local theories ought to be understood as the conjunction of two separate conditions: (1) that the outcome at one end of a so-called Bell experiment[26] could not depend upon the *specific outcome* at the other end, and (2) that the outcome at one end could not depend upon the *kind of measurement* made at the other.[27] Because these two conditions functioned jointly, a violation of just one of them would be enough to produce a contradiction with locality.

Jarrett argued that if it could be shown, first, that quantum theory violated only one condition, and, second, that the condition it violated had no bearing on SR, then a détente with SR could be envisioned in spite of the otherwise nonlocal character of quantum entanglement. Jarrett's central conclusion was that only the first condition was violated by quantum theory and that this violation did not contradict the light-limit, thanks to the randomness of quantum outcomes, which made controlling them (and thus sending superluminal signals) impossible. Had quantum theory violated the second assumption instead or as well, this *would* have entailed a conflict with SR, since a superluminal signal could then be sent by inferring the kind of measurement being made at one end of the experiment from the outcome at the other. With the analysis of locality in terms of these two separate conditions Jarrett appeared to have achieved something of a conceptual détente between quantum theory and SR.

Jarrett's initial analysis has produced an ongoing debate over the best way to understand and characterize the relation between entangled objects. Two options still being considered emerged relatively early in the debate. Some philosophers have adopted a more classical and causally oriented approach, characterizing the relation in terms of an uncontrollable (or at least humanly inaccessible) causal influence between separate and distinct entities across arbitrary distances—an approach commonly associated with the term "nonlocality." Others have adopted a more ontologically adventurous approach, characterizing the relation in terms of physical holism, which problematizes talk of interaction between separate and distinct entities— an approach commonly associated with the term "nonseparability."[28] Still other explanations have been advanced more recently, including scenarios involving tachyons, backward causation, and multiple worlds.[29] I cannot

---

[26] Typically, the particles are made to fly out of an entangling source in opposite directions, so one speaks in terms of "either end" of the experiment.

[27] Jarrett, "Physical Significance of the Locality Conditions," 569–89. A refined version of the argument appears in Jarrett, "Bell's Theorem," 60–79.

[28] For early perspectives, see Cushing and McMullin, *Philosophical Consequences of Quantum Theory*; Redhead, *Incompleteness, Nonlocality, and Realism*. For more recent perspectives, see Albert, *Quantum Mechanics and Experience*; Berkovitz, "The Nature of Causality," 87–122; Chang and Cartwright, "Causality and Realism," 169–90; Dickson, *Quantum Chance and Non-Locality*; Healey, "Holism and Nonseparability in Physics"; Maudlin, *Quantum Non-Locality and Relativity*.

[29] For a discussion of tachyon and many-worlds proposals, see Dickson, *Quantum Chance and Non-Locality,* 48–52, 177–78; Maudlin, *Quantum Non-Locality and Relativity,* 61–80, 212–20. For a well-known account involving backward causation, see Cramer, "Transactional Interpretation," 647–87. It is worth noting as well that the equally old but distinct issue of whether or not quantum theory implies causal

address, let alone pretend to settle, any of these ongoing debates here; what is important in this context is that I attend in what follows to the various ways in which these diverging views and interpretations of quantum entanglement might shape and inform the theological analogy.

3.2 Brief Overview of Spin-½ Quantum Systems

It has become increasingly common to present the basic mathematical principles of quantum theory through a discussion of spin-½ particles (such as electrons) using the relatively simple notation developed by Paul Dirac. Many of the central and counterintuitive features of quantum theory can quickly be introduced in this way without a significant loss of rigor. In this section I will provide only a brief overview of the concepts necessary for understanding the argument developed in the next section regarding the nature of causal relations among entangled objects.[30]

In Dirac's representation of quantum theory, the outcome of any particular measurement on a quantum object is represented by what I will call an "outcome state,"[31] which is identified with the symbol$| \ \rangle$. This symbol becomes meaningful only when it encloses another symbol representing some possible outcome of a measurement. In the case of spin-½ particles there are only two possible outcomes to any spin measurement, "spin-up" and "spin-down" (in units of $\hbar/2$), which are typically represented as$| \uparrow \rangle$ and $| \downarrow \rangle$.[32] According to quantum theory, the state of any spin-½ particle

---

indeterminism intersects with the history of entanglement at one crucial point: Bell's "rediscovery" of entanglement in the 1960s. It was Bell's interest in the deterministic and, as he called it, "grossly nonlocal" version of quantum theory developed by David Bohm that led him to write his seminal 1964 paper; Bell, "Einstein Podolsky Rosen Paradox," 404. Bohm's version can be found in Bohm, "Suggested Interpretation," 369–82.

[30] A basic introduction to this system of notation can be found in the classic text, Dirac, *Principles of Quantum Mechanics*. For a recent general introduction to quantum theory that begins with a discussion of spin-½ systems and employs Dirac's formalism, see Sakurai and Tuan, *Modern Quantum Mechanics*.

[31] The traditional term is *eigenket*.

[32] Quantum spin, or intrinsic angular momentum, is a property of subatomic particles analogous to the classical property of angular momentum, which is a measure of rotational motion. This quantum property was postulated in order to explain the subtle behavior of spectral lines of radiation produced by atoms in a magnetic field. An atomic behavior called the normal Zeeman effect could be accounted for by applying the classical notion of angular momentum to electrons orbiting the nucleus, but a complication of this behavior known as the anomalous Zeeman effect could only be explained by applying the notion of angular momentum to electrons themselves, as if, in addition to orbiting the nucleus, they were also rotating about their own axis. David Bohm notes that early attempts to describe quantum spin as the actual rotation of electrons failed to provide a satisfying explanation of this phenomenon. (Given the observed momentum and the postulated radius of an electron, calculations had the electron's surface spinning at many times the speed of light!) Only later with the advent of Dirac's relativistic wave equation was spin "explained" as a requirement of relativistic invariance. But even in a nonrelativistic treatment of quantum mechanics reference to spin is not illegitimate. As Bohm puts it, "the electron still acts [in the nonrelativistic limit] as if it had an intrinsic angular momentum of $\hbar/2$"; Bohm, *Quantum Theory*, 387. James Cushing has suggested that

after being measured will always correspond to one of these two outcome states, that is, it will always be $y = |\uparrow\rangle$ or $y = |\downarrow\rangle$ (quantum states, generally, are designated "y"). A remarkable consequence of one of the central principles of quantum theory—the principle of linear superposition—is that any linear combination[33] of these two outcome states is itself a realizable state. This idea, though deeply counterintuitive and a serious stumbling block to the visualization of quantum processes, is represented by the following straightforward expression:

$$\psi_S = a|\uparrow\rangle + b|\downarrow\rangle, \tag{1}$$

where a and b are simply numbers called "probability amplitudes" and the subscript S indicates generic combination or "superposition." The physical realizability of this kind of state lies at the heart of debates over what quantum mechanics means, for y_S appears to describe a quantum object that is in some sort of "indefinite" state, neither definitely spin-up nor definitely spin-down, but a fuzzy "combination" of the two.

Although y_S has come to be understood by physicists as a physically realizable state for spin-½ particles, it is *not* regarded as a possible outcome state; that is, according to quantum theory, superpositions like y_S can never be the outcome of a measurement. (Obtaining y_S as the outcome of a measurement would be like walking into a room and finding the light both on and off at the same time, which does not accord well with experience.) But if y_S is forbidden as an outcome state, what *does* happen when a spin-½ particle in state y_S is measured? According to the commonly held view, the state "jumps" into one of the two possible outcome states.[34] Although it is not possible to predict which state will turn up, one can predict the probability of finding either state with the help of its corresponding probability amplitude. The rules of quantum theory dictate that the probability of a particular outcome state is the square of the absolute value of its probability amplitude, so that for y_S the probability of finding $|\uparrow\rangle$ is $|a|^2$ and the probability of finding $|\downarrow\rangle$ is $|b|^2$.[35] Here the random and statistical nature of quantum theory becomes readily apparent. Whether any given measurement will produce $|\uparrow\rangle$ or $|\downarrow\rangle$ cannot be known in advance because the outcome of each measurement is irreducibly random. Furthermore, if we find the outcome to be $|\uparrow\rangle$ in some particular measurement, the only way to establish that the particle was initially in y_S rather than in $y = |\uparrow\rangle$ is to repeat the same measurement on a large

---

for the purposes of understanding Bohm's version of the EPR experiment, it does no real harm to imagine the spin of an atomic system as the spin of a ball or planet rotating about its axis; Cushing, "A Background Essay," 3. However, this property must be understood as only analogous to the classical property because, among other reasons, measurements of quantum spin yield discrete rather than continuous values. Regardless of the axis chosen for measurement, that is, one always finds $\hbar/2$ or $-\hbar/2$ and never any intermediate value. Quantum spin is thus said to be "quantized."

[33] A linear combination of states includes sums only of the states themselves, never of their squares, cubes, etc.

[34] This is one way of describing what is traditionally called the "collapse of the wavefunction" or "projection postulate."

[35] In a complication irrelevant to our discussion, taking the absolute value is necessary because probability amplitudes can be complex numbers.

number of identically prepared particles (called an "ensemble") and then examine the long-term (i.e., statistical) results. If it turns out that $|\uparrow\rangle$ occurs 100 percent of the time, then we were mistaken about the particle(s) being in state ψ_S. However, if $|\uparrow\rangle$ occurs $|a|^2$ percent of the time and $|\downarrow\rangle$ occurs $|b|^2$ percent of the time, then our initial judgment was correct.

Things become slightly more complicated when we allow for the fact that spin measurements always occur along some particular direction or axis. To avoid confusion in our earlier formulation of the basic superposition state, ψ_S, we should also have identified the particular axis of measurement inside the symbol for each outcome state as well. Choosing an arbitrary axis a, for simplicity's sake perpendicular to the line of flight, as the axis of eq. (1), we can rewrite it as

$$\psi_{S_a} = a\left|\uparrow_a\right\rangle + b\left|\downarrow_a\right\rangle. \tag{2}$$

But now we encounter a surprising twist—and one that can be difficult to grasp. It is possible, using the rules of quantum theory, to rewrite eq. (2) in terms of any other axis of measurement. What is crucial to understand here is that rewriting eq. (2) in this way does not constitute a change in the particle's state, only its representation! To understand what this means, one must recognize that the quantum state of an object is always expressed *in terms of* some kind of potential measurement. There is, so to speak, no "naked" representation of a quantum state. Any particular representation of a state is just that, a representation *in terms of* some measurement relevant to the object's properties (spin along a, energy, momentum, position, etc.—all of which are commonly referred to as "observables"). So a transformation from one representation to another corresponds not to a change in the particle's own state but to a change in what we want to know about it. This turns out to be important when we have a spin-½ particle in a state whose representation we know in terms of one axis but whose spin we want to measure along a different axis.[36] Given two arbitrary axes, a and b, the rules of quantum theory dictate that the relation between the set of spin-½ outcome states for a and the set for b can be expressed in terms of the following relations:

$$\left|\uparrow_a\right\rangle = \cos(\tfrac{\theta}{2})\left|\uparrow_b\right\rangle + \sin(\tfrac{\theta}{2})\left|\downarrow_b\right\rangle \tag{3}$$

and

$$\left|\downarrow_a\right\rangle = \sin(\tfrac{\theta}{2})\left|\uparrow_b\right\rangle - \cos(\tfrac{\theta}{2})\left|\downarrow_b\right\rangle, \tag{4}$$

where q is the angle between a and b. As an example, we can use these transformations to rewrite ψ_{S_a} in terms of b-axis outcome states. Substituting eqs. (3) and (4) into eq. (2), we obtain

$$\psi_{S_b} = a\cos(\tfrac{\theta}{2})\left|\uparrow_b\right\rangle + a\sin(\tfrac{\theta}{2})\left|\downarrow_b\right\rangle + b\sin(\tfrac{\theta}{2})\left|\uparrow_b\right\rangle - b\cos(\tfrac{\theta}{2})\left|\downarrow_b\right\rangle,$$

which, after combining terms, simplifies to

$$\psi_{S_b} = \left(a\cos(\tfrac{\theta}{2}) + b\sin(\tfrac{\theta}{2})\right)\left|\uparrow_b\right\rangle + \left(a\sin(\tfrac{\theta}{2}) - b\cos(\tfrac{\theta}{2})\right)\left|\downarrow_b\right\rangle.$$

---

[36] Yes, spin is spin, but technically, spins along different axes count as different "observables."

So, for example, the probability of finding $|\uparrow_b\rangle$ when performing a b-axis measurement on a particle whose state we knew, initially, in terms of the a-representation (eq. 2) is given by the more complicated expression, $|\, c_\uparrow\,^2\cos^2(q/2) + 2\, c_\uparrow c_\downarrow \sin(q/2)\cos(q/2) + c_\downarrow\,^2\sin^2(q/2)|$, which is just the square of the probability amplitude associated with this outcome.

With these details under our belts, we are now ready to see what entanglement looks like in Dirac's notation.

Entanglement occurs when the superposition of each particle of a multi-particle system is intertwined with the superpositions of the other particles in the system. In brief, entanglement is "linked multi-particle superposition." The classic example of entanglement is the so-called spin-singlet state, which is a particular state of a system composed of two spin-½ particles. If we write this state in terms of performing measurements of both particles along the same axis (say, a), it is represented as follows (dropping the superposition subscript S from here on):

$$\psi_{\alpha\alpha} = \tfrac{1}{\sqrt{2}}|\uparrow_\alpha\rangle_1|\downarrow_\alpha\rangle_2 - \tfrac{1}{\sqrt{2}}|\downarrow_\alpha\rangle_1|\uparrow_\alpha\rangle_2 .^{37} \qquad (5)$$

For the sake of clarity, a new subscript appears to the right of each individual outcome state indicating whether it refers to particle 1 or 2 (pairs are ordered from left to right). A number of important features of this representation of the singlet state need to be noted. First, because the quantum object now under consideration is a two-particle system, a complete measurement of this system will involve two separate measurements, one on each particle. This corresponds to the fact that the basic terms of eq. (5) are now *pairs* of outcome states rather than individual outcome states. Second, the basic terms $|\uparrow_a\rangle_1|\uparrow_a\rangle_2$ and $|\downarrow_a\rangle_1|\downarrow_a\rangle_2$ are missing from this superposition. Though there is nothing logically or physically impossible about these outcomes, their absence from this state simply means that, given ψ_{aa} as the initial state, the probability of finding either one is zero. One important implication of this is that same-axis measurements of spin-singlet systems will always yield anti-correlated outcomes; if one measurement leads to $|\uparrow\rangle$ then the other must give $|\downarrow\rangle$ and vice versa. And finally, although the spin-singlet state is represented in eq. (5) in terms of a-axis measurements for both particles, it need not be. In fact, it was Bell's genius to realize that while quantum theory's same-axis predictions for the spin-singlet state did not produce an empirical violation of the locality principle, its different-axis predictions did. (EPR and Bohm had previously only considered the equivalent of same-axis measurements.) Taking our cue from Bell, we can construct a general representation of the spin-singlet state for different-axis measurements by substituting eqs. (3) and (4) into just those individual outcome states associated with particle 2 in eq. (5). This leads to the following lengthy but in fact rather simple expression:

$$\psi_{\alpha\beta} = \tfrac{1}{\sqrt{2}}\Big[\sin\!\big(\tfrac{\theta}{2}\big)|\uparrow_\alpha\rangle_1|\uparrow_\beta\rangle_2 - \cos\!\big(\tfrac{\theta}{2}\big)|\uparrow_\alpha\rangle_1|\downarrow_\beta\rangle_2 - \cos\!\big(\tfrac{\theta}{2}\big)|\downarrow_\alpha\rangle_1|\uparrow_\beta\rangle_2 - \sin\!\big(\tfrac{\theta}{2}\big)|\downarrow_\alpha\rangle_1|\downarrow_\beta\rangle_2\Big] .(6)$$

---

[37] The minus sign makes the spin of the overall system equal to zero, which is one of the reasons for the special properties of this state.

This representation of the spin-singlet state can be used to generate predictions for an experimental setup in which the measurement for each particle in the pair occurs along a different axis. As one example of this kind of prediction, the probability of finding the outcome$| \uparrow_a \rangle_1 | \uparrow_a \rangle_2$ is just $\frac{1}{2}\sin^2(q/2)$.

At this point our brief tour of spin-$\frac{1}{2}$ systems can come to a close, for we now have at our disposal all the tools needed to understand the thought experiment of the next section.

3.3 A Quantum Thought Experiment

Recall that John Bell demonstrated the inability of local theories to account for the behavior of a pair of particles in a spin-singlet state as predicted by quantum theory. More technically, he showed that the statistical "correlation" pattern[38] predicted by quantum theory for different-axis measurements cannot be explained by appealing to any sort of "common cause."[39] Entanglement, then, can make an empirical difference to outcomes from the "global" perspective of correlations. But does it also alter the individual or "local" behavior of an entangled particle? This is the question I wish to explore in the following thought experiment.

We can begin by imagining a simple situation in which we have a spin-$\frac{1}{2}$ particle prepared in the basic superposition of eq. (2) along some arbitrary axis b and with the probability amplitudes $a = -b = 1/\sqrt{2}$ (the reason for designating the axis as b rather than a will become apparent below):

$$\psi_\beta = \frac{1}{\sqrt{2}} \left| \uparrow_\beta \right\rangle - \frac{1}{\sqrt{2}} \left| \downarrow_\beta \right\rangle. \tag{7}$$

For this particle the probability of finding$| \uparrow_b \rangle$ is $\frac{1}{2}$ and the probability of finding$| \downarrow_b \rangle$ is $\frac{1}{2}$. Thus a measurement of this particle could lead to either $| \uparrow_b \rangle$ or$| \downarrow_b \rangle$; we can't say ahead of time which outcome state will occur, but we can say that the probability of each is equal. Now we ask ourselves the following question: Would our prediction have been different if, in fact, the particle had not been in the basic superposition of eq. (7) but instead

---

[38] Mathematically speaking, the act of correlating the two outcomes means assigning each of them a number and then multiplying these numbers together. For example, if we assign +1 to$| \uparrow \rangle$ and –1 to$| \downarrow \rangle$, then the correlated result of any experimental trial might be +1 or –1, depending on whether the outcomes are the same or different, respectively. A long-term average approaching, say, –1 would indicate that$| \uparrow \rangle_1 | \downarrow \rangle_2$ and$| \downarrow \rangle_1 | \uparrow \rangle_2$ pairs predominate.

[39] Imagine that I tear a playing card in half, give one half to you, and keep the other half for myself. You look at your piece and discover that you hold one half of a spade. This means, of course, that when I look, I will also discover one half of a spade. Our outcomes are correlated, but does the outcome of your discovery "cause" the outcome of mine? No. There is a "common cause" at work: the two halves were correlated at the start by virtue of being part of the same card. One way of stating the result of Bell's theorem is to say that it identifies the maximal amount of common-cause correlation that can be present between outcomes for *different* types of measurements made on a two-particle system (e.g., on the percentage of times that upon dividing and "measuring" a card drawn randomly from a full 52-card deck, my "measurement" of suit reveals, say, a spade and your "measurement" of number reveals, say, a six). The quantum mechanical prediction exceeds the limit identified by Bell, and thus quantum correlations cannot be accounted for by appealing to a common cause.

had been entangled with another particle as described by the general (i.e., different-axis) spin-singlet state of eq. (6)? In other words, would the presence of entanglement have made any difference to our original prediction?

To answer to this question, we need to find a way of comparing the outcomes predicted in the entangled case with those that would have been predicted in the unentangled case. As we proceed, it will be important to be clear about which particle we are referring to when discussing a particular measurement event. I will employ the following designations: e_0 refers to a measurement event relating to the original (unentangled) particle; e_1 refers to the a measurement event relating to an additional (entangling) particle; and e_2 refers to a measurement event relating to the original (but now entangled) particle. It is crucial to keep in mind that e_0 and e_2 refer to measurements of the *same* particle, the only difference being that e_0 corresponds to no entanglement while e_2 corresponds to entanglement.

With regard to the two measurements that occur under the presumption of entanglement, e_1 and e_2, I will stipulate that they are time-like separated rather than space-like separated, and that e_1 occurs before e_2. This stipulation does nothing to lessen the degree the entanglement between the two particles, but it does allow us to clarify the direction of causal relation between the corresponding measurements by making their temporal order unambiguous.[40] This simplifies our task of assessing whether or not entanglement "makes a difference locally" by allowing us to invoke the notion of conditional probability and to pose the following question: "Given a particular outcome of e_1, what is the probability of some particular and subsequent outcome of e_2?"[41]

We can begin our examination of the predictions for the entanglement scenario by confirming that the mere existence of entanglement does not lead to a prediction (for e_2) that diverges from the prediction made under the assumption of no entanglement (for e_0). We can use eq. (6) to calculate, for example, the "marginal" probability—the probability of an outcome independent of any other consideration—for$|\uparrow_b\rangle_2$ in the case of e_2. We do

---

[40] Space-like separation is the guarantor of true isolation between events (from the perspective of SR, at least), and it is one of the most difficult things to achieve in laboratory tests of Bell's argument. It poses a difficulty for analyses of the causal relations present in entanglement, however, because space-like separated events have no definite temporal ordering: according to some inertial reference frames e_1 occurs first; according to others e_2 occurs first; and according to still others they occur simultaneously. In the case of space-like separation between e_1 and e_2, then, the question "Given a particular outcome at e_1, what is the probability of some particular outcome at e_2?" cannot be answered unambiguously.

[41] Of course, a time-like scenario like the one I have proposed could not be used to prove the existence of entanglement. The point of this thought experiment is not to prove entanglement but to explore the difference entanglement does or does not make to the local behavior of entangled particles. I should also note that my way of laying things out here in terms of causal connection makes a nod toward "nonlocality" in the nonlocality vs. nonseparability debate. The concept of nonseparability is rich with interpretive possibilities, and I have explored it appreciatively elsewhere; Wegter-McNelly, "World, Entanglement, and God." Although the analogy developed here is easier to visualize in terms of the causal model, my presentation could perhaps be recast from the perspective of nonseparability (which I leave to another time).

this by summing over (ignoring) each of the outcomes in eq. (6) that include$|\uparrow_b\rangle_2$ for e_2:

$$P\left(\left|\uparrow_\beta\right\rangle_2\right) = \tfrac{1}{2}\sin^2\left(\tfrac{\theta}{2}\right)+\tfrac{1}{2}\cos^2\left(\tfrac{\theta}{2}\right)$$
$$= \tfrac{1}{2}\left[\sin^2\left(\tfrac{\theta}{2}\right)+\cos^2\left(\tfrac{\theta}{2}\right)\right]$$
$$= \tfrac{1}{2}.$$

Earlier, we arrived at exactly the same prediction for$|\uparrow_b\rangle$ for e_0, confirming that the mere presence of entanglement makes no difference to local probability predictions.[42]

But what about the situation in which e_1 has already taken place? Does this fact make any difference to our prediction for e_2? For the sake of analysis, let's make this question specific: Given the outcome$|\downarrow_a\rangle_1$ for e_1, what is the probability of finding$|\uparrow_b\rangle_2$ for e_2? The standard formula for calculating "conditional" probability, which can be brought into the quantum context without alteration[43] is

$$P(X/Y) = \frac{P(X\,\&\,Y)}{P(Y)},$$

which can be read as follows: the probability of X given Y is equal to the probability of X and Y occurring together divided by the probability of Y. Looking back at eq. (6), we can locate each of the items needed to obtain the relevant prediction:

$$P\left(\left|\uparrow_\beta\right\rangle_2 / \left|\downarrow_\alpha\right\rangle_1\right) = \frac{P\left(\left|\uparrow_\beta\right\rangle_2 \& \left|\downarrow_\alpha\right\rangle_1\right)}{P\left(\left|\downarrow_\alpha\right\rangle_1\right)}$$
$$= \frac{\tfrac{1}{2}\cos^2\left(\tfrac{\theta}{2}\right)}{\tfrac{1}{2}\cos^2\left(\tfrac{\theta}{2}\right)+\tfrac{1}{2}\sin^2\left(\tfrac{\theta}{2}\right)}$$
$$= \frac{\tfrac{1}{2}\cos^2\left(\tfrac{\theta}{2}\right)}{\tfrac{1}{2}\left[\cos^2\left(\tfrac{\theta}{2}\right)+\sin^2\left(\tfrac{\theta}{2}\right)\right]}$$
$$= \cos^2\left(\tfrac{\theta}{2}\right).$$

By a similar calculation, the other "anti-parallel" (i.e., opposite spins) conditional probability, $P(\downarrow_b\rangle_2 / \uparrow_a\rangle_1)$, is also $\cos^2(\theta/2)$. The conditional probability of each of the two "parallel" conditional probabilities is just $\sin^2(\theta/2)$.

This analysis yields the intriguing result that the conditional probabilities for the entanglement scenario (i.e., the probability of a specific outcome of e_2 given a specific outcome of e_1) are, in general, not equal to the

---

[42] Moreover, the same result obtains if we use the same-axis representation of the singlet state in eq. (5) instead, which nicely illustrates Jarrett's point that the *kind* of measurement prepared at the distant end of a Bell experiment has no bearing on the outcome at the near end.

[43] I am indebted to both Abner Shimony and Alisa Bokulich for confirming the appropriateness of this method of calculating conditional probabilities in the quantum context (personal communication).

probabilities for the no-entanglement scenario (e_0) and depend crucially on q, the angle between a and b. In general, the different possible values for P can be divided into three distinct cases:

Case 1 ($q = 0°$): $P(antiparallel) = \cos^2(0°) = 1$

$P(parallel) = \sin^2(0°) = 0;$

Case 2 ($0 > q > 90°$): $1/2 > P(antiparallel) = \cos^2(q/2) > 1$

$0 > P(parallel) = \sin^2(q/2) > 1/2;$

Case 3 ($q = 90°$): $P(antiparallel) = \cos^2(45°) = 1/2$

$P(parallel) = \sin^2(45°) = 1/2.$[44]

In the first case, our knowledge of an outcome of e_1 allows us to do something we could not do in the absence of entanglement: it allows us to predict with certainty the corresponding outcome of e_2. For example, if we find$|\uparrow_a\rangle_1$ for e_1 then we know we must find$|\downarrow_b\rangle_2$ for e_2. In the second case, the outcome of e_1 also leads to a prediction for e_2 that differs from the one we would have made on the assumption of no entanglement, but here we cannot know for certain which outcome state will appear. In the third case, the outcome of e_1 makes no difference at all to our ability to predict what will happen, since the prediction based on the assumption of entanglement matches exactly the prediction based on the assumption of no entanglement. To summarize, the predicted probabilities of$|\uparrow_b\rangle_2$ and$|\downarrow_b\rangle_2$ for e_2 (i.e., under the assumption of entanglement) vary with q and disagree with the predicted probabilities for e_0 (i.e., under the assumption of no entanglement), *except* when $q = 90°$.

From this analysis it is clear that, in general, entanglement does sometimes make a difference locally to the behavior of an individual particle, at least in the sense that the predicted probabilities for each outcome state differ depending on whether or not one assumes entanglement (at least, in two out of three cases). But prediction is an epistemic issue. The question with which we began this section, "Does entanglement alter the individual or 'local' behavior of an entangled particle?" has to do with what actually happens, not with what we predict will happen. To get at this question more directly, we need to examine the actual impact that the switch from being unentangled to being entangled has on the behavior of our original particle. The ideal way to do this would be to compare the outcomes of e_0 and e_2 in light of some particular outcome at e_1, but this simply cannot be done for a quantum particle. One could measure the spin of the original unentangled particle (e_0), reset the experiment, entangle it with another particle, measure the spin of the second particle (e_1), and then measure the original particle again (now e_2). But the randomness of quantum outcomes means that we cannot say that the outcome of e_2 is what it would have been if our original particle had been entangled from the start. The best we can do (and we can still hope that it will be enough) is to compare our predictions for e_0 and e_2 in as much detail as possible. Let

---

[44] The pattern simply continues in sinusoidal fashion as q increases to 180° and then reverses itself.

us consider, then, the switch from e_0 to e_2 in light of some particular outcome, say $| \uparrow_a \rangle_1$, for e_1.

Turning first to the predictions for our original particle (e_0), all we can say on the assumption of no entanglement is that the outcomes $| \uparrow_b \rangle$ and $| \downarrow_b \rangle$ are equiprobable. We cannot say any more than this, again thanks to quantum randomness. Now we need to identify the outcome of e_2 that would obtain on the assumption of entanglement, given our knowledge of the specific outcome $| \uparrow_a \rangle_1$ of e_1. The subsequent outcome of e_2 will vary depending on q, as we determined above, so again we need to divide the result into three cases. This leads to the following "e_0 vs. e_2" comparisons:

Case 1 ($q = 0°$): $| \uparrow_b \rangle_0, | \downarrow_b \rangle_0$ equiprobable $vs. | \uparrow_b \rangle_2$;

Case 2 ($0 > q > 90°$): $| \uparrow_b \rangle_0, | \downarrow_b \rangle_0$ equiprobable $vs. | \uparrow_b \rangle_2, | \downarrow_b \rangle_2$ not equiprobable;

Case 3 ($q = 90°$): $| \uparrow_b \rangle_0, | \downarrow_b \rangle_0$ equiprobable $vs. | \uparrow_b \rangle_2, | \downarrow_b \rangle_2$ equiprobable.

In each case, we need to decide whether the outcome of e_2 is the same as, or different from, the outcome of e_0. In the first case, the best we can do is to note that the difference, if there is one, is between either of the two possible outcomes occurring equiprobably and one of these outcomes occurring with certainty—after all, the set $\{ | \uparrow_b \rangle, | \downarrow_b \rangle \}$ includes $| \downarrow_b \rangle$. This might or might not constitute an actual difference, depending on the specific outcome that would have obtained without entanglement (e_0).[45] In the second case, it is not clear whether we can say that the outcome of e_2 is different from the outcome of e_0, since both outcomes are still possible, although with differing probabilities. Were we to repeat this entangled measurement a large (technically, infinite) number of times, we could at least say that the outcome of e_2 would necessarily be different from the outcome of e_0 at least once, since the overall statistics (as reflected in the probabilities) are different.[46] And what about the third case? Is the outcome of e_2 here the same as, or different from, what the outcome of e_0 would have been? All we can say in this case is that the outcome of e_2 might be the same as, or might be different from, what the outcome of e_0 would have been, and that the probabilities of each possible outcome remain unchanged.[47]

---

[45] The specific outcome that obtains in this case with regard to e_2 follows from the fact that when $q = 0°$, the outcome of e_1 forces the outcome of e_2 into an anti-parallel spin.

[46] In this case, "$| \uparrow_b \rangle_2, | \downarrow_b \rangle_2$ not equiprobable" follows from the fact that the probabilities for antiparallel and parallel spins lie between 0 and ½, and between ½ and 1, respectively, ½ itself not included; given these ranges, the probabilities of $| \uparrow_b \rangle$ and $| \downarrow_b \rangle$ cannot be identical.

[47] Here, "$| \uparrow_b \rangle_2, | \downarrow_b \rangle_2$ equiprobable" follows from the fact that the probability prediction is unchanged from e_0.

It is difficult to know how best to interpret these results regarding the causal dependence of the outcome of e_2 on the outcome of e_1, but let me suggest a way forward. In the first and third cases quantum entanglement does not *necessarily* alter what happens at e_2. In both cases, the specific outcome of e_2 that obtains might be the very outcome that would have obtained for e_0. What one *cannot* say, then, is that the specific outcome of e_2 would not have obtained if the specific outcome of e_1 had not previously obtained. Assuming for the sake of argument a simple counterfactual definition of causal dependence, the outcome of e_1 *cannot* be said to have caused the outcome of e_2, despite the fact that the two particles were entangled with one another. In these two cases entanglement does not lead to any "locally" identifiable causal relation between e_1 and e_2—there is no causal dependence of the second outcome on the first.[48]

In the second case, entanglement *does* alter what happens at e_2, at least if one takes the long-run view (or, alternatively, if one takes probabilities to apply meaningfully to individual events); so in this case entanglement does manifest itself "locally."[49] The specific instances of $e_1 \rightarrow e_2$ causation cannot be pinned on any particular event, however, since there is no way of performing an event-by-event comparison of what did actually happen at e_2 and what would have happened at e_0. Summarizing my analysis of these three cases, we might say that two different kinds of entanglement between two spin-$\frac{1}{2}$ particles in the spin-singlet state appear to be possible: one which includes $e_1 \rightarrow e_2$ causal dependence and one which does not. I will refer to these as "directive" and "nondirective" entanglement, respectively.

One possible objection to my argument for the possibility of nondirective entanglement could arise with regard to my analysis of the first case. There I argued, in essence, that the claim of counterfactual causation could not be sustained because the outcome of e_2 might have been the same as it would have been without entanglement (since$| \downarrow_b \rangle$ is a member of the set $\{ \uparrow_b \rangle, | \downarrow_b \rangle \}$)—one is not entitled, in other words, to make the counterfactual claim that it definitely "would have been" different. But the following objection might arise: since it is impossible to know what the actual outcome of e_0 would have been, the claim that the outcomes of e_2 and e_0 could have been the same is an empty assertion. What really counts, one might argue, is the narrowing of possible outcomes that occurs in the shift from e_0 to e_2. Both$| \uparrow_b \rangle$ and$| \downarrow_b \rangle$ were possible outcomes for e_0, but$| \downarrow_b \rangle$ was the only possible outcome for e_2. Thus, goes the argument, there *is* a significant difference in the first case that sustains the counterfactual claim, namely, the possible outcomes being narrowed from two to one; this difference means that entanglement in this case is directive rather than nondirective. I think it is sufficient to point out in response that this objection leads to the opposite result in the third case: there, the lack of any nar-

---

[48] It is perhaps no coincidence that the global (correlation) pattern produced when $q = 0°$, $90°$ in Bell experiments on spin-$\frac{1}{2}$ particles does not differ from the result predicted on the basis of the locality assumption. Entanglement does not disappear in these cases, but its global effect mimics the result that would obtain without entanglement.

[49] Of course, this instance of causal dependence cannot be exploited for superluminal signaling because the inherent randomness of the outcome for e_1 makes it uncontrollable.

rowing of the options forces the opposite conclusion: no relevant difference exists between the outcomes of e_0 and e_2. So, even allowing this objection, the third case would *still* be an example of the lack an $e_1 \rightarrow e_2$ causal implication, leaving intact my original claim that there are two kinds of entanglement, causally speaking: directive and nondirective.

Is there a causal manifestation of spin-singlet entanglement from the local perspective in addition to the remarkable global manifestation Bell discovered with regard to correlated behavior? Is there, to put the question more precisely now that we have worked through the details of the thought experiment, causal dependence of one outcome of an entangled pair on another, in addition to causal dependence of correlation patterns on joint outcomes? The answer would appear to be, in some instances, yes, though the causal connection cannot be pinned to any specific particle-pair (only to some pair in an infinite collection of such pairs). In other instances, though, the answer appears to be that there is no causal dependence, at least in the sense that there is no counterfactual alteration of individual (i.e., local) outcomes. This particular aspect of the analysis is rather surprising. Who would have expected, prior to the advent of quantum physics and the subsequent discoveries of John Bell, to find an example in the physical world of an empirically motivated claim for the existence of a causal relation between two physical objects that manifests itself relationally (i.e., globally) but not in the individual behavior of the related objects (i.e., locally)?

In presenting this thought experiment I have gone to some length to explore a particular aspect of quantum entanglement. The point of working through this imaginary (but in principle experimentally realizable) scenario at length was to render plausible the idea of a causal relation that can manifest itself relationally without also manifesting itself individually in the behavior of the related entities. If the argument I have made for the coherence of nondirective entanglement has merit, then it is interesting to consider possible extensions of this idea beyond its original context for the sake of philosophical and theological reflection. The argument I have presented here, to be clear, is not about what *must* obtain with entanglement generally but about what *can* obtain if conditions are "just so." The particularity of the example may seem to lessen the analogical potential of the idea, but the analogy I wish to develop in the following section turns only on the coherence and potential fruitfulness of the idea, not on the ubiquity of entanglement (whether nondirective or directive) in the world's physical processes.

4 God–World Entanglement and Natural Evil

Within quantum physics, entanglement expresses the general idea that two separated objects can be related in such a way that truths about the relation between the objects do not supervene on truths about the objects themselves. That is, there are things that can be said meaningfully of the system that are not reducible to statements about the system's parts. The thought experiment presented above exposes one radical though hidden implication of this general idea, namely, that in certain circumstances the presence of an entangled relation can be said to make no counterfactual difference to the individual behavior of the objects. The relation is detectable from the "global" perspective of correlated behavior but not from the "local" perspective of individual behavior. In light of this intriguing result, I now turn to an examination of the idea that God might be thought of as non-directively entangled with the physical world.

There is a certain theological appeal to the idea that a relational (i.e., trinitarian) God who creates an entangled world (i.e., entangled in the scientific sense) would be entangled (i.e., entangled in a theological and analogical sense) with the world God has chosen to create.[50] One key aspect of a trinitarian account of creation is that any notion of a firm distinction between God and the world must be developed with great care. Such a distinction can be affirmed with regard to the doctrine of *creatio ex nihilo* (in the sense that God creates the world, not out of God's own being, but literally from nothing), but it must be nuanced to allow for the claim that creation is the product of a trinitarian act. Specifically, if the second person of the Trinity is not only the formative (i.e., structural) principle but also the substantive (i.e., material) principle of creation—the mediator between God and creation, as Pannenberg puts it—then it will be possible, even necessary, to affirm that the world exists in intimate relation to the divine life. Such an assertion is difficult to maintain apart from a specifically trinitarian understanding of the act of creation.[51]

If the second person of the Trinity were only the formal principle of distinction, and not materially linked to all of creation through the incarnation, there would be no compelling reason to consider the entanglement of God with creation. One could still speak of the entanglement of creation itself as the product of the creative act of God, but there would be no grounds for speaking in the same manner of an entangled relation between God and creation. On this point, I think, the incarnation changes everything—it not only allows for such a connection but demands it. Through the incarnation, the second person becomes the exemplar of all that is distinct from God.[52] God, precisely as other-than-the-world is found in the incarnation to be the formal and material principle of the world's existence. From the creaturely perspective, God may be perceived as standing apart from or against the seemingly autonomous processes of the physical world. But from the divine, incarnational perspective, the physical world is no further away from God than God's own decision to turn out from God's self in the act of creation. The image of a physical world nondirectively entangled with God takes its cue from the richness of this trinitarian, incarnational perspective on the divine act of creation. A physical world nondirectively entangled with God has God as its source and sustainer,

---

[50] One important though not immediately relevant implication of this approach is that it supports and encourages an outlook of methodological naturalism among the sciences: no amount of empirical investigation would ever uncover evidence of a God who is nondirectively entangled with the physical world because the physical sciences confine themselves to an empirical (i.e., "local") examination of the physical world, i.e., they intentionally set aside the possibility of God's existence and activity. Theology, in contrast, attempts to transcend this limitation in order to assess the behavior and character of the physical world as a part of the larger (i.e., "global") God–world relationship. This way of characterizing the distinction between science and theology blocks the traditional approach of natural theology.

[51] From a trinitarian perspective, the freedom of the world to be what it is never exists apart from the fact that it is in relation to God; Gunton, *Triune Creator,* 10.

[52] Pannenberg, *Systematic Theology,* 2:31.

even of that aspect of its existence one might call its "relative causal autonomy."

Pannenberg characterizes the trinitarian God–world relation primarily in terms of the category of "distinction," noting that because of the incarnation "creatures are related to their Creator by their distinction from God and to one another by their distinctions from one another."[53] But in light of Pannenberg's own insistence on the undiminished unity of the trinitarian persons in the second person's free decision to cede divinity to the first, a description of creation solely in terms of distinction falls short. One can say that creation's experience of its relation to the divine in terms of distinction—most pertinently, in the fact that it is not divine—results from the work of the second person. But creation is not the work of the second person alone. It is also the work of the third person. One must also say, therefore, that the third person sustains creation's ongoing relation to God, just as the third person sustains the ongoing communion of the trinitarian life. Because the cooperative work of the second and third persons of the Trinity constitutes both the formal and material principles of creation, a trinitarian notion of the relation between God and creation must be expressed both in terms of distinction and union. The idea of a physical world nondirectively entangled with God blends these notions, making the physical world's distinction from God a product of its union with God through the entangling reality of incarnation. God's gift of being to the physical world comes precisely in the form of a relation that grants this world its relative causal independence.

Could it be, then, that there is ultimately no great divide between the natural world's self-inflicted suffering and God's ongoing relation to it as its creator? Consider the following comment from Karl Rahner:

> Who can say how far [the end of the world] is the very "running-itself-to-death" of the course of the world itself (which is happening in accordance with its eternal laws), how far a halt is called by the creative and restraining word of God, how far both of these things ultimately come to the same thing![54]

This enigmatic statement (made in the context of a reflection on the meaning of bodily resurrection) appears to hint at, or perhaps better, hope for the conceivability and coherence of affirming both realities at once. To integrate these claims would be to offer a kind of answer to the problem of natural evil—not one that explains in any conclusive way why God chooses to allow natural evil in the act of creation but one that shows the compatibility of natural evil with God's active presence in the natural world. If we are to transform Rahner's query into something more than a hope for the appearance of coherence amidst the swirl of apparent self-contradiction, we will need an image that can guide us toward a clearer sense of how these two claims might be mutually affirmed.

What would it mean for the problem of natural evil, then, to think of God and the physical world as nondirectively entangled with one another through God's creative activity? From a nondirectively entangled relation brought about by God would come a physical world characterized by rela-

---

[53] Ibid.

[54] Rahner, *Content of Faith,* 654.

tive causal independence, a physical world free to run its own course and ultimately free to evolve in ways that might lead to suffering on the part of life within it. This world would be intimately related to God, and the character of its own existence would be a product of its relation to God—yet its causal processes would not be counterfactually directed by God. Its relative independence would be a sign of the coalescence of the efficacy of the divine presence entirely on the relation between the physical world and its Creator. Natural evil, in such a world, would be a sign not of God's absence but rather of God's peculiarly empowering and nondirective presence. The physical world could be seen to take its own causal integrity and independence from the divinely established and maintained relation between it and God rather than in contradiction to this relation.

What could it mean for the efficacy of God's presence in the physical world to coalesce in the relation rather than in the physical world itself? In the quantum basis for this analogy, the meaning and justification are clear (if surprising) enough. The presence of entanglement between two particles can lead to a situation in which the behavior of each particle differs only from the perspective of correlated behavior. No difference appears from the perspective of each particle's individual behavior. The two particles act jointly and in relation to one another in a way that sustains the relation but at the same time makes it possible for each to act independently. In the theological analogy, it would thus be wrong to infer from the apparent integrity and relative independence of physical processes that a God nondirectively entangled with the world either just goes along with, or is a victim of, whatever happens. Rather, the relation God creates with the physical world is one in which God is actively present to the world in a manner that grants the world its own freedom.

In light of these comments, some final caveats are in order. First, a world entangled at the physical level with God would not have to be a place in which God related to all of its aspects in a nondirectively entangled manner. Though God might have a nondirectively entangled relation to human beings or other forms of life *qua* enmattered beings, this would not preclude God from having a distinctively personal relation to us *qua* persons. The simultaneous strength and limit of this analogy is that it aims to provide an image solely for God's relation to physical processes apart from any considerations of life or personhood. Relating these different considerations once they have been distinguished in this way is a crucial issue that would need further attention, but I cannot pursue it here.

Second, the analogy with nondirective quantum entanglement works only on an indeterministic interpretation of the quantum formalism. Both directive and nondirective quantum entanglement collapse into strict causal determinism if a framework such as the one developed by David Bohm is assumed. In Bohm's case, the randomness of individual events is lost, except at the level of prediction, and the causal dependence between individual outcomes is present in all three q-cases above. On this particular issue, I adopt a "what if" strategy similar to the one developed by Robert Russell in his own exploration of the concept of quantum divine action.[55]

---

[55] Among Russell's numerous publications on this topic, see, for example, Russell, "Divine Action and Quantum Mechanics." I have surveyed Russell's engagement with quantum mechanics in Wegter-McNelly, "Atoms May Be Small."

We share the presumption of an indeterministic interpretation of the theory, not because either of us thinks it is the best or only possible interpretation, but because each of us wishes to press our arguments as far as they will go for the sake of seeing what they can and cannot do.

There is, however, an important difference between Russell's proposal and mine. Whereas his aims to transcend the realm of analogy by pointing to a specific locus of divine action (i.e., the "event" of wavefunction collapse), my proposal remains within the analogical realm and thus can speak only of "mode," not of "locus." Mine is an analogical description that gets at the *how* of divine action. The *where* of divine action remains undefined on my analogy, except to say that the efficacy of God's active presence in the physical world is located in the entangled relationship between God and the physical world. This limitation might be construed as an advantage, however, at least in the sense that my proposal maintains an arguably greater sense of the difference between divine and creaturely agency. If analogies fail to convince when they presume too much similarity, this analogy, at least with respect to Russell's proposal, presumes *less* similarity by virtue of its being an analogy rather than a concrete identification of the (a)causal nexus within which God is presumed to act. Even with regard to the *how* of divine action there is an important conceptual dissimilarity in my proposal between the "randomness" of quantum events and the "freedom" with which God is presumed to act. God would be in control of God's own activity in a way that random quantum events are not "in control" of their own outcomes, again pressing the otherness of God with regard to physical processes.[56]

Third, it is perhaps obvious, but still bears mentioning, that an analogy with *directive* entanglement leads to a different understanding of the relation between God's presence in physical processes and natural evil. Although it still could be said on the directive analogy that God's act of establishing a relation to the physical world serves as the basis for its independence generally, this could no longer be said to be true at every moment. There would be times when God could be said to "steer" the course of events counterfactually, that is, God would push at least some of them in a direction they would not have otherwise gone. But with God engaged in this kind of activity at least some of the time, the question returns: Why is God not doing this all or at least more of the time to prevent suffering? And so the compatibility of God's active presence in the world and natural evil would be weakened in the directive version of the analogy (cf. the article by Clayton and Knapp in this volume, which makes the same point but in the context of a different argument).

Finally, let me make two constructive points regarding the analogy. First, on the point of God's otherness, the addition of a nonpersonal analogy to the traditional collection of mostly personal analogies for divine action provides a way of reemphasizing God's otherness vis-à-vis our own human perspective on the world. That a personal God might choose to interact with the physical world in a way that mirrors or accommodates its own mode of existence implies not that God's personhood is left behind in such interaction but that God accommodates Godself in relation to different parts of creation in order to faithfully reflect and engage the particular

---

[56] Of course, this difference destroys the analogy if pushed too far.

character of those parts. Second, although I stipulated a time-like relation between entangled events, the fact that entanglement also occurs between space-like separated events, for which there is no unambiguous temporal ordering, indicates that at least some relations within creation transcend the temporality of "local" existence. Though not an absolute transcendence of time, this phenomenon might be thought of as a reflection of divine eternity, closing the conceptual distance between quantum entanglement and God–world entanglement and increasing the plausibility of the analogy.

In this essay I have developed a model of the God–world relation in the form of an analogy from quantum physics. The basis of this analogy, what I have called nondirective quantum entanglement, suggests the possibility of seeing God's active presence in the physical world as compatible with the relative autonomy of its processes and thus with the existence of natural evils. A physical world nondirectively entangled with God would be a place in which God's presence, though apparently ineffective, could be seen as the very basis for creation's relative independence and causal autonomy. The promise of this proposal depends, in the first instance, on the coherence of the argument for global causal dependence without local causal dependence. Ultimately, however, the proposal must be judged in terms of its ability to offer those who continue to be harmed by nature's physical processes a way of overcoming the perceived contradiction between natural evil and God's presence in creation.

WHY IS GOD DOING THIS?
SUFFERING, THE UNIVERSE, AND CHRISTIAN ESCHATOLOGY

Denis Edwards

The South Asian tsunami of December 26, 2004 brought intense suffering to the peoples of Indonesia, Sri Lanka, India, and Thailand. It left more than 221,000 dead, and many more injured. Millions had their homes and livelihoods destroyed and were caught up in long-term trauma and grief. In Australia, as in other parts of the international community, the tsunami called forth an immediate response of compassion and generosity. It also raised theological questions: If God is a God of love, why is this happening? Why is God doing this? Newspapers responded to these questions with a number of articles from various religious leaders. In the face of such devastating loss, the answers offered seemed inadequate. Some failed to respect the pain of the victims and their families and others distorted the Christian gospel.

Scientists explained the cause of the tsunami, pointing to the dynamic movement of the 13 tectonic plates that make up the crust of the Earth and to the earthquakes that occur when these plates collide with one other. The South Asian tsunami was the result of an earthquake near the meeting point of the Australian, Indian, and Burmese plates. The Australian plate is rotating into the Indian plate creating a region of seismic activity in the Indian Ocean. It appears that what happened in December 2004 was a subduction earthquake, where one plate passes under the edge of another. The Indian plate slipped past the Burmese plate, causing the vertical displacement of water that became the destructive tsunami.

There is a close relationship between the movement of plates and the evolutionary history of life on Earth. The constantly changing dynamic system of tectonic plates allows for the emergence of mountain ranges, rivers, rain forests, and fertile plains, providing habitats that allow life to evolve in new ways. There is an inner connection between the physical processes that drive the geology of Earth and the diversity of life. The evolution of life, with its abundance and beauty, is accompanied by terrible costs to human beings and to other species. Here, as elsewhere in the natural world, the costs are built into the system. They are built into the geology and the underlying physics of a dynamic life-bearing planet.

1 The Issue: A Universe with Suffering Built In

The foregoing description of the tsunami and its causes brings into focus the specific issue of this book. The emergence of life on Earth over the last 3.8 billion years can be understood only in the context of the geological dynamics of our planet; Earth can be understood only as part of the planetary system that formed around the young Sun in the Milky Way Galaxy about 4.5 billion years ago; our galaxy can be understood only within the context of the expanding universe that began from an extremely dense and hot state 14 billion years ago. The laws and constants that govern the universe allow it to be life-bearing, but the costs of emergence are built in at the level of these same laws and constants.

In the intense questioning of God that followed the tsunami, my own theological instinct was to say nothing in the public discussion. It seemed a time for grief, for lament, and for human solidarity. Theological comment ran the risk of trivializing the ongoing pain. In the context of a local Christian community, however, there was no choice but to say something about the religious issues that were being discussed so widely. It seemed important to ask whether it is appropriate to think of God as sending disasters to some people and as saving others from them. This, of course, is to call into question the widespread Christian assumption of an interventionist view of divine action. It seemed necessary to offer an alternative view of God working in and through the natural world. And in this natural world, tsunamis, with all their costs, are deeply connected to what brings life to the Earth and its creatures. The Christian sources do not offer any kind of adequate intellectual answer to the question that asks *why* God creates in such a way. What Christianity can offer, based on the Christ-event, is a God who embraces the pain of the world and promises that all things will be healed and transformed in the power of resurrection. Christians live as people of the promise. Only in trusting in the promise do they have any kind of response to offer to suffering.

These inadequate reflections are recalled here because implicit in them is the agenda that faces Christian theology in the light of the suffering built into the universe. The theological task is to respond, however inadequately, to the idea that so much that is beautiful and good arises by way of increasing complexity through emergent processes that involve tragic loss. The costs are evident in the 3.8 billion year history of life with its patterns of predation, death, and extinction. We know, as no generation has known before us, that these costs are intrinsic to the processes that give rise to life on Earth in all its wonderful diversity.

Responding to this issue will require contributions from a variety of disciplines, including those of science, philosophy, and philosophical theology. My own limited response will be one that stands in the tradition of Anselm's *fides quarens intellectum* and Ian Barbour's *theology of nature*. Beside the fact that this is my own discipline, two further considerations suggest this line of approach. The first is encapsulated in comments made by Robert John Russell and Thomas Tracy in their contributions to this volume: Russell proposes that a response to natural evil must go beyond both science and creation theology to involve Christian eschatological theology; Tracy's philosophical analysis arrives at an "epistemic stalemate," and he points beyond this to the eschatological promise of Christianity, insisting on the need to offer hope for *all* victims of natural suffering, non-human and human. The second consideration relates to Wesley Wildman's argument in this volume; I will not respond directly to his argument that creativity and suffering are co-primal and both spring from God, but find myself challenged by his work to try to articulate an alternative Christian theology of divine action that deals honestly with the pain of the world and that takes some responsibility for the view of divine action that is communicated in and by the church.

My proposal is that while a Christian theological notion of divine action cannot offer a full explanation of suffering, it can remove common misunderstandings that spring from traditional Christian notions of divine action, and it can offer an eschatological vision that sees suffering in the context of hope based on the resurrection. Such a theology would need to be eschatological from the ground up. It would need to be in creative dialogue with sciences such as cosmology and biological evolution. A helpful notion of divine action would need to offer some alternative to the popular view of an interventionist and arbitrary God. It would need to offer hope not just to human beings but to the whole of creation. It can be argued that this theology of divine action ought to bear some relation to the extensive dialogue on divine action that took place in the 1990s. Russell describes the consensus that emerged from this work as a notion of divine action that is not only *general* but also *special,* that is not only *subjective* but also *objective,* and that is *noninterventionist.*[57] In this consensus, noninterventionist divine action refers to the view that God acts in nature without breaking or suspending the laws of nature. The position I develop below is situated within this consensus, but takes the specific form of a theology that sees God acting consistently through secondary causes.

I believe that a number of theologians might offer a "thick" christological, trinitarian, and eschatological rethinking of divine action. They include Wolfhart Pannenberg, Jürgen Moltmann, and the one I have chosen to work with, Karl Rahner. Before turning to Rahner, I will attempt to clear the ground by discussing the way a Christ-centered theology might locate itself vis-à-vis scientific cosmology and the way it might see itself in relation to the limits of what it ought to say about suffering and about God.

2 Theology's Location in Relation to Scientific Cosmology

The kind of Christian theology I have in mind is committed to certain ideas about the universe: that God creates *ex nihilo;* that God is present and engaged in every aspect of the universe in *creatio continua;* that God has purposes in creating the universe that include not only the emergence of life and of human beings but also the Christ-event and the eschatological fulfillment of all things. On the basis of its own sources it does not possess special information about the origin or emergence of the universe. It does not compete with science, but engages with what scientific cosmology has to offer, because it sees this as the best information currently available about the way God's creative act takes effect in our world. When science puts forward a view of the universe as expanding from a compressed hot state over the last 14 billion years, theology can dialogue with this worldview, recognizing that theology done in the light of science will always have a tentative and revisable character because it is always open to further scientific developments.

Over the last few years, astronomers such as Martin Rees have popularized the idea that the observable universe may be part of a much larger "multiverse," or ensemble of universes.[58] As William Stoeger, George Ellis, and Ulrich Kirchner point out, in a recent article, there are two distinct reasons why the idea of a multiverse has received increasing attention: on the one hand it has been seen by some scientists as intrinsic to the origi-

---

[57] Russell, introduction to *QM,* ii–iv.

[58] Rees, *Before the Beginning,* 177–86.

nating process that generated our own universe; on the other, it has been offered as an explanation of why our universe seems to be fine-tuned for life and consciousness.[59] I will consider these separately.

2.1 The Multidomain Universe

A number of research programs in early universe cosmology, such as the chaotic inflation associated with Andrei Linde's work, suggest that the processes that would have brought our universe into existence from a primordial quantum configuration would also have generated other universes or universe regions. I follow Stoeger, Ellis, and Kirchner in calling this kind of situation a multidomain universe as distinct from a multiverse in the strict sense, which refers to a collection of causally disconnected universes. How does theology relate to the idea of a multidomain universe? In my view, Christian theology has nothing to say from its own resources about whether the universe as a whole consists simply of the observable universe or of a universe of many domains, as some cosmological models suggest. Of course, it does matter to Christian theology if we belong to a multidomain universe. Christianity would need to ask about how such a greatly expanded universe might be thought of in terms of God's saving love given in Christ's life, death, and resurrection. The size of the observable universe has already forced theologians to consider whether there may be other economies of salvation at work in other parts of this universe. This becomes far more of an issue when one thinks of the possibility of causally disconnected domains of a greatly expanded universe. While this certainly raises genuine questions for theology, it is not a reason for theology to think it can know from its own resources the limits of God's creation. Christian theology has no reason to think it knows the mind of God on this issue and no reason to limit the bounty of God. Whether our universe is part of a multidomain universe is a question for further research in cosmology, and Christian theology is open to either outcome.

2.2 The Fine-Tuned Universe

The second reason for the popularity of the multiverse is that it provides a way of avoiding the theological issue raised by the fine-tuning of the universe. If any of the parameters that govern our universe, including its constants and initial conditions, were even slightly different, then life and human consciousness could not have emerged. The universe is made in such a way that it is ordered to complexity, to life, and to humanity. It is a life-bearing universe. This leads to the question: Why is our universe so fine-tuned to the emergence of life? Of course, one possible answer to this question is the existence of a creator. Some scientific thinkers avoid the theological implications of fine-tuning by invoking the idea of the existence of range of diverse universes, covering all possible parameters. In this case, they argue, it is not surprising that one universe, ours, turns out to be congenial to the emergence of life.

While this argument appears in scientific texts, it is not strictly a scientific argument. Rees, for example, writes in support of the multiverse

---

[59] Stoeger, Ellis, and Kirchner, "Multiverses and Cosmology," 2.

proposal: "If one does not believe in providential design, but still thinks the fine-tuning needs some explanation, there is another perspective—a highly speculative one, so I should reiterate my health warning at this stage. It is the one I much prefer, however, even though in our present state of knowledge any such preference can be no more than a hunch."[60] Rees is rightly cautious about the lack of evidence for this form of multiverse proposal, and acknowledges that it seems in tension with Ockam's Razor, but says that he still prefers the idea of the multiverse to providential design. Clearly this is, at least in part, a *theological* argument. It is an argument based on avoiding the idea that fine-tuning could be caused by a creative providence. Of course, theological convictions have often motivated scientific research in the past and there is no reason why they should not do so in the future. But Christian theology has no need to oppose the idea of providential design. It sees the fine-tuning of the universe as expressing the will and action of the Creator. It is not inclined to embrace the multiverse concept simply in order to avoid the theological implications of fine-tuning.

Christian theology, however, does not have any special information about how God brings about this fine-tuning. There is every reason to assume that it occurs through secondary causes that are open to scientific exploration. And if a case were to be made for a multiverse on scientific grounds, then Christian theology would need to be open to this massively enlarged picture of the creation. While theology can and should reflect on the interrelationship between fine-tuning and the emergence of life, mind, and spirituality, I am not inclined to attempt to make a case for the existence of God dependent on fine-tuning. Such an argument could make sense only if the observable universe is all that there is. It is important for Christianity to keep an open mind on further scientific developments that may support the idea of a multidomain universe and even the possibility of some form of multiverse. The biblical God is a God of boundless fecundity, endless creativity, and infinite mystery. There is no theological reason to limit such a God's creative action to the observable universe.

3 The Limits of Theology in Relation to Suffering and to God

Theologians who are committed to thinking from the perspective of the living memory of Jesus of Nazareth—his life, death, and resurrection—find that there are limits to what they can say about both suffering and God: suffering does not find any kind of full explanation in the Christian tradition, but is understood as a critical memory that calls for liberating and healing praxis and that opens in hope to a future in God; God is not subject to our limited human concepts but confronts us as incomprehensible and uncontrollable Mystery.

3.1 The Critical Memory of Suffering

The theologian Edward Schillebeeckx recognizes that certain kinds of suffering can be transformative. But he finds what he calls a "barbarous excess" of suffering and evil in our history: "There is too much *unmerited* and *senseless* suffering for us to be able to give an ethical, hermeneutical and

---

[60] Rees, *Our Cosmic Habitat,* 164.

ontological analysis of our disaster."[61] Auschwitz cannot be rationalized. Nor, he says, can the quiet suffering of a grieving parent in one's own neighborhood. This excess of suffering provokes protest and resistance. The appropriate human response to suffering is action that brings liberation and healing. But death sets a radical limit to all human responses and shows that far more is needed. Schillebeeckx speaks of suffering as a "negative experience of contrast." The human "no" to the excess of suffering opens out in hope to a better world that can claim our "yes." This "open yes" becomes the basis for healing and liberating action. And from time to time there are experiences of meaning and happiness that sustain and nourish this "open yes." Schillebeeckx sees this human experience of longing for liberation as finding transcendent depth in the various religious traditions. He understands Christians as finding in Jesus the human face of this transcendence and the one in whom the yearning of humanity becomes a well-founded hope.[62]

Christianity contains both a challenge to respond to suffering and a vision of a God who is with creation in its travail and who promises to bring it to its transfigured fulfillment (Rom 8:19–22). The cross of Jesus stands as an abiding challenge to complacency before the suffering of others and to ideological justification of the misery of the poor. It brings the suffering of the poor to the center of Christian faith. The resurrection offers a dynamic vision of hope for the suffering of the world, but it does not dull the memory of the suffering ones. They are always present, forever imaged in the wounds of the risen one.

Johannes Metz speaks of the *memoria passionis* as "the dangerous memory of freedom in the social systems of our technological civilization."[63] This dangerous and critical memory provides an alternative way of seeing. It breaks through ideological commitments. It can lead to solidarity, to alternative lifestyles and to personal and political action. The Christian gospel does not explain suffering, but presents it as a constant challenge to liberating praxis.

3.2 The Critical Memory of the Shattering Otherness of God

In one of his essays, David Tracy speaks of the "shattering otherness" of God.[64] The Christian theological and mystical traditions carry a sense of God's otherness that is far more radical, far more "shattering" than is assumed in everyday discourse both inside and outside the church. The Greek patristic theologians developed the idea of an apophatic or negative theology, based upon the infinite distance between God and any human concept of God. The great medieval theologians acknowledged the analogical nature of all language about God and carefully incorporated a negative moment in all their positive affirmations. The Fourth Lateran Council (1215) declared something that theology takes as foundational: "No similarity can be said to hold between Creator and creature which does not imply a greater dissimilarity between the two." The fourteenth-century English mystical clas-

---

[61] Schillebeeckx, *Christ: The Human Story,* 725.

[62] Schillebeeckx, *Church: The Christian Experience,* 6.

[63] Metz, *Faith in History and Society,* 109.

[64] Tracy, "Hidden God," 5–6.

sic speaks of meeting God in the "Cloud of Unknowing." John of the Cross's beautiful poem images the union of love with the Beloved as a "Dark Night." The radical incomprehensibility of God is the beginning point and end of all theology. What we can comprehend is not God. All of our concepts and all of our words come from our everyday experiences of things in this world. They cannot be used of God in any univocal way. We speak truly of God only in stumbling ways and only within the limits of analogies and metaphors from everyday experience.

As Rahner has often pointed out, God is *abiding* mystery to us. God is precisely what we cannot comprehend. While Christians believe that this uncontrollable and incomprehensible mystery has come close to us and is revealed to us in Jesus of Nazareth and in the outpouring of the Holy Spirit, this does not do away with the mystery.[65] In Christ and the Spirit God is revealed as unthinkable love, love beyond all comprehension, a love that is closer to us than we are to ourselves. And in the light of this revelation, God is understood as a dynamic trinitarian Communion, as shared life in which otherness is embraced. But the otherness of this Mystery is not something theology can comprehend or control.

In this section I have been pointing to what might be seen as limits or parameters for a theology based on the Christ-event that attempts to respond to the issue of suffering. From within its own tradition, such a theology has no resources to explain suffering. It is called to memory, solidarity, liberating practice, and hope. And its deepest tradition keeps it in full recognition of God's shattering otherness. God is a God before whom, like Job, we finally fall silent. These two parameters suggest that theology on its own resources will never be able to offer a full rational explanation for suffering in relationship to God, but this is not to argue that nothing can be said. Important things can be said within these parameters, and one of them concerns the way theology and the church understand and image God acting in the world.

4 Rahner's Theology of Divine Action

With these limits in mind, I will explore Rahner's concept of divine action, asking whether it has anything to offer at both the pastoral level and the level of the theology-science dialogue, in response to the suffering that is built into creation. Much of Rahner's work is found in short, often topical, papers in the twenty-three volumes of his *Theological Investigations*. While he nowhere offers an extended treatment of divine action as such, I will attempt to show that he has a theology of divine action, that this forms a systematic whole, and that it has something to offer in reflection on the suffering that is built in to the universe. I will gather his work into a short synthesis, built up around six characteristics of his view of divine action, and will argue that the first two, God's self-bestowal and creation's self-transcendence, form the twin systematic foundations for the rest.

---

[65] Rahner, *Foundations of Christian Faith,* 44–89.

4.1 God's Self-Bestowal

According to Rahner, the central insight of Christian revelation is that *God gives God's self to us* in the Word made flesh and in the Spirit poured out in grace. The self-giving of God defines every aspect of God's action. Because of this, the story of the universe, and everything that science can tell us about its long history, is part of a *larger* story, the story of divine self-bestowal. The creation of the universe is an element in the radical decision of God to give God's self in love to that which in not divine.[66] This means that the story of salvation is the real ground of the history of nature, and not simply something unfolding against the background of nature.[67] The story of the universe exists *within* a larger vision of the divine purpose.

God creates in order to give God's self to creation as its final eschatological fulfillment. This fulfillment will be the salvation not only of human beings but also of the whole creation. God wills to bestow God's very self in love, and creation comes to be as the addressee of this self-bestowal. Creation, incarnation, and final fulfillment are united in one great act of divine self-communication. The incarnation is not thought of as an add-on to creation. It is not an afterthought. It is not a corrective for a creation that went wrong. It does not come about simply as a remedy for sin, although it is this. With the Franciscan school of theology, exemplified in Duns Scotus (1266–1308), Rahner holds that God freely chooses, from the beginning, to create a world in which the Word would be made flesh and the Spirit poured out.[68] The Christ-event is the irreversible beginning of God's self-giving to creation that will find its fulfillment only when the whole of creation is transformed in Christ.

Rahner insists that what is most specific to the Christian view of God is the idea of a God who bestows God's very self to creation.[69] This is a God who creates creatures that are *capax infiniti,* who without being consumed in the fire of divinity, are able to receive God's life as their own fulfillment. Christianity's insistence, against pantheism, on the difference between God and the world does not mean that there is a distance between God and the creature. Rather, God in God's being *is* this difference.[70] What is truly Christian is a theology, which while maintaining the radical distinction between God and the world, understands God's self-giving in such a way that God is the very core of the world's reality and the world is truly the fate of God. The history of the universe attains its ultimate meaning from the fact that this history is directed to the self-bestowal of God, which will be the

---

[66] Rahner, "Christology in the Setting of Modern Man's Understanding," 219. I am building here on a forthcoming article in *Philosophy and Theology,* "Resurrection of the Body and Transformation of the Universe in the Theology of Karl Rahner."

[67] Rahner, "Resurrection: D. Theology," 1442.

[68] This means that creation and incarnation are "two moments and two phases of the *one* process of God's self-giving and self-expression, although it is an intrinsically differentiated process." *Foundations of Christian Faith,* 197.

[69] Rahner, "Specific Character of the Christian Concept of God," 185–95.

[70] "God is not merely the one who as creator establishes a world distant from himself as something different, but rather he is the one who gives himself away to this world and who has his own fate in and with this world. God is not only himself the giver, but he is also the gift." Rahner, "Specific Character of the Christian Concept of God," 191. See also his "Christology in the Setting," 224.

final fulfillment not just of human beings but the whole universe in God, and this future in not only promised but already begun in the life, death, and resurrection of Jesus Christ.[71]

How is God's creative presence to creatures to be understood? Rahner rejects the notion that the divine immanence can be understood simply on the model of efficient causality. This model is based on the relationship between finite beings that are distinct from one another prior to the causal relationship. This is an inadequate model for a God who creates in a process of self-bestowal. Rahner finds a better model in the theology of grace, and its final fulfillment in glory. This cannot be understood simply as efficient causality whereby one finite entity achieves an effect in another. It is better understood as a kind of formal causality, by which the indwelling God, while remaining radically transcendent, really determines a creature's being. In formal causality a principle of being becomes a constitutive element in another being by communicating itself to the other.[72] God communicates God's self to us and we are transformed in God. In grace, and in glory, God freely gives God's self to us as our fulfillment and we are divinized. Because God is God and not a created cause, neither divine transcendence nor creaturely freedom is compromised.

On the basis that creation is one, and that the one history of the *whole* creation is directed to divine self-bestowal, Rahner proposes that the divine indwelling characteristic of grace is an appropriate analogy for the fundamental relationship that God has with the whole universe and all its creatures[73] This means that Rahner can say that the creative immanence of God to the world is of such a kind that "the reality of God himself is imparted to the world as its supreme specification."[74] Creation is intrinsically directed towards self-bestowal. It is not simply that God creates something other over against God's self, but that God freely communicates God's own reality to the other. The universe emerges in the process of God's self-bestowal. God is always immanent to the world in self-giving love. Rahner sees this self-bestowal of God, as the absolutely transcendent, as "the most immanent factor in the creature."[75] This concept of creation is eschatological from the ground up.

4.2 Creation's Self-Transcendence

The concept of divine self-bestowal looks at creation from the side of God, from the perspective of the divine purposes. A second fundamental concept looks at this same divine action from a creaturely perspective. It points to the effect of God's immanent presence: *creation has the capacity for self-*

---

[71] Rahner, "Book of God—Book of Human Beings," 223.

[72] Rahner, *Foundations of Christian Faith,* 121. In some earlier works, Rahner calls this *quasi*-formal causality, with the *quasi* indicating the uniqueness of this kind of formal causality, in which both divine transcendence and creaturely integrity are fully maintained.

[73] What is true of grace is always valid "in an analogous way for the relationship between God's absolute being and being which originates from him." Rahner, "Natural Science and Christian Faith," 36.

[74] Rahner, "Christology in the Setting," 225

[75] Rahner, "Immanent and Transcendent Consummation," 281.

transcendence. This concept is worked out in Rahner's anthropology[76] and in his evolutionary Christology,[77] but it functions throughout many aspects of his work. Along with the concept of divine self-bestowal, it provides a way of grasping the radical unity of God's one and undivided act, an act that involves creation, redemption, and final fulfillment.

The theological tradition has understood continuous creation as God sustaining creatures in being *(conservatio)* and enabling them to act *(concursus).* Rahner transforms this into a theology of becoming, a theology of self-transcendence. In his evolutionary Christology, he begins from the fundamental *unity* he finds in creation. All of creation is united in its one origin in God, in its self-realization as one united world, and in its one future in God. In this context, he reflects on the transitions to the *new* in the history of the universe, particularly when matter becomes life, and when life becomes self-conscious spirit. The emergence of the new requires explanation, not only at the level of science, but also at the level of theology, which needs to understand God's creative act as enabling the universe to become. Rahner argues for a theology of the active self-transcendence of creation, by which he means a dynamism that is truly intrinsic to creation, but which occurs through the creative power of the immanent God. He sees it as the constant "pressure" of the divine being that enables creation to become more than it is in itself.[78] This "pressure" does not belong to the essence of the finite being and it cannot be discerned by the natural sciences. It is understood as the interior, dynamic relationship of all things in the evolving universe to their Creator.

The material universe transcends itself in the emergence of life, and life transcends itself in the human. In human beings the universe becomes open to self-consciousness, freedom, and a truly personal response to God in grace. Within this context, Rahner sees the Christ-event as the definitive self-transcendence of the created universe into God.[79] Jesus in his humanity is a part of evolutionary history, a part that is radically open to the divine bestowal. If the Christ-event is considered from below, it can be seen as the self-transcendence of the evolving universe into God. If it is considered from above, it can be seen as God's irreversible self-bestowal to creation. In this one person, we find the irreversible self-communication of God to creatures and the definitive human acceptance of this communication. This is what makes Jesus the savior.[80]

In one of his essays, Rahner writes of the *two* ways in which the theology of God's immanence needs to be developed for our time. What he points to are the two aspects of divine action discussed in these first two sections, God's self-bestowal and creation's self-transcendence.[81] I think it can be concluded that, for Rahner, these are *two consistent and fundamental*

---

[76] Rahner, *Hominisation,* 98–101.

[77] Rahner, "Christology within an Evolutionary View," 157–92; *Foundations of Christian Faith,* 178–203.

[78] Rahner, "Natural Science and Christian Faith," 37.

[79] In Jesus, we find the "initial beginning and definitive triumph of the movement of the world's self-transcendence into absolute closeness to the mystery of God." Rahner, *Foundations of Christian Faith,* 181.

[80] Ibid., 193.

[81] Rahner, "Christology in the Setting," 223–26.

characteristics of divine action. These two aspects of the one divine action are mutually interrelated: It is God present in self-bestowal who enables creaturely self-transcendence. Divine self-bestowal and creaturely self-transcendence characterize not only creation, grace, and incarnation, but also the final consummation of all things in Christ.

4.3 Enabling Creaturely Autonomy to Flourish

Rahner sees the act of creation as an absolutely unique relationship. It is not an extrapolation from, or an intensification of, causal or functional relationships between things in the world.[82] Creation is a relationship in which the world is always totally from God and always dependent on God. In this relationship God establishes the creature and its genuine *difference* from God's self. The creature has its own otherness, integrity, and autonomy.

A fundamental principle in Rahner's theology of the God-world relation, one that he often repeats, is expressed in the axiom: *The radical dependence and the genuine reality of the existent coming from God vary in direct and not in inverse proportion.*[83] In ordinary experiences of causality, the more an entity depends upon something else, the less it possesses its own reality and autonomy. There is an *inverse* relationship between dependence and autonomy. But the relationship of creation is radically different. It is a incomparable relationship, one that does not suppose a pre-existing other, but creates the other as other, constantly maintaining it as creation while setting it free in it own autonomy. In this relationship, dependence on God and creaturely freedom and autonomy exist in *direct* relationship to one another.

This claim is finally grounded for Rahner in the human experience of grace. The person who has experienced freedom and responsibility in the depth of their being, and has known this before God and as grounded in God, is in a position to understand something of this relationship between God and creature. In this kind of experience, a created person experiences his or her own freedom as a reality. It is experienced as "a freedom coming from God and a freedom for God."[84] The more we are grounded in God, the more free we are. This leads to Rahner's conviction that, in relation to God, "radical dependence grounds autonomy."[85] This provides a glimpse of the nature of all genuine creaturely autonomy vis-à-vis the Creator. Creaturely integrity is not diminished because the creature's existence is radically dependent on God, but flourishes precisely in this dependence.

It is clear, in this view, that divine action is understood as entirely compatible with creaturely integrity. It is important to note, however, that one who holds this view may take one of two possible further positions: either that God acts in creation with absolutely unlimited power or that God's power is a power-in-love that waits upon human freedom and the integrity of natural processes. Rahner's view of the compatibility between divine action and creaturely autonomy is best seen in conjunction with a view of divine power as power-in-love. Walter Kasper, among others, makes this

---

[82] Rahner, *Foundations of Christian Faith,* 77.

[83] Ibid., 79.

[84] Ibid.

[85] Ibid.

view of divine power explicit. He points out that the cross and resurrection of Jesus reveal that divine omnipotence is the transcendent power to give oneself in love. It is radically power-in-love. It is not that God strips God's self of power to reveal God's love on the cross: "On the contrary, it requires omnipotence to be able to surrender oneself and give oneself away."[86] Against all notions of divine power based upon the power of absolute human monarchs, the cross and resurrection *redefine* divine power. It is redefined as the infinite capacity for self-giving love and for enabling the integrity of the other.

It has always been understood that God can only act in accordance with the divine nature. This nature is revealed in the Christ-event as radical love. The divine nature is revealed as transcendently vulnerable in love. The God revealed on the cross is a God whose nature it is to respect the integrity of creatures, to wait on them patiently, to work through them and to bring them to fulfillment. The power of God revealed in the cross and resurrection is a *cruciform* power, not the despotic power of a tyrant. This means that while God's creative and redeeming action is compatible with human and creaturely integrity, and enables them to flourish, there may be times when *God's nature,* as lovingly respectful of both human freedom and the finite limits of creation, sets limits on what God can do at any one stage in the history of the universe. The love that defines the divine nature is a love that lives with the process, a love that accompanies creation, sometimes suffering with it, promising healing and liberation.

Rahner sees the resurrection of Jesus as *ontologically* the beginning of the divinization of the universe.[87] In it, the final destiny of the world is decided and already begun. This is the most powerful thing imaginable. But it will occur only in and through all the creativity and all the finitude, all the setbacks and all the failures of creaturely processes. Nothing will undo the divine promise. But this promise will be realized in and through God working patiently and lovingly with human beings and with the whole creation and bringing it through the process to its divinizing fulfillment.

4.4 Immanent Rather than Interventionist

While Rahner's early work does not explicitly address the issue of whether God acts in an interventionist way, his theology is based from the beginning on the idea of a God who acts from *within* creation in a way that cannot be thought of as interventionist. I will mention three examples. First, he sees *revelation* as occurring universally as human beings experience the always present mystery of God in grace and struggle to articulate this experience explicitly in concepts, words, and symbols. All of this culminates in Christ. A second example is in his *Christology,* where Jesus is seen as the self-transcendence of creation into God and as God's irrevocable self-giving to creation. In his *anthropology,* Rahner works with the Roman Catholic teaching of the immediate creation of the human soul, and interprets this in terms of God's one act of continuous creation that, by means of a process of self-transcendence from within creation, enables the emergence

---

[86] Kasper, *God of Jesus Christ,* 195.

[87] The resurrection is "the embryonically final beginning of the glorification and divinization of the whole of reality." Rahner, "Dogmatic Questions on Easter," 129.

of unique and diverse spiritual beings made in the image of God. In each case the self-bestowal of God and the self-transcendence of the creature are not interventions from without, but the deepest meaning of God's one creative and redeeming act.

Rahner discusses the issue of divine "intervention" briefly in his *Foundations of Christian Faith,* first published in German in 1976. He argues that if we are to experience God in the world, then it cannot be simply as one element in the world, but as the ground of the world. If we are to find God in the openness to mystery that occurs in our experience of created realities, God must be "embedded" in this world to begin with.[88] What is sometimes seen as a special "intervention" of God is to be understood as the historical expression of God's self-communicating presence that it intrinsic to the world. Every so-called intervention, although it is free and unpredictable, is a becoming concrete and historical of that one intervention by which God has embedded God's self in the world from the beginning as its "self-communicating ground."[89] This means, of course, that God never becomes a cause amongst other causes within the world. God acts from within creation in the form of primary causality. This primary causality finds expression in, and is mediated through, a range of secondary, created causes.

Rahner holds for *special* acts of God, seeing them as "objectifications" of God's one self-bestowing action. This occurs when a created reality mediates and expresses the immanent presence and love of God. A created reality has the role of giving expression to the divine action. Because this role really does belong to the created events themselves, Rahner is among those who speak of *objective,* special divine action. He holds that such events are capable of being recognized as special only within the context of subjective experience of grace.[90]

This means that God's one act of self-bestowing love (primary causality), an act that embraces both creation and salvation, can find expression in a variety of created, secondary causes, and these can be seen as objective and special divine acts. Included amongst these would be the graced experience of meeting God in prayer, the Jewish experience of knowing and responding to God in the events of Exodus and Covenant and the Christian experience of meeting God in the created humanity of Jesus of Nazareth. In such divine acts, the one self-bestowing act of God finds objective expression in and through a range of created secondary causes. These include words, persons, and events. To those with eyes to see, these become symbolic mediations of the divine. In this way, Christians understand Jesus of Nazareth as the radical symbol, or sacrament, of God. They see him as the radical expression of, and mediation of, the divine self-bestowal.

Rahner offers a "modest" example of a special divine act. A person has a "good idea" that proves effective and is experienced as a gift from God. Can this be understood as genuinely inspired by God? There may well be a natural explanation for the good idea, in psychological or neurological terms. But this does not mean that it cannot be seen as an act of God—in the sense that in this experience one encounters the God who is present in every dimension of creation and is really mediated in this event of a good

---

[88] Rahner, *Foundations of Christian Faith,* 87.

[89] Ibid.

[90] Ibid., 88.

idea. When a good idea is experienced subjectively by one who sees the event as the objectification and place of encounter with the ground of all reality, it can be understood as willed by God, as God-given, and hence inspired. It becomes, in Rahner's view, a genuine experience of God's special providence.

In a later (1980) book, coauthored with Karl-Heinz Weger, a book that sets out to deal honestly with difficult questions of the day, Rahner makes his most developed and explicit comments on interventionist and noninterventionist approaches to divine action. He speaks of a "fundamental change" in the move away from an interventionist view of God.[91] He insists that, even for traditional theology, God was not one object amongst others in the world, nor a being outside the world who intervenes from time to time, but the immediately present, immanent, all-embracing, and ultimate ground of being. But he acknowledges that this traditional theology also envisaged, as a matter of course, interventions of God that could be located at certain points of space and time. The traditional idea of the history of salvation was "based mainly on the model of interventions by God." This interventionist model coexisted with a more universal view of God as the "deepest energy of the world."[92] The universal and interventionist models of divine action were never completely reconciled in Christian theology.

Rahner does not condemn the older approach, because he too holds for a genuine history of salvation and for a particular and special revelation, above all in Jesus Christ. But he argues for the emergence of a "universalist basic model in which God in his free grace, from the very beginning and always and everywhere, has communicated himself to his creation as its innermost energy and works in the world from the inside out."[93] He wants to show is that it is possible "with all due caution and modesty, to do without a particularist model of external intervention by God into his world at particular points of space and time, without having to interpret Christianity 'naturalistically.'"[94]

Within this context, Rahner sees Jesus as the one who makes God's deepest promise historically accessible and irreversible. This promise is already the fundamental energy at work in all things in the universe, and because of this Rahner believes that it is possible to understand the event of Jesus without the image of intervention.[95] It is the resurrection of Jesus that gives expression to this promise. The resurrection is an event of revelation, but not one coming from "outside."[96] This important claim is simply

---

[91] Rahner and Weger, *Our Christian Faith,* 57.

[92] Ibid., 77.

[93] Ibid., 78–79.

[94] Ibid., 84.

[95] "The event of God's promise of himself in Jesus makes that deepest promise by God of himself to the world historically accessible and irreversible. It is always and everywhere the fundamental energy and force of the world and its history. It is therefore perfectly possible to understand the event of Jesus without the aid of images of an intervention in the world from outside. In doing without such an image, however, we must let history really be history and clearly realise that this deepest energy and power of the world and its history is God in his sovereign freedom, who, by his free promise of himself, has made himself this deepest energy and force of the world." Ibid., 103–4.

[96] Ibid., 111.

stated by Rahner. In my view, it is one that needs further development in contemporary theology.

In the light of this discussion, I will attempt to make more precise the notion of immanent or noninterventionist divine action that I am proposing: (1) It sees God's action as one differentiated act of self-bestowal, involving creation, salvation, and final fulfillment, rather than as a number of acts. (2) It thinks of God's action as working from within the processes of an emergent universe rather than as breaking in upon the universe from a divine realm beyond it. (3) Because it understands God to act consistently through secondary causes, it sees God as working in and through the laws and constants of nature rather than as violating, suspending, or bypassing them. (4) It sees God's action as enabling creaturely integrity and autonomy to exist and flourish rather than as limiting it. (5) It understands God as respecting and waiting upon the integrity of the processes of nature rather than as sending natural disasters to some while saving others. (6) It understands special divine acts, including words, persons, and events, as ways in which God's one ever-present act of self-bestowing love finds new symbolic and effective expression in our lives and history rather than as separate divine interventions.

4.5 The Hiddenness of the Future

Rahner's insistence on the promise of the resurrection is accompanied by an insistence that we cannot picture its content. To think we might picture our final and definitive state "would be still more absurd than to suppose that the caterpillar could imagine what it would be like to be a butterfly."[97] He takes up this theme in a systematic way in his essay on the hermeneutics of eschatological statements.[98] At the center of his approach to the interpretation of the biblical and traditional depictions of the future, as, for example the great wedding feast, or Christ judging between sheep and goats, is the thesis that *the future of the world in God remains radically hidden to us.*

This future is the coming of the incomprehensible God. It is announced and revealed in Christ, but it is revealed only as "the dawn and the approach of mystery as such."[99] This conviction provides Rahner with a basis for distinguishing between genuine interpretations and false interpretations of eschatological predictions and images. A genuine interpretation will preserve the mystery, while a false one will present the future as if it were the literal report of a spectator. This criterion is based on Rahner's conviction that the future is known as an inner moment of the present. Genuine knowledge of the future in God is knowledge of the eschatological present. It is a projection based on what is already experienced of God. The future will be the fulfillment of the salvation already given in God's self-communication through grace, in the Spirit, and through Jesus Christ, the Word made flesh.

---

[97] Rahner, "Hidden Victory," 156.

[98] Rahner, "Hermeneutics of Eschatological Assertions," 323–46. He incorporates much of this into his *Foundations of Christian Faith,* 431–37.

[99] Rahner, "Hermeneutics of Eschatological Assertions," 330.

What we encounter in the experience of grace is the God whom Rahner calls *absolute future*. He distinguishes *absolute* future from all *this-worldly* futures.[100] The absolute future is God's self-bestowal: it is the consummation of creation and redemption already promised and initiated in the life, death, and resurrection of Jesus. By contrast, all "this-worldly" futures occur in the ordinary dimensions of time and space as particular events or states of this world. Each of them, by definition remains open to a further future. The Christian claim is that the evolution of the universe will end, not in emptiness, but in the divine self-bestowal. This absolute future is already a constitutive element of the unfolding of the universe. It has already found irrevocable expression and been made visible in Jesus Christ. His resurrection is the promise and the beginning of the absolute future, the transformation of all things in Christ. Absolute future is another name for God. It not only comes towards us but is also the sustaining ground of the movement towards the future.[101]

God, then, is known not as one object among others that we might plan for the future, but as the ground of this whole projection towards the future. This absolute future has definitively promised itself to us in Jesus Christ.[102] The absolute future comes to each person in grace. It is offered to each and its acceptance is the ultimate task for each. The content of Christian preaching consists of the question of the absolute future and, properly speaking, *of nothing else*.[103] Because of this, it is essential to the church and a fundamental task of theology to act as the guardian of the "*docta ignorantia futuri*."[104] Rahner uses the language of learned ignorance, associated with Nicholas of Cusa (1401–1464), to point to theology's critical role of resisting closure with regard to the future. Keeping open the question of the future is theology's most basic task. Rahner holds that even in final blessedness, God will remain the eternal mystery to which human beings commit themselves in the ecstasy of love. Christianity in its essence is openness to the question of the absolute future.

4.6 New Creation: Self-Transcendence of the History of This Universe

What is the relationship between the coming of God as our absolute future and the world that we attempt to construct? Rahner's response is in terms of his fundamental principles of divine self-bestowal and creaturely self-transcendence.[105] The coming Reign of God will not simply be the outcome of the history planned and accomplished by humans. Nor will it simply come upon us from outside. It will be the deed of God. It will be the self-bestowal of God. But it will also be the *self*-transcendence of history.

Rahner argues that there is a dialectical tension between the two statements that *history will endure* and that *history will be radically trans-*

---

[100] Rahner, "Marxist Utopia and the Christian Future," 59–68; "Fragmentary Aspect of a Theological Evaluation," 235–41; "Question of the Future," 181–201.

[101] Rahner, "Marxist Utopia and the Christian Future," 62.

[102] Rahner "Question of the Future," 190.

[103] Ibid., 188–89.

[104] Ibid., 181, 198.

[105] Rahner, "Theological Problems Entailed," 260–72.

formed. This tension maintains an openness to the future while still giving a radical importance to the present. He makes the strong claim that "history itself constructs its own final and definitive state." What endures, he says, is the *work* of love, as it is expressed in the concrete in human history. It is not merely some distillation of our works of love but our history itself that passes into its definitive consummation in God.[106] History is embraced by God in the Christ-event as having eternal meaning. Human freedom and the moral character of our actions have final and definitive significance because they are taken up into God and transformed in Christ.[107]

This line of thought is related to the idea articulated more recently by John Polkinghorne,[108] and built upon by Robert John Russell,[109] that new creation will come from the old *(ex vetere)*, in both *continuity* and *discontinuity* with it. Rahner embraces these positions in his theology of the self-transcendence of the history of the universe in new creation. His thought moves in the same direction when he asks, in another article, whether final consummation will come from within creation or from beyond: Will our final consummation be immanent or transcendent?[110] Will our future be continuous with our present experience (immanent)? Or will it be discontinuous with our present experience (transcendent)? Again, Rahner insists, this future consummation *is* the self-bestowal of God. And this self-bestowal is not only our fulfillment, but also *the principle of the movement* towards this fulfillment. The present is sustained by the future. It is the immanent God who constitutes this future and enables the self-transcendence of the creature into this future. This means that our final consummation is *both* immanent and transcendent.

What of the consummation of the universe as a whole? As always, Rahner insists that matter and spirit constitute a unity. If one were to consider matter as such, in the abstract, it would never reach its consummation. But the material world has always been sustained by the creative impulse of self-transcendence by which it tends towards the spirit. This impetus is nothing other than the impetus of the divine self-bestowal. The incarnation of the Word is present in this creative impulse from the beginning. The material world is not something to be cast aside as a transitory stage in the journey of the spirit. The material world itself is to be transformed in Christ. From the beginning, this movement of the universe towards consummation has been one of self-bestowing love. Because this self-bestowal in love is the most immanent element in every creature, Rahner concludes: "It is not mere pious lyricism when Dante regards even the sun and the other planets as being moved by that love which is God himself as he who bestows himself."[111]

---

[106] Ibid., 270.

[107] Ibid., 271.

[108] Polkinghorne, "Eschatology: Some Questions," 38–41; "Eschatological Credibility," 48–55; *Faith of a Physicist,* 162–70.

[109] Russell, "Eschatology and Physical Cosmology," 283; "Bodily Resurrection," 14–28.

[110] Rahner, "Immanent and Transcendent Consummation," 273–89.

[111] Ibid., 289.

5 Conclusion: Suffering Built In to the Universe in the Light of Christ

I have been proposing that one partial response to the suffering that is built into our life-bearing universe can be found in a theology of divine action based upon the Christ event. At the center of this theology is the Christian claim that the resurrection of Jesus Christ is the promise and the beginning of the final healing and divinization of the whole of creation. How is this theology of divine action related to the loss, suffering, and grief caused by the South Asian tsunami? What does it have to say in the light of the natural evil that is built into the universe from its origin? What it offers is an account of a God whose nature is to love and respect the natural world in its emergence and in its integrity, who does not overrule the natural world, but works in and through its processes bringing all things finally to healing and fulfillment in God's self.

Some key characteristics of this theology of divine action can be summed up in the following points:

1. It is an act of self-bestowing love that reaches its culmination only when all things are taken up, healed, and divinized in the trinitarian life of God.
2. It takes effect in creation's God-given capacity for self-transcendence.
3. It is one differentiated act rather than many discrete acts.
4. It is better envisaged as springing from within creation rather than coming from outside.
5. God's one act (primary causality) embraces creation, salvation, and fulfillment, and it finds expression in a variety of secondary causes, including created persons, words, and events, that can be seen as objective and special divine acts.
6. It is compatible with creaturely autonomy and enables it to flourish.
7. It is a power-in-love, a love that waits with infinite patience on human freedom and on creaturely integrity in achieving the divine saving purposes.

Such a view of divine action can offer a partial response to natural evil. The costs of evolution can be understood in the context of a God who acts in fidelity to the divine nature, as self-bestowing love; who is present to each creature, embracing each in love, delighting in each and suffering with each; who is radically respectful of both human freedom and the integrity of natural processes; who works through the self-transcendence of creaturely processes; who will bring each creature, in a way that is appropriate to each, to redemption and fulfillment in the divine Communion.

This approach raises at least three fundamental questions that cannot be addressed here. First, how precisely can resurrection be thought of in noninterventionist terms? This question clearly sets an important agenda for further theological work. Second, how can Christian eschatological hope be reconciled with cosmology's bleak predictions about the future of the universe? This is an issue facing all Christian eschatologies, one that has been taken up directly by authors such as Russell and Polkinghorne in recent work.[112] Third, how do we go beyond Rahner's thought about the material universe being transformed in Christ to think about the eschatological fulfillment of nonhuman biological creatures? Does God really care about every sparrow that falls to the ground? Elsewhere I have attempted some response to this question,[113] arguing that we can think of the Spirit of God not only accompanying every sparrow with love in its life and its death, delighting in it and suffering with it, but also as bringing it to a redemptive fulfillment in the trinitarian life of God.

In my view, this concept of divine action is one partial but important response, on the basis of the Christian sources, to the suffering built in to the universe. It has something to say at both at the level of the science-theology dialogue of this volume and at a pastoral level. At the level of the science-theology dialogue, the theology of divine action developed here is in creative dialogue with the science-religion discussion of divine action in the 1990s, where divine action is understood as both general and special, as noninterventionist, and as not only subjective but also objective. It adds to the discussion a dynamic notion of God's creative act as eschatological self-bestowal and as working in and through creaturely self-transcendence. This theological view of divine action can be seen in relation to the views of other contributors to this volume. In their own distinctive and often more philosophical approaches, a number of contributors identify the goods that are achieved in the emergence of a universe in which suffering is built in. The goods they identify include the emergence of intellectual and moral agents capable of being taken into loving communion with God (Tracy, Murphy, Clayton, and Knapp), the evolution of a universe capable of eschatological transformation (Russell, Tracy), and creation's participation in creativity (Tilley). As is obvious from the outline of divine action offered above, the first two of these, communion with God and the transformation of the universe are affirmed and developed in the theological notion of divine self-bestowal. The third, creation's participation in creativity, is expressed and developed in the theological notion of creation's God-given capacity for self-transcendence. Divine self-bestowal and creation's self-transcendence are the two particular goods that this theology of divine action specifies.

This approach offers a further contribution to the general discussion with the argument that God's omnipotence is redefined by the cross and resurrection: divine omnipotence is understood not as the capacity to do absolutely anything, but as the divine capacity to act in fidelity to the divine nature. This nature is one that lovingly respects and waits on the integrity of human freedom and the autonomy of nature and will bring all to fulfillment in and through the process. This view of divine action offers a thoroughly eschatological theology in the notion of divine bestowal that works in and through every aspect of creation, grace, incarnation, and final fulfillment. It involves a promise not just to human beings but to the whole of creation as one creation. All will be taken up into God and transformed in Christ. All will be divinized.

---

[112] Russell, "Bodily Resurrection," 3–30; "Eschatology and Physical Cosmology," 66 315; Polkinghorne, "Eschatological Credibility," 43–55; "Eschatology: Some Questions," 29–41.

[113] Edwards, "Every Sparrow," 103–23.

At a pastoral level, this theology of divine action provides a viable alternative to an interventionist God who arbitrarily brings suffering to some and healing to others. It sees God as working in and through natural processes in a way that waits upon and fully respects the integrity of the processes and the freedom of human beings. It sees God as lovingly accompanying each creature in its life and its death; it sees the cross as revealing a God who enters into the starkest pain of the world and suffers with suffering creation; it sees the resurrection as a promise that creaturely suffering and death will be redeemed and healed as each creature finds its meaning and fulfillment in God's self-bestowing love. In this kind of theology, standing as it does in the Christian tradition, what is argued for is the plausibility of holding that while we do not fully understand why suffering is built into the universe, we can still entrust ourselves, our universe, and all its creatures in faith, hope, and in love, to the absolute mystery of self-bestowing love.

INCONGRUOUS GOODNESS, PERILOUS BEAUTY, DISCONCERTING TRUTH: ULTIMATE REALITY AND SUFFERING IN NATURE

Wesley J. Wildman

Introduction

Speaking of ultimate reality is bizarre. Most theologians and a few philosophers are captivated by such speech, however, and choose it even while understanding its final futility. Rejecting the advice of probably wiser but possibly timid souls, theologians go ahead and dream their dreams, spinning them into great tomes of wisdom and ritualized traditions of debate. It is an act of defiance in the face of inevitable human ignorance and an act of simple trust in the receptive grace of ultimate reality. Speaking of ultimate reality is only simple for the ignorant or the young. For theologians who know their labors are the very stuff of dreams, the impossibility of the task is captivating. They bypass the crisp argument to get to the elegant paradox that reflects their warring instincts. They refuse the novel for the sake of ancient insight. They turn from conventional wisdom in the name of loyalty to truth as hard as diamonds. Good theologians never forget the knife-edge character of their journey, strung out between betraying the sacred by saying too much and abandoning their duty by surrendering to the beckoning silence. Theirs is a wax-winged thrusting of the soul heavenwards to the sun.

Nothing conjures theological dreams more than the reality of suffering in nature. Suffering, in one way, is not a problem for theologians; it is the very lifeblood of their book-bent bodies. Every theological tradition draws life from the relentless power of suffering to make us seek dreams. Judaism staggers under the weight of being chosen to suffer, Christianity has a crucifix at its center, and Islam has the most profound construal of surrender imaginable. Hinduism's samsaric vision frames suffering in serenity, Buddhism teaches us how to escape it, and Chinese religion tells us how to organize our lives to manage it. Occasionally we see theologians valiantly defending their faith in face of suffering, but this act of loyalty means something only to needy believers and nasty detractors; such defenses are essentially a public service. Less often we see theologians refusing the question, brushing it aside with a story or a joke, or aggressively attacking defenders of the faith as somehow betraying the very thing they seek to honor and protect. Bless us all. In the ominous light of suffering, all theology is a kind of agonized writhing. But there is nothing to be done for us. Again and again we return to the streets of suffering, scouring the sidewalks for the penny we know must be there, the coin that will make sense of everything.

Cursed to wander in search of secrets, then, let us not waste energy on defending the universe or its divine heart, at least not when we are praying or talking to one another. We will do our spirited public service defenses and our dutiful institutional rationalizations when we must, of course. And there really is no problem with any of that; obviously humanly recognizable divine goodness can be defended in the face of suffering if we are deter mined enough. We have all the standard tools ready to hand, from best-of-all-possible-worlds arguments to unavoidable-side-effect-of-overriding-good arguments. If we get desperate, we can fall back on "God can't stop it," "God has a bigger plan," "God suffers with us," "It is not really real," or "Just wait until it is all put right." And if the arguments are not quite as compelling as we would like them to be, neither are the assaults on the overall goodness and meaning of life, the universe, and everything, so there is no need to panic. Our acts of institutional maintenance and defense are certainly successful enough to keep the hostile hordes at bay.

So let's take a deep breath, set all that aside, and return to our first obsession, our calling and our curse: the theological task of speaking of ultimate reality in face of the proximate reality of suffering in nature. The varied phenomena of suffering in nature press hard upon all theological theories of ultimate reality. It is impossible to approach such an interpretative task without appreciation for the creativity displayed in intellectual efforts past and present, within Western cultures and outside them. Likewise, it is impossible to step so sure-footedly among the shards of past efforts that we forget that before which we wander and play with our theological concepts, that greatest of all realities which drives us to silence when we are most attentive to it and yet suffers our speculations without reproach. Humility is the theologian's byword on pain of irrelevance.

2 A Reverent Competition among Three Views of God

It is for these reasons that I set out to bypass theodicies and defenses for now and to speak of ultimate realities instead. Rather than presuming a theory of ultimate reality for the sake of a theological meditation on suffering, however, I suggest we take less for granted and use suffering as a source of selective pressure on God ideas. I am particularly interested in the effects of a full awareness of the reality of suffering in nature on a reverent competition among three theological approaches to God, two of which are also accounts of ultimate reality, and all of which are important in the contemporary theological scene: determinate-entity theism, process theism, and ground-of-being theism. In what follows I shall try not merely to describe these three approaches to speaking of God but also to express each one's beating heart and its fundamental attraction, thereby to discern the resulting theological criteria that each marshals to guide its strategy for making sense of suffering in nature. This reverent competition will allow me to articulate how ground-of-being theism handles the problem of suffering in nature. This theological view is an awkward partner for common human moral expectations but deeply attuned to the ways of nature and resonant with the wisdom about suffering that is encoded in many of the world's religious and philosophical traditions.

The intended audience for this argument is qualified experts interested in the topic, and it includes broadly theistic theologians, religious naturalists, metaphysicians, and intellectuals from a variety of religious traditions. I fully realize that I cannot completely satisfy the legitimate intellectual demands of such a complex audience, but I am motivated to try because it is the proper audience to evaluate a view of ultimate reality and its relation to the world. Some will object to my attempt to produce a general theological interpretation of suffering as intellectually futile and will urge me to pick a smaller audience within which I can realize their favorite theoretical virtue, namely, high confidence in the theological model relative to a religious group that can make use of it. It follows that not everyone is interested in theology as a kind of general inquiry, as I am. While I do not fault such objectors for their less general and more practical interests, I do think they should not prejudge the possible results of theological inquiry in a broader audience, particularly at the dawn of an unfamiliar era of comparative metaphysics.

I pause to comment on a terminological difficulty. I will treat these three views of God as types of theism. Some would argue that the word "theism" has become so closely associated with the personalist elements of determinate-entity and process views that it is misleading as a description of ground-of-being theologies. To put my response compactly, I contend that this personalistic focus in the doctrine of God is excessive. It is a post-Reformation distortion, deeply linked to the "turn to the subject" in modern philosophy and theology. It also tends to reflect the economic and social values of middle-class suburban white Protestants. I am unwilling to allow the general category of theism to be held hostage to what I hope amounts to a passing trend in the history of Western Protestant theology. But I acknowledge the terminological difficulty and do not want it to cloud or mask my view of ultimate reality as ground of being.

2.1 Determinate-Entity Theism

The first view in our reverent competition is determinate-entity theism, which conceives God as an eternal, all-powerful being with a compassionate awareness of every circumstance and moment of suffering, the ability to act in history and nature, and the moral quality of humanly recognizable goodness to a supreme degree. This view can be called "personal theism" because its initial moves are to analogize the divine nature using the intentional and agential capacities of human persons, but I shall continue to refer to it as determinate-entity theism, which I think is a more accurate term. This is because the best of the determinate-entity views are not slaves to personal analogies. They make allowance for the ill-fitting aspects of personal analogies for God by introducing balancing symbols. For example, the common metaphor of blinding light applied to the divine presence suggests that the divine wisdom would necessarily sometimes have to be incomprehensible and the divine will occasionally inscrutable. These balancing mechanisms resist the ever-present danger of anthropomorphism in God-talk. Indeed, the flatly anthropomorphic versions of these views are the object of aggressive attacks both externally from skeptics and internally in the sacred texts of all major theistic religious traditions. From both directions, the critiques regard excessively anthropomorphic views as superstitious, and thus (like all superstition) constantly in need of special pleading in face of the contraindicating evidence of worldly experience. By contrast, subtler anthropomorphic symbolism, balanced with nonanthropomorphic symbols for the sake of empirical adequacy, has played an essential role in all theistic religions.

The inevitability of anthropomorphism is due to the particularity of the human imagination. To this extent the projection analyses of religion from Feuerbach to Freud are correct: human beings must picture ultimate reality in terms of the highest and most profound reality they know, which is themselves, or at least their parents, their rulers, their warriors, their shamans, and their priests. And they are also obviously correct that this casts doubt upon all anthropomorphic picturing of ultimate reality. We can allow this and yet insist that nothing about projection strategies for interpreting ultimate reality entails atheism—neither the sheer fact that they occur nor the reflexivity of their occurrence, neither their pervasiveness nor their moral dangers. This is where some of the projection theorists allowed the enthusiasm of new realization to lead them into theoretical distortion. As grappling with these critiques for more than a century has made clear, the world still demands theological interpretation in terms of ultimate realities even though projection reflexes are an inevitable accompaniment of such interpretations. The theological and practical religious challenge is to manage projection-driven anthropomorphism, not to eschew it altogether, and by and large the great theistic traditions have done this fairly well.

What kind of determinate entity is the divine reality? Theologians giddy with the joy of speech may lose track of the fact that they construct answers to this question. They may attempt to evade responsibility by saying that they merely faithfully and thoughtfully update a tradition that bears forward an answer whose origins lie hidden in untold revelatory transactions, or by self-consciously submitting themselves to a purported revelation whose veracity lies beyond question on pain of banishment from one's beloved community of theological companions. It makes no difference: we construct in our hermeneutical glosses, in our conceptual rearrangements, in our ignorant rediscoveries, in our claims of allegiance, in our self-righteous criticisms, and in our charges of improper novelty. And we incur responsibility as we do so. Let us skip past the clever subterfuges whereby we pretend at merely receiving and handling sacred ideas. We should embrace the fate of theologians, which is to speak into life ideas of ultimate realities for each other and on behalf of the great traditions of spiritual practice and social organization that orient the living, including us, as we wander the tangled paths joining living memory to a future unknown, save for the certainty of our own deaths. If we are lucky, our ideas of ultimate realities will make contact with the peculiar cosmic harmonies and noisy dissonances that test and constrain our speculations. If we are very lucky, our theological ideas will bring a measure of contentment and conviction, even though we know we will gladly surrender them as our deaths bring us near to the ultimate reality itself, which always looms before and hovers beneath us. And this is really what we are up to on this view: we are trying to make sense of our lives, of the lives of those we love, even of the lives of those we hate or whose experience we cannot comprehend. To this end, we take unto ourselves the power to speak into reality the dreams we dream, to conjure and weave, to construct and argue. We will not shrink from our anthropomorphic instincts and their associated limitations, nor from our simple need to make sense of life in this vale of joy and tears.

With no evasions, then, what will we say is the determinate character of the ultimate reality we call God? We will assert, simply, what we most long to be true: that ultimate reality makes final, beautiful sense of everything; that in God there lies meaningful hope for the downtrodden, and even for our own sorry hearts; that God will make our souls live on in a perfectly purified realm whose proportions we can almost imagine but never realize in this world; that God can be the object of our love and will respond to us personally, knowingly, and graciously. Thus, God will be good in a humanly recognizable way, powerful in a way that is relevant to fulfilling our longings, all-knowing in a way that unmasks those longings, and benevolent in a way that forgives our stupid hatred and fear and self-destructiveness. If God needs to take action in our history and nature to secure the possibility of these unendingly desired outcomes, then God's mercy will bend to the task.

In short, God's determinate nature is known in our longings. Everything else we say theologically must serve this overridingly important vision of ultimate reality, and this becomes the *crucial criterion of determinate-entity theism*. It is anthropomorphic, yes. But our humanly shaped imaginative capacities open this reality up for us, so there should be no shame about constructing a theological environment for human life that matches our deepest longings. Our great hope is answered in the booming resonances of cosmic space and time, and the answer is yes! This is good news of the stunning, life-changing, shockingly apt sort.

Determinate-entity theism is fundamentally optimistic, magnificently anthropomorphic, and existentially thrilling. It stays close to the hopes and fears of ordinary human beings. It communicates to most people effectively and inspires sublime art and music and architecture. It requires that sacred texts be submitted to no more than a courteous minimum of demythologization, which really amounts to mere cultural updating. It has relatively few variants, and their battles with one another for credibility do not seriously threaten the big picture. So God may judge and damn the wicked, as our own vengeful hearts demand, or God may run the scythe of judgment through every heart and draw us all into the divine presence with grief-tinged joy and thankfulness; God may send suffering to advance an inscrutable plan or tolerate unwanted suffering for the sake of nurturing souls that can richly long for divine fellowship. Such disagreements do not wreck the coherent and attractive flavor of determinate-entity theism, with its central ideal of perfect personal goodness, so long as we are willing to regard them as subordinate details. Determinate-entity theism's portrayal of the divine personality mutates over cultures and eras, reflecting different fashions and needs. Yet the same basic idea of God persists through all of these variations.

The crucial theological criterion associated with this blessed draft is uncompromising. Theologians devoted to a God that miraculously answers human longings are intensely sensitive to intellectual moves that dilute the strength of the brew or import alien elements that confuse its flavor. They find the most perfect recipes for their preferred drink in the sages of the past and are suspicious of newfangled ideas that always and only seem to wreck perfection in the name of short-sighted, arrogant innovation. Yet they are also unafraid to translate their favorite recipe in the language of the day; there is not a trace of secret gnosis here. Theirs is a welcome and healing draft, after all, and compassionate souls can only commend it to their self-deluded and self-destructive companions on the way. More concretely, they unfailingly diagnose theological compromises with alien cultural wisdom as seductions to be resisted, and they typically see demythologization programs as advancing other agendas that finally cannot be rendered compatible with their own. Yet they have no difficulty understanding that real wisdom may be present in religious traditions other than those they know best, for their vision of ultimate reality is not at risk from the realization that there may be many authentic ways. Narrower theologies claiming exclusive prerogatives face an insoluble problem in the face of religious pluralism, but this determinate-entity vision of ultimate reality can be made as capacious as it is perfectly fitted to human longings.

Compassionate forms of determinate-entity theism in our era tend to regard suffering in nature as an unfortunate by-product of otherwise good natural processes. An argument of this sort is common in this volume: if God wanted to create creatures capable of freely entering into loving fellowship with their creator, then God would have no choice but to allow the natural evils that come with finitude and physicality and the moral evils that freedom brings. In an earlier era these same theologians may have regarded suffering as divine punishment or testing, or otherwise sent by God to achieve morally unimpeachable divine purposes. For example, the accidental death of a child was routinely interpreted as God taking the child for special divine companionship, and this could and still can bring genuine comfort to grieving parents willing to surrender to this vision of tragedy consistent with perfect divine love.

Theological fashions aside, all of these determinate-entity views affirm that the world we inhabit is the best of all possible worlds. The crucial criterion absolutely requires this, and without it, the very point of determinate-entity theism in the form discussed here fails utterly. God must be good in a humanly recognizable way and powerful in a humanly relevant way for human longings to find their perfect answer in ultimate reality, for this creation to be a home rather than a hopeless and hostile environment for futile human writhing. Of course, the ultimate divine purpose for this world of suffering and love may remain rationally obscure, or clear only in the sense that confidence in revelation can make theological claims clear, but even then God's goodness, the creation's wondrous purpose, and the sacred meaning of every human life is assured. At root, this is the best of all possible worlds in the same sense that God is the best possible answer to human longings. The associated interpretation of suffering in nature spins itself around this central axis.

2.2 Process Theism

Our reverent competition's second candidate is process theism. Importantly, this is not a theory of ultimate reality. The process God in most of its variations is one actual entity alongside many others that constitute the fluxing process of reality, albeit one with a special role. God's role is to maximize value in the cosmic process, including by making the greatest possible use of every configuration of events, including those involving suffering. God is not creator in this view, so God is not all-powerful. Removing suffering is not a possibility now or at any future time, therefore, but making the most of every welcome occasion and each disastrous event is the perpetual divine responsibility. God does this by constituting the unfolding divine nature with the awareness of all that happens, arranged and related so as to maximize value and goodness. This leads to the first sense in which the process God is good literally by definition: the so-called consequent divine nature is maximally good by construction in and through its prehension of feeling of the world process itself. The other sense in which God is good by definition derives from the process account of causation: God conveys to every moment of the world's process a vision of possibilities that functions as a lure toward beauty, goodness, and truth appropriate to each kind and level of event. The source of this lure is partly the so-called primordial nature of God, which is a wondrous vision of possibilities. God's goodness is postulated at each pole of this dipolar theism, and it is humanly recognizable goodness, at least at our level of complexity.

The goodness of God is assured by definition, therefore, but key questions about ultimate reality remain open, including what it is and whether it is good. Most versions of process theism do not seek a theory of ultimate reality in the form of a God concept. On the contrary, the God concept of process theology—whether in Whitehead's "single actual entity" form or Hartshorne's "series of actual occasions" form—is practical for human beings only to the extent that it does not repeat the mistakes of the past, premier among which is to make of God an ultimate reality. This stance does not make a theory of ultimate reality impossible, but it does entail that a process-based theory of ultimate reality, though it includes the process God, must go well beyond God to encompass the various principles to which the cosmos and God alike are subject. Whitehead's philosophical cosmology, as expressed in *Process and Reality,* is pluralistic: there are many fundamental categories. Ultimate reality is not one thing, the ontological solution to the philosopher's problem of "the one and the many"; rather, it is the pattern by which creativity perpetually produces one from many. It is that which is closest to experience and thus the least abstract and most pervasive feature of reality. In this sense it is what is ultimate about reality. But it is not God, even though, like all processes within reality, God expresses it.

The basic theological instinct of process theism is easy to recognize and deeply moving. It aims to deliver us from an unhealthy obsession with an illusory picture of God. Process theologians are the very few theologians who resist the drug of ultimacy speech in the name of usefulness; they can feel the tug but they just say no. If we surrender our attachment to a God of infinite power who creates everything, makes sense of everything, and finally gives meaning to everything, then we can open ourselves to a more modest but more concretely satisfying picture of God. The plausibility of traditional determinate-entity theism is massively strained by our ordinary experience of life. We see around us not a perfectly good world, and certainly not anything that would recommend the idea of a perfectly good omnipotent creator. Rather, our experience suggests that reality is a morally neutral environment for the interplay of a host of processes, within which we witness both beauty and goodness, on the one hand, and disintegration, and pointless suffering, on the other.

This shows us that traditional determinate-entity theism is anthropomorphism run amok; it is a stubborn assertion of human longings against the unrelenting facts of existence. Our attachment to such a picture of God is self-delusory and, in the long run, distorts our view of everything else. Process theism beckons us to move toward a divine light of a different sort, one in which God is explicitly aligned with the parts of the cosmic process that make for beauty and goodness. Stop coercing the cosmos to fit a God-concept driven by over-expansive human longings! Exchange that rapacious God-concept for one actually fitted to human spiritual longings and moral aspirations, and the theodicy problem evaporates. The price for this invaluable benefit is merely surrendering the futile quest for a morally comprehensible ultimate reality.

The deep intuition here is that anthropomorphic instincts in theology work only when they are limited to the patterns and parts of the world process that are scaled to human longings. To project human longings and expectations beyond this limit is, unsurprisingly, intellectually disastrous and spiritually frustrating. Let the whole of reality be what it is! It is a relief to lower one's eyes and to focus on making concrete sense of our longings and aspirations in relation to the world of our actual experience. In this world many things occur that threaten to destroy us, from accidents to natural disasters, from predators to human stupidity. Instead of calling these things good, as seems to be demanded in some sense when we say they are the creations of a recognizably good deity, just stop trying to make human-friendly meaning out of them. They are threats to the integration of goodness and beauty at human and other levels, even if they are co-conditions of integration and emergence. But nothing stops us from picturing God as unambiguously on the side of maximizing values of goodness and beauty. Taking refuge in this God, we shield ourselves from the harsh light of the entire world process, which is as hostile to life as it is supportive.

For the traditional determinate-entity view, a meteor slamming into the Earth that destroys the human species is an unfortunate side effect of a process that God created and sustains for the sake of higher goods, and the (severe) problem is why God did not intervene to prevent the destruction of an entire ecosphere. On the process view this is a disaster, period, and God is left to integrate the horrible consequences of an event beyond divine control into the divine nature. Whereas we may have difficulty relating humanly to the God who creates a world with such possibilities, we have no trouble relating to a God who does not control such events but is left to pick up the pieces afterwards. This shows why anthropomorphism must be limited to the domain within which process theism operates, and also both why it works well within that domain and why it works badly beyond it.

Nothing stops us speculating on ultimate reality within the process framework, just as Whitehead did. Indeed, we could integrate the pluralistic process conception of ultimate reality into a competing idea of God. This gives us two candidates for the use of the word "God"—the process entity, God_P (P for Process), which is scaled to most human spiritual longings and moral intuitions, and creativity itself, God_C (C for Creativity), which is the source of pain and pleasure, purpose and pointlessness. On the one hand, process theologians hold that God_C is religiously useless because it is morally impenetrable, so most resolutely ignore it and advise others to do likewise. Specifically, the process view alleges that traditional determinate-entity theists deeply mischaracterize God_C when they say it is good in a humanly recognizable way, and wrongly assert that God_C is able to bring meaning to everything that happens in way that reflects a personal center of consciousness. Whatever God_C is, on the process account, it is exceedingly resistant to anthropomorphic modeling, and certainly nothing like the personal God of so many sacred texts and religious pieties. The human-like activity and moral sense of God_P are what matters to most human beings and to the human species thought of as a civilizational, cultural, and moral project.

If God_C is what matters to some theologians, it is because they love what can neither be rationally comprehended nor morally assimilated. This is an unhealthy intellectual obsession with no concrete benefits for the world and only perplexity and despair for the ones so magnificently obsessed. Process theists would gladly liberate their theological companions caught in the conceptual chaos of overreaching anthropomorphism that arrogantly forgets its proper limits. But process theists know from hard experience that not everyone shares their instincts about what is important and useful and valuable. Some willingly enslave themselves to an impossible master, after which the delusion that the master is good and loving causes only the most exquisite agony, the agony of the slave who cannot afford the luxury of seeing things as they manifestly are, but can only survive in a world woven from illusions.

2.3 Ground-of-Being Theism

The third view in our reverent competition is ground-of-being theism. Ground-of-being theisms deny that God is a determinate entity in all or most respects. Their theological advocates are deeply wary of anthropomorphism in theology, and as the metaphor "ground of being" suggests, they tend to look toward universal features of reality rather than to exceptional features such as human beings for imaginative symbolic material. Though ground-of-being theisms are theory building efforts rather than apophatic in themselves, they are often aligned with apophatic traditions in theology because they are stations on the way to refusing theological speech in a kind of mute testimony to that which finally transcends human cognitive abilities. They have a rich heritage in Western and South-Asian theological traditions, and in more naturalistic forms within Chinese philosophy. Indeed, they are strikingly similar across these cultural boundaries, particularly in their refusal to tame ultimate realities with humanly recognizable moral categories and in their rejection of an intentional, agential divine consciousness.

Ground-of-being theists share points of agreement with their rivals. On the one hand, they concur with the process critique of the traditional forms of determinate-entity theism we have been discussing. There surely is a place for anthropomorphic modeling in theology, but ground-of-being theism and process theism alike say that determinate-entity theists slip into a world of illusions when they suppose that the creator God_C is good in a humanly recognizable way. On the other hand, ground-of-being theists share the determinate-entity theists' instinct to bypass the process deity God_P in favor of the creativity deity God_C. These two sorts of theologians are obsessed with the ultimate reality from which the process theist turns away in the name of religious and cultural relevance; they willingly enslave themselves to the complexities and perplexities of thinking of ultimate reality as God.

Ultimate-reality-enslaved ground-of-being and determinate-entity theists do not just docilely accept the process theologians' charge of futile attachment. They retort that process theology is precisely the wrong kind of anthropomorphically woven tapestry of illusion. God_P, they argue, is a condensation of humanly supportive trajectories in the universe that happily skips over the rest in a kind of half-hearted and half-spoken Manichaeism. God$_P$ is a mere invention, with the conceptual level pitched to guarantee divine goodness and ramify religious hopes and beliefs. The process refusal to link God with ultimate reality is an arbitrary constricting of the theological purview and a betrayal of the theologian's Promethean calling.

This mutual recrimination over brazen anthropomorphism and futile attachments is one of the exquisite ironies of contemporary theology. The acceptance in our time of a theological viewpoint that eschews a theology of ultimate reality—a rejection formerly unthinkable in orthodox theological circles within theistic religions, yet deeply resonant with some of the portrayals of God in sacred texts—makes this irony possible. Is ultimate reality our bane or our blessing? Should we flee the theologian's self-appointed calling in the name of moral intelligibility or embrace it no matter what the cost? These questions hint at an existentially profound difference in the instincts of contemporary theologians.

Of course, ground-of-being theologians participate in these recriminations but also claim unique theoretical virtues and are subject to distinctive criticisms. In offering a theory of ultimate reality they share the virtues of determinate-entity theism, while in refusing to say that the creativity deity God$_C$ is good in a humanly recognizable way they share in the empirically robust realism of process theology. But the endpoint—an ultimate reality not personally good in a humanly recognizable way—can this even be called a worship-worthy God? Determinate-entity theists argue that the ground-of-being theists' refusal to see ultimate goodness in the heart of reality deals a killing blow to the human aspiration to feel at home in the universe and to believe that human lives have purpose and meaning. Process theists would not put it that way, of course, but they would echo the criticism by saying that the word "God" has to be aligned with goodness at human scales or else it loses its religious relevance. Ground-of-being theism only confirms the process theologian's suspicion that God$_C$ is religiously useless. At least the determinate-entity theists make an effort to preserve religious relevance by insisting that God$_C$ is good in a humanly recognizable way! Ground-of-being theists basically accept the process theist's analysis of ultimate reality as conceptually incomprehensible and morally impenetrable but then call it "God" anyway. This, say process critics, is more an abuse of terminology than their own alleged misuse of the word "God" for something other than ultimate reality.

Ground-of-being theologians seek an empirically adequate theory of ultimate reality, and this drives their refusal to allow that this ultimate can be unambiguously morally good. These views tend to regard suffering as ontologically co-primal with creativity in the divine life and in the world. Thus, they do not treat suffering as an unwanted side effect of otherwise good natural processes and good divine purposes; to do this is an exercise in futility when the unwanted side effects are not minor but rather, on average, of about the same size and importance as the good events. They do not associate God only with human-scaled and human-focused goodness; to do this would be merely to attribute to God the dubious quality of being convenient for humans. They do not affirm that this is the best of all possible worlds; claiming that all of reality is finally good is a mere clutching at straws as the whirlwind of creativity and suffering spins us around. To what, then, does this interpretation of ultimate reality lead? And why go there?

I shall return to these questions below. For now I note the fundamental instinctive difference between ground-of-being theism and its competitors in this reverent contest. In a supermarket full of potential theological virtues, and being on a tight budget, ground-of-being theists spend their money on plausibility rather than religious appeal, making the most of the latter given what the former allows. They would mistrust some of the prominent items in the human longings aisle that promise to satiate spiritual hunger, and they would scour the top and bottom shelves for affordable nutrition. They would stubbornly refuse to be lured by colorful claims of ultimate intelligibility, ultimate meaning, ultimate purpose, and ultimate justice, and instead stalk the realism and deference aisles, looking for simple, everyday items that deliver these wonders in more modest measure. Perhaps this reflects a kind of disappointed failure of human longings; it certainly expresses a determined effort to accept ultimate and proximate reality.

If our reverent competition reduces to a mere choice among theological personality styles—and surely such considerations are relevant—then this is where we stand. The determinate-entity theists are the optimists and fight against disbelief. They prize reality as ultimately good and human life as ultimately meaningful above all. Those who do not they interpret as stubbornly, self-destructively refusing to accept the wondrous miracle that the character and purposes of ultimate reality are limned in human longings. The process theists are the activists and fight against soporific delusions. They value religious relevance, rational intelligibility, moral clarity, and transformative action above all and so refuse to speculate on ultimate reality while resolutely affirming an alliance between God and human interests. They interpret those who see things differently as in thrall to the illusory mythic sentiment that ultimate reality is a proper object of human religious and intellectual instincts. The ground-of-being theists are the mystics and fight against resentment. They treasure above all the whole of reality as it is without illusions and without limitations to human interests and longings, and they surrender themselves to it, whatever it may be, and without reserve. They interpret those who do not as unable to cope with life as it most truly is and as reserving their love for an idealized, humanized image of God.

3 The Bane of Anthropomorphism

I naturally resonate with and feel some attraction toward all three of these ways of conceiving God in relation to the challenge of suffering in nature. Yet I shall argue that the alternatives to ground-of-being theism face significant conceptual problems in mounting their response and that these problems derive chiefly from trying to make of God a moral agent unambiguously good in a humanly recognizable way.

3.1 The Argument from Neglect

Determinate-entity theism must face what I call the "argument from neglect." This argument contends that the determinate-entity theist's conception of God does not rise even to human standards of goodness and so certainly is not humanly recognizable as perfectly morally good in a way that befits deity. The argument turns on an analogy with human parents, as follows.

A human parent, indeed parents in many species, must constantly balance the need to protect and guide offspring with the need to allow the offspring freedom to learn. Loving parents do not hesitate to intervene in a child's life when they discern that ignorance or mischievousness or wickedness is about to cause serious trouble, and perhaps irreparable disaster. Parents rescue the child, interjecting education, punishment, or encouragement as needed. As time goes on, children need less guidance but parental interference rightly persists until the child is largely independent. Wise parental interference does not limit a child's freedom; on the contrary, it enhances it by protecting the child from freedom-destroying injuries and character defects, and by leading them patiently but surely toward freedom-enhancing independence and moral responsibility. We hold parents negligent, and sometimes criminally negligent, when they fail to intervene when necessary for the sake of their child's safety and well-being.

Human beings are like children in respect of moral and social-civilizational matters. God, on the determinate-entity theist's account, has all of these responsibilities in relation to human beings that human parents have in relation to their children. God should intervene to educate and guide, to punish and redirect. If it is claimed that this does in fact occur, then it certainly does not occur often or effectively enough for God to avoid the same charge of negligence that we would bring upon a human parent acting in similar fashion.

I think creatures besides human beings have a claim on parental protection and nurture from a determinate-entity God. This becomes an especially important point in relation to natural disasters, where the scale of injury and death in other species is frequently far larger than the human losses we most notice. Animals may not be able to raise their voice in complaint, but human beings can do so on their behalf. Together, we feel neglected, exposed to the elements, and left to comfort ourselves with illusions of ultimate love and perfect nurture that experience finally does not support. We get our love and protection, our education and wisdom, not from God's parental activity but from our own good fortune at living in a cosmic era with few meteoric collisions, from our own determination to build stable and rewarding civilizations, and from our own discoveries about the world that we pass along to our children. The idea of God as protective, solicitous parent may make a difference in our lives in the way that a wondrous story can bring comfort and solace, but that is as far as it goes.

3.2 The Argument from Incompetence

Ground-of-being theism is immune to the argument from neglect, but process theism is vulnerable to a variant of it. Process theism properly and predictably replies to the argument from neglect that God is always trying the divine hardest to educate and alleviate suffering in an ultimate reality that is partly hospitable and partly hostile to human interests. But then the well-earned counterreply is that the process God$_P$ may not be negligent but certainly is incompetent. In other words, in what I shall call the "argument from incompetence," God$_P$ is not powerful enough to merit our worship and allegiance and we should go in search of Godc—the only deity that finally matters, even if its moral character is indigestible.

Of course, if a theological interpretation of suffering in nature requires a theory of ultimate reality then process theology was never a candidate anyway. If our aim is less systematic—say, theological support for a religiously relevant response to suffering in nature—then process theology may be the right sort of proposal, but it is, as I have tried to show here, inadequate. Note that the argument from incompetence does not demonstrate the incoherence of the process idea of God but merely its religious ineffectualness specifically in response to suffering, contrary to claims that its supporters typically make on its behalf. In the case of the argument from neglect, the target is the conceptual coherence of determinate-entity theism.

3.3 The Alternative: Religious Indigestion or the Breath of Life?

Ground-of-being theism is immune to the argument from neglect and the argument from incompetence. But at what price? Its rejection of a personal center of divine consciousness and activity is religiously indigestible to many people, and knocking out two of its major competitors is the very opposite of a good outcome for such folk. Ground-of-being theism simply does not meet their basic criteria for acceptability as a theological interpretation of God. So they will understandably continue the struggle on behalf of determinate-entity and process theisms, representing the interests of their religious constituencies.

Meanwhile, ground-of-being theism is the very breath of spiritual life for some other people. It has a lot to commend it, particularly in relation to the problem of suffering in nature, so long as the ruling theological criteria allow both suffering and blessing to flow from the divine nature itself. This is simply too much for many and yet simply perfect for some. Recognizing this apparently unbridgeable chasm between conflicting theological and religious intuitions, I become less eager to persuade others to become ground-of-being theists. Rather, I want to argue that ground-of-being theism, so often neglected in contemporary theology and in science-religion discussions, should be taken with great seriousness as an intellectually compelling account of ultimate reality, even if it is seen as a threat to determinate-entity and process forms of theism. Given the conceptual structure of the problem of God in relation to suffering in nature, if ground-of-being theism is seen as a threat at all, then it should be seen as a major threat rather than a negligible one.

4 The Blessing of Ground-of-Being Theism

I am arguing that ground-of-being theism should be taken seriously as a theological interpretation of ultimate reality in relation to the challenge of suffering in nature. The reasons go well beyond its immunity to the argument from neglect and the argument from incompetence. In fact, there are two types of reasons. On the one hand, ground-of-being theism possesses native strengths. Some of these are theoretical and derive from placing ultimate reality close to the world of nature as its ontological ground. Other strengths are practical and draw on the advantages for any authenticity-based spiritual quest of accepting the world as it is without evasion or dreaming. On the other hand, ground-of-being theologies highlight the weaknesses of alternatives. Moral and theological interpretations of suffering in nature rely on a range of argumentative resources that function as vital strategies in theodicies, defenses, and the like. Determinate-entity and sometimes process theists use some of these resources to deflect criticisms of their ideas of God based on suffering in nature, including criticisms such as the argument from neglect and the argument from incompetence. In the searing light of ground-of-being theism, many of these strategic resources are not as useful as they may appear at first, and they seem to be little more than shrouded repetitions of the fundamental criteria that determine what counts as an acceptable theological approach.

In what follows, I shall consider both types of reasons, sometimes together. Each of the themes below deserves consideration at length, but I hope it is enough in this context merely to sketch each point quickly.

4.1 Historical and Economic Awareness

Most people in our time living in so-called developed nations typically have high expectations for comfortable and satisfying lives. They expect to avoid many illnesses and to recover from the illnesses they must endure. They expect children to be born healthy, mothers to survive childbirth, and children to grow up and live long lives. They expect that nutritious and tasty food will never be a problem for them and that they can live in comfort with a wide range of pleasurable activities to fill their days in a fundamentally stable society. If they get bored or sad, they expect to be able to make lifestyle decisions that mitigate the problem. They expect to be well educated, spiritually nurtured, legally defended, and militarily protected. They expect that technology will shield them and those they love from the ravages of nature in all its forms.

Such people differ on ideals of distributive justice, with some wanting to share the wealth and others insisting that the poor and afflicted figure out how to solve their own problems, "just like we did." The religious among them tend to believe that life under this description is a divine blessing even if they take it for granted. Their experience of God typically includes involvement in a religious group where most people experience life much the same way they do, ritual rites of passage with cameras flashing and video recorders preserving memories, and moments of spiritual insight involving God's gracious forgiveness and acceptance and support. Untimely death is rare and so God is seen more as the comforter in face of death than as the bringer of death. Disasters are uncommon and tend to be far off and so God is seen more in the generous and caring responses than in the disasters themselves.

Everyone with any historical sense and any cross-cultural experience understands that this lifestyle and the expectations it creates are exceptional, not typical. It is more difficult to grasp the exceptional nature of the view of God that most people living this way hold. Theistically minded people in other places and eras tended and still tend to see God behind all life events, both the satisfying and the tragic, both the comforting and the discomfiting, both the welcome and the terrifying. They have few expectations that God will make their life circumstances safe and happy, understanding from their own experience even in childhood that they will have to accept unwelcome life events very often in their many fewer than three-score-and-ten-year lifespan. They tend to see everything that happens as part of the manifestation of divine glory, as a result, and expect that happiness will have to come from an internal spiritual connection to God rather than from cultural and economic conditions.

This case really requires comparative survey data for its synchronic, economic wing and historical analysis of theological opinions among ordinary people for its diachronic wing. I can present neither here, but such data do exist both in the form of international surveys and historical studies of sermons at times of tragedy or popular religious opinions as reflected in the press—see, for example, the extensive literature on the Lisbon Earthquake of 1755. Saying "God sent this disaster for inscrutable reasons" used to work better than it does now as a comforting, realistic response to unexpected tragedy. In many places today this response causes outrage, tainting the picture of God's perfect "wouldn't hurt a flea" goodness. I am arguing that the ideal of humanly recognizable and humanly relevant goodness is a cultural construction and that it can deeply influence what we are prepared to accept as a plausible theological interpretation of God in relation to suffering.

Perhaps some determinate-entity theists will say that our comparatively magnificent cultural circumstances have allowed us both to realize that God truly is good in just this way and to throw off the comforting but heavy cloak of God standing behind every event, no matter how painful for human and other sentient beings. Perhaps some process theists will point out that their theological view has become a mainstream option for the first time in an era and within cultures where human control over suffering is such that surrendering the omnipotence of God can seem sensible rather than ridiculous, thus disclosing the way things really are. Seeing this cultural and historical variability in ideas of divine goodness, the ground-of-being theist will continue to praise the divine glory rather than vaunt the divine goodness. They will feel a great sense of solidarity with less comfortable life companions past and present. And they will feel that, in one respect at least, their theological interpretation of God in relation to suffering in nature is not a slave to cultural fashions and economic circumstances.

4.2 Theological Significance of the Ground of Being

Speaking of God as the ground of being removes the possibility of proposing a divine character that is profoundly different from the character of the world. This is its chief theological difference from its competitors. Determinate-entity theism requires a divine goodness that our best scientific vision of the cosmos does not easily support and so positively requires some ontological distance between God and the world and a layer of theological explanation for why the world is the way it appears to be despite the purported impeccability of God's moral character. Process theism associates God's moral character with some but not all aspects of worldly events, thereby framing God as supportive and similar to the good aspects of natural events but resisting and unlike the bad. Ground-of-being theism needs neither to explain a discrepancy nor to distinguish among events to articulate the divine nature. The ground of being is the fecund source of all events, regardless of whether human standards in play at a particular time and place would classify them as good, bad, or indifferent.

But does this not merely sanctify the world as it is?! Indeed, it does, in the particular sense that all structures and possibilities of reality express the divine character. But those expressions include causal patterns and necessities, which also manifest the determinateness of the divine character. Some of these patterns make for life and meaning and some for pointless annihilation, and both possibilities are grounded in the divine nature. The moral conundrum for human beings is not the process theist's bracing Manichaean or Zoroastrian challenge to join the divine side against the chaos of anarchic freedom, but rather the challenge to decide which part of the divine nature we truly wish to engage. Can we choose purposelessness, violence, and cruelty? Yes, and God awaits us along that path as self-destruction and nihility. Should we choose to create meaning, nurture children, and spread justice? If we do, the possibility itself is a divinely grounded one. Does God care which way we choose? God is not in the caring business, on this view. The divine particularity is expressed in the structured possibilities and interconnections of worldly existence; wanting and choosing is the human role.

Determinate-entity theists and process theists alike will feel deeply worried about this refusal to align God with a particular moral path. But this merely repeats the quick (but I hope well-earned) caricatures of the former as the optimists and the latter as the activists. By contrast, ground-of-being theists are the mystics who see divine depth in the way things are. This places ground-of-being theism simultaneously close to some forms of religious naturalism, to some forms of apophatic mysticism, to some forms of Hindu and Buddhist philosophy, to some forms of Chinese philosophy, and to some forms of atheism. The distinctions among these views are only crisp when God is a determinate entity; otherwise they merge in a way that is profoundly satisfying to the ground-of-being theist. This merging of apparently different views is one of the powerful theoretical virtues of ground-of-being theism.

4.3 Spiritual Significance of the Ground of Being

There is a spiritual corollary of the theological decision to place God close to the world as the ground of its being: existentially authentic acceptance of the world's fundamental character is equivalent to love of and submission to God. Despite the mention of God here, the underlying concept is such that there are obvious resonances with the South Asian spiritual instinct that "samsara is nirvana," including in its atheistic Buddhist form. Does this spiritual posture entail that one does nothing about the world as one finds it? No, and here the resonances are more with Friedrich Nietzsche. Nietzsche declared the death of God in much the way that ground-of-being theism rejects determinate-entity theisms.

What this means morally and spiritually, for individuals and entire civilizations alike, is that human beings must accept not only the world with its structured possibilities but also their capacity to choose; indeed, choice is one of those structured possibilities. On this view, the fulcrum for human moral action lies where it always has, despite all theological obfuscations: with human decision. God does not advocate or resist the decisions we make because God ontologically supports all decisions. Improving the world and making it more just is one of the choices before us. This is an invigorating challenge but also terrifying, as Nietzsche pointed out. He praised the spiritual vivaciousness of the one who can take full responsibility for choosing, without any hint of evasion and without any pretense that God favors one choice or another. This is spiritual maturity in the ground-of-being framework.

4.4 Is This the Best of All Possible Worlds?

We can love an optimally good world even if it causes us suffering and even if we feel wide-hearted compassion for other suffering creatures. We may forget our love in a moment of pain, but the habits of trust and acceptance formed in those times when we feel that the world is optimally good bring courage and resolve when we need it most, under the burden of suffering. Some may think that cultivating love for a suffering-filled world is a conceptual trick whose essential purpose, regardless of how we rationalize it, is to cope. Indeed, Buddhist phenomenological and psychotherapeutic analyses of responses to suffering disclose a maze of deflections and projections, attachments and distortions that loom over our noble arguments about the world's optimal goodness and press hard the question of sincerity. But in this vale of tears there are many valuable paths, and some take us into the perpetually uneasy realm of conceptual reasoning and moral judgments about the optimal goodness of the world. We can go there, warily perhaps, but not in total despair.

Is this world optimally good? More precisely, let "suffering landscape" refer to the forms and extent and intensity of suffering relative to the potential for emergent complexity, all within a particular ecosphere. Is the suffering landscape of Earth's natural history and present ecosystem the best of all possible worlds? I think it is feasible to argue that the possibility of alternative cosmic, geophysical, biological, and nutritional arrangements with varying suffering landscapes is difficult to block from a scientific point of view. In fact, there is significant scientific evidence that these alternatives should be possible, particularly alternative geologies (e.g., fewer tectonic plates) and alternative biologies (e.g., ecologies with different nutrition profiles). We cannot show this decisively, because it turns on historical accidents or requires different physical or cosmic contexts. But we can know with some confidence that we are not entitled to assume that Earth's suffering landscape is the best of all possible worlds.

This is an important negative conclusion. It strongly suggests that speaking theologically of "the best of all possible worlds" drives against empirical data for the sake of defending a particular conception of divine goodness. But determinate-entity theists might attempt to rehabilitate the "best of all possible worlds" claim by making it empirically more robust. They can enlarge the scope of this claim to include not only the Earth's suffering landscape, past and present and future, but also the suffering landscapes in every physically (not logically!) possible world. Indeed, if we take the multiverse possibility as a physical reality, then we need to include even suffering landscapes in other universes, relatively few of which would seem hospitable to life and suffering. In that case, the multiverse, of which there is only one, may indeed be the best of all possible worlds. We have rid ourselves of invidious comparisons to other real scenarios because there can only be one multiverse.

Unfortunately, this may not help us to love the world we are in. In the traditional form of the "best of all possible worlds" claim, all beings suffer to greater and lesser degrees. We feel sympathy for those who suffer greatly and perhaps begrudge the luck of those who barely taste the bile of misery, but these variations are part of accepting that this is the best of all possible worlds. A principle of plenitude embraces the Earth's ecosphere: every kind of suffering and elation will come to light in this strange place, and this necessity for fullness means that there can be no complaint against the Earth and its Great Mother if you happen to dwell in the dark lands of suffering; it is all still good. There is an analogy here with the re-habilitated "best of all possible worlds" claim: there may be one multiverse, but there are many suffering landscapes, and a principle of plenitude gov-erns the whole. Earth may happen to have a middling suffering landscape, a high-suffering one, or a low-suffering one. Regardless of our good fortune or cursed bad luck, this is still the best of all possible multiverses—so we tell ourselves. Yet, just as a poverty-stricken family suffering from needless disease and brutal government suppression and raging tsunamis might long to exchange places with a more fortunate family, so we might long to exchange our entire planet's suffering landscape for that of a more fortu-nate locale, where the lyrical theme of emergence is less flawed by the dis-cordant accompaniment of intense and pervasive suffering. Just as the family finds it hard to love its context, so we might understandably find it difficult to love the inferior world in which we find ourselves. And just as it is no comfort to the family to point out that "this is the way the Earth works; it is all good," so it is no comfort to us when we hear ourselves say-ing, with tremulous voice, that this is the way the multiverse works—the fact that things are less than optimal here is merely the price paid for the other fact that somewhere (but not here!) there exists a blessedness that makes this the best of all possible multiverses. And it is all good; Mother Multiverse is beyond reproach.

I think determinate-entity theism is in desperate difficulty in relation to this empirically realistic version of the "best of all possible worlds" claim. It involves allowing that a compassionate, personally interested and active divine being tolerates entire worlds with unfortunate suffering landscapes while also beholding worlds with optimal conditions. That is tough to di-gest. Despite its starkness, however, perhaps this difficulty is not much more severe than that of the traditional form of the "best of all possible worlds" claim, in which the hands of broken-hearted God are tied when it comes to individual moments of suffering—restraint for the sake of the greater systemic good—and yet we still affirm humanly recognizable divine goodness. These considerations show how poor the "best of all possible worlds" claim is as a strategy for articulating an interpretation of suffering in nature relative to determinate-entity theism. In my view, determinate-entity theists may be wisest to postulate that, despite the suggestions of contemporary science, there is no variability in suffering landscapes, per-haps because there is no other life-supporting place in the multiverse, so that the landscape of this and every world is and would be the same, and that this universal suffering landscape is also the best possible suffering landscape. The anthropocentrism of this view is bitter medicine, but this is what is required to leave the "best of all possible worlds" claim in tolerable health.

Of course, neither process theism nor ground-of-being theism needs to resort to such artifices. Both can take the world just as it is in all its variations and disagree only on how to picture God's relation to it all.

4.5 Is Suffering an Inevitable By-product of Good Divine Intentions?

A related strategy is to argue that suffering is the necessary by-product of one or another overridingly valuable state of affairs, such as self-consciousness, moral freedom, spirituality, creativity, relationality, love, the capacity for a relationship with God, the incarnation of Christ, the realization of bodhisattvas, or the ability to attain enlightenment. I think it is incontestable that some degree of suffering is indeed an inevitable by-product of such virtues. The connection between desired outcome and suffering is indissoluble and rooted in the evolutionary realities of emergent complexity.

Neither process theists nor ground-of-being theists have to worry about the puzzles this fact may pose for a view of God, but determinate-entity theists have a lot at stake. Determinate-entity theism needs to establish the "best of all possible worlds" claim for the sake of its vision of divine goodness, and the "inevitable by-product" argument is only meaningful if it serves that end. With that in mind, more is required if the "suffering as inevitable by-product" strategy is to be theologically useful for determinate-entity theists. It is not enough that desirable virtues entail *some* degree of suffering. We need to have reason to believe that they entail the *particular* suffering landscape of Earth, or of the cosmos, or of the multiverse. Might not the same virtues emerge in a slightly different biological or geophysical setting with less actual suffering than we see here? Thus, this "suffering as inevitable by-product" strategy does not appreciably assist the determinate-entity theist in establishing that this is the best of all possible worlds.

4.6 Is Suffering Fundamental?

Rather than trying to interpret suffering as one component in a complex emergent world, appearing only with the emergence of life, it is possible to broaden the ordinary meaning of the world and make suffering fundamental to the whole of reality. Buddhist philosophical cosmology routinely makes this move, particularly in its South Asian rather than its traditional Chinese forms. The *pratı\tya samutpa\da* account of reality specifically denies that objects and entities have "own being" or essential individuation. Rather, all realities are bundles of relations that emerge from the web of interdependent connectedness. By the traditional standards of Western philosophy, this Buddhist approach to philosophical cosmology seems out of balance, overstressing relationality and failing to register the intrinsic elements of the entities that, admittedly, always stand in relation to one another. But Buddhists take great heart from the fact that advancing science, especially in the quantum world of the very small, is making it increasingly difficult to say what a substance is independently of its relations. They regard this as confirming their contention that everything arises in dependent correlation with everything else, and even their particularly intense form of this affirmation that denies independent subsistence to entities of all kinds.

Buddhists are especially interested in human life, as are most religions, but this philosophical cosmology has significance for suffering in nature, also. Thinking through the lens of the First Noble Truth, which asserts that all is suffering, Buddhist philosophers came to view *dukkha,* or suffering, as the immediate and inevitable consequence of *pratı\tya samutpa\da.* There is nothing without suffering because suffering is rooted in relationality and change *(viparina\ma-dukkha)* and simply in being-conditioned *(sañkha\ra-dukkha).* In its particularly complex forms, as in human life, suffering can be analyzed as attachment to the apparently-but-not-actually real, a problem that can be solved through following the Noble Eightfold Path toward *moksha* or liberation. But suffering already arises in nonhuman settings as a correlate of change and being-conditioned. Suffering is universal. Buddhists fight over what liberation means, not least because certain understandings of it, such as suffering-free compassionate presence, suggest a greater distinction between *pratı\tya samutpa\da* and *dukkha* than the philosophical cosmology allows. This is why the doctrine of emptiness, or *s;u\nya\ta,* is so important to Buddhist philosophy. As difficult as it is to speak about the liberated state as *s;u\nya\ta,* this is the deep entailment of any philosophical cosmology that places *pratı\tya samutpa\da* and *dukkha* in such tight connection.

I see no intrinsic problems with this broad and general usage of suffering and the metaphysical visions it sponsors. Importantly, it is easily compatible with a ground-of-being theism, while being more of a stretch for process theology, which has a different account of what is fundamental at much the same level as a *pratı\tya samutpa\da* cosmology. But we need more. A philosophical cosmology boasting an ontologically fundamental concept of suffering needs also to explain suffering in complex emergent forms, such as physical injury, conscious pain, emotional distress, and existential anxiety. So I prefer to use terms such as "change" and "being-conditioned" for the inanimate realm, to reserve the word "suffering" for the biological realm, and to demand a satisfying theory of emergence situated in a philosophical cosmology that links one to the other. This adjustment to suffering-is-fundamental schemes works well with ground-of-being theologies and suits process theology much better.

4.7 Is Suffering Illusory?

The suffering-is-fundamental strategies, at least in their Buddhist and Hindu forms, paradoxically also affirm that suffering is illusion. This is obvious in the case of human life: suffering is caused by attachment, which is rooted in misperception and misunderstanding of the world, a state of affairs correctable through enlightenment. It is not as evident in the philosophical cosmology of *pratı\tya samutpa\da* and *dukkha,* where suffering seems to be as real as the processes of emergent complexity. But the overall goal of *moksha* reframes *pratı\tya samutpa\da* and *dukkha* alike as a kind of deceptive conjuring that finally is unreal.

The main Western version of the suffering-is-illusory strategy goes by the name of privation theory and is famously associated with Saint Augustine. The really bad part of suffering is the evil that causes it, according to Augustine, but evil is merely a privation of good, a distortion with no reality of its own. Privation theory solves a problem in the doctrine of creation *ex nihilo,* which was just firming up in Augustine's time, because it appears to place the explanation for evil on some other doorstep than God's. Finally, of course, shifting responsibility for evil cannot avoid tainting the moral character of God, so Augustine, and more elaborately Gregory of Nyssa, were forced to propose a philosophy of history to bring a dynamic temporal dimension to the story of evil. Evil may seem real now, according to this story, but it is in fact nonbeing and will be shown to be nonbeing, in the sense of sheer nothing, with the unfolding of God's will in history and nature. Thinking of the ontological-historical destiny of evil as nothingness is the basis for our calling it a privation of good in our presently conflicted and suffering-filled circumstances. Interpreting privation theory in connection with philosophy of history helps to deflect the great weakness of illusionist theories of suffering, namely, their failure to take with due seriousness the practical reality of suffering.

Privation theory is theologically quite useful. It has cross-cultural resonances and an important historical-eschatological dimension, and it is a serious attempt to understand reality as good despite the prevalence and intensity of suffering within it. Yet its implications for the doctrine of God as creator remain as problematic in our time as they were in Augustine's. Relative to an eternal and omnipotent divine entity, the historical texture of the flowering and ontological obliteration of evil does nothing to shield the divine reality from the reality of evil. If evil and suffering ever were experienced, then an eternal and omnipotent God must bear the marks of this possibility within the divine being itself. This is why the privation view actually suits ground-of-being more than determinate-entity accounts of ultimate reality (process theism is, of course, not affected by any of this). This is also why the deepest articulation of privation theory within the determinate-entity framework has to be fundamentally incarnational as well as eschatological. But the incarnation does not produce a picture of God as a decisively good and powerful entity. Rather it suggests either that God is to be identified with the moral ambiguity of nature and history, surrendering the divine goodness to divine power as in ground-of-being theism; or else it invites the opposite response, the surrender of power to goodness in the manner of contemporary process theology. This is merely to replay the traditional tri-lemma argument of theodicy—God cannot be all-good and all-powerful if evil is real—with the added observation that not even privation theory can effectively secure the nonreality of evil.

It is impossible not to admire Augustine's and Gregory of Nyssa's intricate attempts to preserve the goodness and power of God in the face of the apparent reality of evil, but I suspect theirs is a vain struggle on the terms in which they framed the problem. In the ground-of-being framework, by contrast, the motivation for a privation theory disappears, but its fruit—divine incarnation and participation in the world's moral ambiguity—have a welcome place.

4.8 Is Suffering Our Fate?

Another strategy for interpreting suffering is to emphasize its inescapability. In one ancient version of this view, it is human fate, and the fate of all plants and animals, to suffer at the whim of the gods. The ancient Near Eastern Epic of Gilgamesh ends when the bizarre hero, having won the secret of everlasting life through a monumental effort of self-assertion against all odds, has the life-giving plant stolen by a serpent while he is momentarily distracted. The story's point is unmistakable: there is no evading the will of the gods, no matter how strong and creative we are. We will suffer and die because it pleases the gods.

This picture of suffering as the fated lot of human beings has more and less personal versions. In Gilgamesh things are intensely personal: the gods witness Gilgamesh's agonies and deliberately thwart his attempt to escape them. In the course of Stoic philosophy's development, fate gradually became less a matter of the imposition of divine will and more a matter of causal determinism. Fate was a recapitulation in the life of every individual being of the simple fact that everything in nature is rigidly determined, not so much by divine whim as by the very nature of the world, which became more and more the same thing. In twentieth-century existentialism, there were representatives of both views. But on all sides the moral benefits of understanding suffering as the result of determinism or fate are supposed to be courage to face the world as it is, free of self-deceptive illusions and artificial comforts. Bear up! Face misery and death with dignity!

It has long been understood that, despite appearances, doctrines of determinism do not interfere as much as one might think with practical action aimed at improving the circumstances of life. After all, according to determinism, we necessarily do whatever we do, whether it is sitting around paralyzed by despair or actively transforming social conditions and alleviating suffering. So I do not urge that clichéd criticism as a genuine practical problem with the "suffering as fate" strategy for interpreting suffering in nature. Moreover, Western philosophical compatibilism offers ways of affirming nature's freedom and human moral responsibility in the presence of divine and physical determinism, just as Chinese philosophy offers ways of understanding human action as free within the highly structured flow of power within nature. Potential contradictions of this sort do not amount to much of a problem either.

Process theism cannot support this picture of suffering as fate. In the context of determinate-entity theism, the most obvious theological difficulty with the "suffering as fate" idea is that it requires us to picture God personally inflicting suffering on creatures. As I suggested earlier, awareness of our historical and economic location will help us see that there have been times and there are still places where such a picture of divine power is reassuring—a God who can afflict us surely has the power to save us from our enemies! But it is an unfashionable view when technology and political economy combine to make life conditions for human beings mostly comfortable and safe, thereby reducing moments of suffering, at least in the sense of injury and pain, to the level of inconvenient interruptions from which we are entitled to expect rapid deliverance through medicine or the justice system. In such blessed cultural havens, to think of God as wielder of weal *and* woe serves merely to insult the socially accepted view of the perfectly good divine character.

Process theology surrenders divine power to guarantee divine goodness, in response, so the process deity (Godp) inflicts no suffering even if ultimate reality (God$_C$) does. Most determinate-entity theists affirm the humanly recognizable goodness of God by rejecting divine infliction of suffering while maintaining divine omnipotence. Ground-of-being theism offers more flexibility in relation to the suffering-as-fate interpretation, and can take it or leave it depending on other factors. In general, I think the idea of God as source of worldly fate is the most under-explored strategy in our time for theologically articulating the meaning of suffering in nature. There is more potential in this idea, even for the less anthropomorphic versions of determinate-entity theism, than is usually assumed in comfortable Western cultural contexts. The best example may be Islam, which continues to preserve the idea of a personal God of unlimited power, merciful and compassionate yet decisive in action, and beyond human moral judgment absolutely.

4.9 Kenosis Strategies

This takes us directly to kenosis strategies. These explicitly theological proposals are supposed to explain why God does not intervene to deliver us and the rest of nature from needless suffering. God enters into a covenant with nature in which divine power is self-limited to enable complexity to emerge in nature, whereby moral and spiritual life can flower. I think that such kenosis strategies reflect the cultural and economic conditioning I have described; in fact, kenosis can be quite indigestible as an explanation of suffering in other cultural settings because it defangs a deity whose fierce intervention is desperately needed. Moreover, I do not think that the biblical concept of kenosis (Phil 2:1–11, especially v. 7) has anything to do with this kind of covenantal divine self-limitation; biblical kenosis is explicitly and solely Christological in character, and Paul's aim in introducing it is to inspire humility among arrogant and fighting church factions. Yet we can liberate recent kenosis proposals from their mistaken claim to biblical authorization based on this Philippians passage, and we can also set aside cultural conditioning factors as genuine but not determinative of the truth of the matter. Then we must consider the theological import of the kenosis strategy on its own terms.

The more humanly recognizable we want our model of God's goodness to be, the more we need some explanation of the apparent absence of effective divine intervention to alleviate needless suffering, and thus the more important it is that we make the kenosis strategy stick. Theologies that regard humanly recognizable goodness in God as a mistaken or dangerous criterion for theological adequacy, such as ground-of-being theism and some forms of determinate-entity theism, can take or leave kenosis. But kenosis is vital to determinate-entity theisms affirming God as a caring personal being. Thus, it is in relation to such variants of determinate-entity theism that we must consider the effectiveness of the kenosis strategy. In this context, however, kenotic interpretations of God's personal goodness in relation to suffering face a severe conceptual problem, as follows.

Either the kenotic God retains omnipotence or not. In the first case, kenosis seems artificial and unconvincing because it is merely a reversible divine decision rather than a fundamental and inescapable feature of divine creation. That is, if God deliberately embraces self-limitation for the sake of allowing moral and spiritual life to flower, then there ought to be nothing to prevent God from making exceptions as needed. Kenosis in this sense does nothing to protect the humanly recognizable goodness of God. The second case requires a reflexive and automatic kind of kenosis. God must be determined in some primordial creative event as already and eternally self-limited, so that no decision to be any other way in any subsequent context is possible for God. Process theology illustrates this, in one sense: a process theogony could have ultimate reality (Godc) creating the world in such a way that God can be present to the world always and only as the process deity Godp. (This is only a fantasy, of course: the process God is an actual entity that plays the hand it is dealt just as every actual entity does, and no theologically relevant theory of ultimate reality is possible. But it is worth noting that such a view is possible in the theological metaphysics of Robert Cummings Neville, though in practice he does not exploit the possibility.)

So it appears that kenosis is most convincing as a part-explanation of suffering in nature when it is ontologically forced, but then it is not exactly embraced willingly and thus is not truly kenosis. If kenosis is a matter of deliberate divine self-limitation through which omnipotence perseveres, then it is no longer truly binding and thus unconvincing as an explanation of divine silence in response to needless suffering in nature. The upshot is that kenosis does much less than its proponents claim to preserve the humanly recognizable goodness of God. It is merely a hidden repetition of the fundamental (and otherwise understandable) criterion of theological adequacy active for determinate-entity theists and, despite appearances, offers no materially new conceptual resources.

4.10 Will Everything Be All Right in the End?

We saw earlier that privation theory is at its best when it incorporates a historical dimension whereby the essential nonbeing of evil, though not evident now, will become evident at some point in the future. Unsurprisingly, determinate-entity theists typically deploy an eschatological vision in which the ontological destiny of suffering is *nihil,* nothing. Suffering will fall away when the consummation of all things comes, finally confirming God's personal moral perfection and power. The most interesting aspect of this to me is that eschatological deployments from the camp of determinate-entity theism show an awareness of the empirical difficulties that this view must manage.

Ground-of-being theists can also deploy eschatological visions of suffering, but in slightly different terms and to a very different end. In one such vision, suffering will plunge back into the abysmal creative surge from which it sprang, along with life and spirit, leaving nothing at all or else only the glorious fruits of the vast cosmic-historic process to that point. This cosmological vision can take or leave the hypothesis of an overall direction to the cosmic process; this is always and only an empirical question about the way the divine character expresses itself determinately in the structured possibilities of reality. Interestingly, one of the best-known ground-of-being theists, Paul Tillich, uses many pages in the third volume of his *Systematic Theology* to say that there is no overall direction of progress in the universe as a whole even though there is abundant meaning.

At this point he shows his affinity for the view of cosmic history in Schelling rather than Hegel, two other ground-of-being theists. But perhaps the question really ought to be settled empirically rather than speculatively.

The samsaric version of this view is particularly impressive but deeply perplexing. *Karma,* a kind of universal moral law of cause and effect that lies beneath the experience of suffering, gives birth to life and spirit, goodness and bliss, and delivers souls through a tortuous path into the arms of Brahman via enlightenment and *moksha.* Though the Vedas do speak of cosmic epochs and Siva's resetting of all of creation, there appears to be no post-samsaric recreation of reality in this vision but only deliverance from it. This is a one-by-one personal eschatology rather than an eschatology of nature and all of creation; suffering is the means of liberation to something else, but is not in itself defeated. The *"samsa\ra is nirva\na"* option in this context is akin to realized eschatology within Christian theology: liberation is not in some other world but here and now in the midst of suffering. The samsaric vision is probably the most positive of all interpretations of suffering in nature because it renders suffering *in itself* as productive and nurturing, the path along which all beings trudge on their way to a blessed goal. This seems quite different from regarding suffering as a necessary but unfortunate side effect of an otherwise valuable process. We can appreciate all this, but we must also note that there is no solution to the raw challenge of suffering in nature here. There is reframing, akin to reframing suffering as privation or illusion, but there is no easing, no overcoming. But perhaps there should not be any overcoming of suffering or of the samsaric reality that structures existence.

It falls to more history-minded religions and philosophies to articulate the possibility of an eschaton that transforms nature itself, not merely individuals within it. Keeping in mind the argumentative stakes—the determinate-entity theist deploys eschatology to give evidence of God's personal goodness in relation to suffering in nature—let us trace the argument about the possibility of such an eschaton.

Beginning from the scientific point of view, can there be an alternative embodiment of life and spirit that is free of the travails of nature that we know and see about us? I have hinted that it is possible to imagine alternative biological and ecological circumstances that might be freer of suffering than our Earth is. But in all of those biology-based scenarios, we cannot get along without glucose, or some substitute biological fuel, among other things, and thus suffering in nature might be reduced relative to Earth's suffering landscape but it cannot be eliminated. Thus, even if Isaiah's vision of the lion eating straw like the ox were to come true, we would still have suffering in nature, especially in the form of plant injury and accidents. What about leaving biology behind altogether and imagining a form of life that uses light as the sole energy source? If feasible, this would eliminate predation altogether, but it would still leave accident and injury and possibly disease to worry about. So there is no science-supported basis for picturing a new heaven and a new earth that is free of suffering. Well, perhaps that is to be expected. A new heaven and a new earth without suffering would be so spectacular a transformation that it would have to be supernatural in character, so we should probably assume that our scientific knowledge should be ignored as irrelevant.

But this is too hasty. Consider the matter from the angle of eschatological lifestyle. In Judaism, Christianity, and Islam, eschatological visions
of a new heaven and new earth typically involve moral and spiritual beings
living in relation to one another, worshipping God, able to recognize loved
ones and to remember life, able to grow and learn and change, but all this
without any trace of suffering either in these blessed beings or in their
natural environment. But I doubt that we can picture embodiment, change,
and growth without the use of natural resources, even if light is our food.
Moreover, if the "suffering as inevitable by-product" arguments have any
credibility, then they apply to these pictures of the new heaven and the
new earth just as much as they apply to the world we know. I freely admit
to feeling the allure of idealized pictures of a suffering-free afterlife, and I
have no difficulty admiring the boldness with which the history-minded religions artfully wield their supernatural eschatological resources to justify
their affirmation of God's perfect personal goodness—in fact, within this
framework, nothing less would take suffering as seriously as it ought to be
taken. Yet I cannot see how the proposals are coherent, no matter how
badly they are needed.

Furthermore, I suspect that the fruit of theoretical success in articulating a coherent eschatology would only be theological disaster for determinate-entity theism. It would only reinforce skeptical questions about
God's humanly recognizable moral goodness by introducing an embodiment
scheme that boasts growth and change and relationality yet no suffering.
In other words, *that* world and *not this* world would be the best of all possible worlds. Such a God would be flagrantly morally inconsistent.

5 Conclusion

This has been a typologically driven argument rather than one rooted in
specific theological positions, so many complexities and subtleties have
been neglected. Yet typologically driven arguments can still produce robust
conclusions. My conclusion is that there is no adequate reply to the argument from neglect, as it is directed against the coherence of determinate-
entity theism. Neither is there any satisfying answer to the associated argument from divine incompetence, as it is directed against the much
trumpeted efficacy of process theism in supporting a religious response to
suffering.

It is good, some say, to be in the middle of theoretical disputes, participating in the virtues of all sides. When the middle proves to be unstable
due to underlying incoherence, however, there is real cause to look around
for some other place to stand, there to reconfigure alternatives so as to understand oneself, once again, as standing in a new middle, but now upon
firmer ground amidst more fruitful theoretical disputes. Determinate-
entity theism of many kinds requires God to be a compassionate entity with
personal knowledge of suffering, the power to act in history and nature,
and all the while to be perfectly good in ways that human beings can grasp.
This places determinate-entity theism in the middle of a host of daunting
theoretical difficulties. There has been no shortage of attempts to defend it
but the resources for the main defense efforts face serious difficulties, as I
have tried to show. The problem, in a nutshell, is that this idea of God is an
admirable but finally ineffective attempt to deal with the empirical gap be-

tween life as we experience it and the goodness we long to affirm in God as ultimate reality. Nothing shows this more clearly than the problem of suffering in nature, as leveraged by the argument from neglect.

Yet there are other central places to stand, theologically, and other ways to configure the surrounding theoretical landscape. I urge battle-weary determinate-entity theists to look over at the intellectuals gathered around the ancient idea of God as ground of being, the power and creativity of the structured flows of nature, the ontological spring of matter and value. This God is not good in a humanly recognizable way, nor personal in character, yet when we assert God's goodness despite its incongruity with our anthropocentric ways of thinking, our minds are led higher to larger patterns and wider virtues in which suffering is no longer merely an unwanted side effect of otherwise wondrous physical processes but a creative source in its own right. This God is beautiful from a distance in the way that a rain forest is beautiful, but just as it is unpleasant for humans to live unprotected in a rain forest, so it is perilous to be in the direct presence of divine glory. We suffer there as well as surrender in bliss. The truth about this God is deeply disconcerting, not easily assimilated into our humanly configured cultural worlds and religious habits of thinking. Yet this is the truth that sears our souls, that awakens us again and again from our anthropomorphic theological slumbering, and that drives us to love that which destroys even as it creates.

This image of God is a less-perplexing concept in many types of Hinduism, or even in most forms of Islam, past and present. It is built into the heart of Chinese religious cosmology and is amply present in most tribal religions. It was less difficult in ancient and medieval Judaism and Christianity than it has become in modern times and in first-world cultures. The paucity of contemporary Christian imagery and symbolism corresponding to dancing destroyer-creator Siva shows just how difficult this idea might be in the context of the Christian religious and theological imagination—and this despite unmistakable biblical imagery of this type in Job and in apocalyptic literature, especially the book of Revelation. The destroyer-creator ground-of-being insight survives most clearly in the Western epic tradition, thanks to Blake and Dante and others, and in medieval art depicting hell as punishment. But the incarnational-sacrificial Christological rubric utterly reframes this insight in the terrifying narratives and imagery of suffering saints and martyrs. Christianity has always had an idiosyncratic approach to suffering because of its Christological lens. But for precisely this reason, Christian theologians might take a more positive view of suffering in nature than they have typically taken. Despite my appreciation for the Greek metaphysical tradition, at this point the early Christian theologians seem to have underestimated the philosophical resources of their primary narrative, rashly subjecting it to Greek intuitions about suffering as incompatible with fully realized being. The result is an insoluble conundrum in which suffering in nature has to be framed as a foreseen but unintended side effect of a creation that is good because of its other virtues, and this creation is the gift of an omnipotent, wise, compassionate, and agential God who does not intervene as often as moral obligations demand because it would ruin the beauty and moral independence of the creation. This is a theological conundrum, if ever there was one.

Suffering in nature is neither evil nor a by-product of the good. It is part of the wellspring of divine creativity in nature, flowing up and out of the abysmal divine depths like molten rock from the yawning mouth of a volcano, searing and burning—maybe with ecological benefits, maybe with no discernable redemptive elements whatsoever. Luminescent creativity and abysmal suffering are co-primal in the divine nature as they are in our experience. To acknowledge the ground of our being in these terms is to accept suffering as our fate and the fate of all creatures, and also to do what we will with these circumstances, whether that means sitting idly by or launching into the world with banners waving and guns blazing. All things testify to the divine glory. All things without exception.

IV. CHANGING THE TERMS OF THE DEBATE

Brad J. Kallenberg

Don Howard

THE DESCRIPTIVE PROBLEM OF EVIL

Brad J. Kallenberg

Earth's cramm'd wi' heav'n
And every bush and tree with fire of God;
But only he who sees
Takes off his shoes.

—Elizabeth Barrett Browning

1 Introduction

Language is like the cane in the hand of the blind person. The better one becomes at getting around with the cane, the more he or she is apt to forget the cane but *through* the cane perceive the objects scraped and tapped by the other end. A defective cane may distort the world perceived by the blind person. So too, defective use of language threatens to muddy our understanding of the things we talk about. When discussing something as difficult as natural evils, a frequently undetected defect in our language-use is overly attenuated description. In what follows, I will sketch three conditions under which attenuated description multiplies confusion in general conversation. I will then describe ways in which the lexical shortcuts taken in discussion about "natural evils" can be corrected. Although, it remains to be seen whether conversants are willing to pay the cost involved. For, in order to talk most clearly about "natural evil," and thus understand the problem most deeply, those doing the talking must employ descriptions that require correlative practical actions in order to be intelligible. I give an example of how the juxtaposition of two components, rich descriptions and appropriate action, makes possible the trained eye to perceive a pattern that, while falling short of an explanation per se, serves as a satisfactory response to natural evils. I conclude my essay by proposing a protocol for advancing the conversation about natural evils.

2 Humans Respond to Evil by Talking

It is a matter of historical record and empirical fact that natural phenomena can bring about intense suffering. The fact that suffering, which is entailed by tsunamis, wildfires, earthquakes, tornadoes, floods, lightening strikes, etc., is unwanted, unnecessary, undeserved, or overkill is bound up with the very concept "suffering."[114]

Humans respond to instances of suffering in their environment with a variety of reflexive behaviors. Ludwig Wittgenstein called all such behavioral responses "primitive reactions." This term is helpful for expressing the fact that human beings, like other animals, respond to their physical environment in physical ways.[115] In the face of relatively minor pain we brace

---

[114] For a catalogue of types of suffering, see Wildman's essay in this volume.

[115] The term is a little misleading if we think that *primitive* means something like "presocial" or even "prerational." For Wittgenstein, primitive reactions could be socially- or rationally-conditioned, though they need not be. See Kallenberg, *Ethics as Grammar*, 161–214.

ourselves, we wince, we cry out, we pout. As biological creatures, these reactions are primitive and reflexive. Just as primitive and reflexive are our
reactions to the suffering of others: we weep, we touch, we empathize, we
hug, we offer aid, we ward off others from danger, we shelter the children,
and we attempt to avoid future instances of similar dangers. All such habitual reactions (and more) constitute the complicated form of our life as
human animals.

One set of behaviors among all those that human beings perform in response to pain and suffering, is that of verbal behaviors. We are the animals, who more than any other, respond to our surroundings audibly. Human beings respond to their world *by talking*. We call for help, we offer
words of comfort, we recount the event to others, we educate the children,
we discuss blame for past events, we strategize for the future. And so on.

One tiny fraction of all the conversations we have about any particular
tragedy is the puzzling we do together over events that seem beyond human control. Of course, sometimes when we ask Why? we are not looking
for an explanation but simply sighing. Yet on occasion, perhaps when we
are at a relatively safe distance from tragedy and suffering, we talk in order to make the best sense we can of the offending event. We instinctively
hope that our conversation will produce an adequate *explanation* of evil.
While an explanation may put an end to our questions (as explanations
may occasionally do), more often than not, what we are really after are not
explanations but rather ways of telling the story of our lives in a way that
encompasses tragedy in a *satisfactory* way.

I use the term "satisfactory" as a way of reminding us that the vast
majority of our lives as physical critters is governed by the metric of "satisfactoriness."[116] For example, engineers compete among themselves to offer
solutions to a particular design problem. Some proffered designs may be
clearly unsatisfactory. But obviously, there may be *many* "satisfactory" designs. Granted, some designs may be more satisfying than others with respect to a given set of contextual constraints. But on no grounds is one warranted in concluding that there is a single, logically-compelled "correct"
design.

If verbal responses to evil fall under the class of physical behaviors
humans perform to cope with a particular physical environment, then it
makes sense to evaluate these verbal behaviors in a way similar to the way
we assess engineering projects: is it "satisfactory?" For, conversations
around the Why? question are themselves a form of verbal design. The constraints against which a "satisfactory" description of suffering is assessed
includes, among other criteria, its fit with the complicated form of life
shared by speakers. But whose form of life is in view? For the purposes of
this present essay, I am presuming a reading audience composed of scien-
tifically-minded religious believers. So, on the one side, a "satisfactory" verbal response must comport with the best scientific practices and models
available. On the other side, the "satisfactory" answer must fit with the
contours of a religious form of life.[117] Just as it is true that not every

---

[116] Kenny, "Practical Reasoning."

[117] "Satisfactory" does not reduce to something that merely coheres with other
propositional statements, since the religious form of life itself cannot be reduced to
assent to some set of propositions.

natural drive ought to be fulfilled nor every primitive reaction be deemed trustworthy, so too, for religious believers not every askable question is worth asking. Truth is, in the communal practice of living well that goes by the name "religion," some questions *ought not* be asked, since the artful habit of resisting the asking of some questions is taken by the group to be itself a crucial part of the larger project of shaping this group's moral habits (virtues), particularly their *habitual manner of seeing the world*.

Iris Murdoch once observed that we can only act in the world that we see.[118] One way to think about "vision" is to list the variety of aspects under which (or "lenses" through which) we view our environment. For example, a bridge can be viewed under the aspect of structural integrity or it can be viewed under the aspect of artwork (or romance or politics or theology; i.e., bridges themselves may be speech-acts). Each aspect corresponds to a distinct manner of interrogating the bridge. Not every person is equally skilled in considering the bridge under a given aspect. In this example, the aspects of artistry and engineering are not mutually exclusive, they are complementary. In other cases, skills for one way of observing may interfere with skills required by the complementary aspect. Balancing these demands is part and parcel of what counts as "satisfactory."

I hasten to point out that despite the priority we tend to assign to some batches of words over others, every "satisfactory" verbal design will have to vie for acceptability against the field of other satisfactory designs. Within communities who view their environment under *both* the aspect of science and that of religion, no verbal responses to the Why? that surrounds suffering has an *automatically* privileged status in the common life. Like all other speech acts, sentences are spoken not to the wind, but to and with each other. As such, our sentences are measured for their felicity by how well they are taken up by our conversation partners and grounded by subsequent behavior.[119] As is always the case, any given speech act is open to lexical ambiguity, semantic misunderstanding, and grammatical confusion. But the most satisfactory speech act is the verbal design that minimizes such difficulties.

But there is a linguistic gopher that, because it is so often overlooked, cuts off fruitful conversation at the root. This pest is the habit of using the phrase "natural evil" as an adequate description for naming what is being talked about. In this essay, I challenge this designation as overly simple and the breeding ground for grave confusions that preempt the satisfactoriness that our conversations about suffering might otherwise achieve. I will show that just as attenuated descriptions in ethics render moral obligations invisible, so too attenuated descriptions of "natural evils" make the plausibility of religious belief virtually disappear. Additionally, in this paper I argue that broadening the description with respect to narrative time and space helps us perceive the description for what it is: a speaker-involving speech act. What I propose amounts to a shift in focus from the talked-about-subject ("natural evil") to the formerly transparent linguistic behaviors themselves. The upshot of this shift will be the emergence of

---

[118] Murdoch, "Vision and Choice in Morality."

[119] I take as a benchmark for intelligibility, Wittgenstein's dictum: "*Practice gives the words their sense.*" Wittgenstein, *Culture and Value,* 85e.

formerly unnoticed relevant patterns from the now rich descriptions to the end that a fitting response to the so-called natural evil can be disclosed.

3 Attenuated Descriptions of "Natural Evil" Hinder Conversations about Suffering

Attenuated description is a common occurrence in conversations between persons who are adequately fluent. But attenuation is a bad idea under three conditions.

3.1 When Conversing with the Nonfluent

Taking shortcuts is human nature. It is not more precise to say "hand to me the four-foot pole with the 2,907 bristles bound together and attached to the one end" than to say simply, "hand me the broom." We instinctively conserve our energy in speech as well as in labor. "Broom" enters the English language as the name of a family of artifacts ranging from whisk brooms to broom trees and everything in between. If on this occasion, there happens to be an array of brooms to choose from, the listener would likely "read" the context instantaneously (e.g., we are standing in a garage) and select the proper broom (i.e., the push broom) for such an occasion. The usefulness of the word "broom" depends both upon the fluency of the listener and the listener's aptitude for reading the surroundings.

 Naming is the most common form of attenuated description. But names can be misleading and may even hinder communication if fluency is in short supply. For example, the sentence "F equals m multiplied by a" is only recognized as true for those who are fluent in the terms. It may not help the uninitiated to learn that F, m, and a go shorthand for "force," "mass," and "acceleration." After all, the following sentence, while grammatical, is sheer nonsense: "The force of love equals the mass of the cheesecake multiplied by the acceleration of the economy." Clearly, "What force?" "What mass?" and "Which acceleration?" are questions that cannot be answered simply by affixing more labels. What is needed in this case is the *tutoring* of the novice by means of a progressively complex series of stories the response to which cultivates the requisite skill of similarity recognition by the novice to the end that he or she becomes able to carry on conversations about "$F = ma$."

3.2 When Facing a Referential Impasse in the Conversation

A second instance in which attenuated description can perpetuate confusion is when the referent is ambiguous. If one overhears a conversation about a famous dead Greek person named Aristotle, one is likely to enter a conversation on the assumption that Plato's most famous pupil is the referent to the name "Aristotle." But if the next topic to arise is the referent's personal wealth, perplexity would abound until it became clear that the "Aristotle" in question was not a philosopher, but the famous dead Greek who married John F. Kennedy's widow in 1968 and who himself subsequently died in 1974. Of course, the speaker can get some of the details wrong (Aristotle Onnassis actually died in 1975) and still achieve understanding simply by filling in *enough* details to disclose the referent's identity. How much detail is "enough" is of course a function who the listener is and of how much skill is required to read the surroundings.

3.3 When How to "Go On" Is Unclear

A third instance in which attenuated description muddies the water is related to the second. Attenuated description not only breeds perplexity about the identity of the referent, it fosters confusion about how to go on. Consider the following example. Francis of Assisi is championed as the paradigm of charity. Refusing to take over his father's prosperous enterprise, Francis disavowed his family wealth by stripping naked and swapping his rich man's tunic for the flea-ridden rough shirt of a local beggar. Thereafter becoming the most famous of the mendicant preachers, Francis's self-induced poverty is taken by some to be a morally supererogatory habit. If however, I model my own life after St. Francis and give away my fortune (ha!), I would *not* be acting in imitation of St. Francis. Why? Because his life and mine are similar only under vastly attenuated descriptions: "religiously minded males intent on growth in personal holiness." If the descriptions are expanded only slightly to include marital status (Francis never married, Kallenberg is married and father of three children), then the voluntary poverty that is supererogatory in Francis's case may prove to be downright immoral in my own.[120]

With simple illustrations in hand, it is now possible to see how conversations about "natural evil" have been infected by overly attenuated descriptions in ways that multiply confusion.

4 Recent Descriptions of Natural Evil Have Been Attenuated

There are at least three ways in which descriptions of "natural evil" have been unwisely attenuated in contemporary conversations about theodicy. Confusion enters the conversation when (1) occasions of suffering are described in self-distancing ways, when (2) the narrative context is overly compressed, and when (3) the range of acceptable causal explanation excludes nonefficient forms of causality.

4.1 Overly Restricting the Involvement of Human Agents in the Description

Objectivity is a virtue in the humanities as it is in the sciences. For example, in historiography, theologically-minded historians presuppose a stance of "methodological atheism" in order to do their practice well.[121] This is as it should be. Yet both scientists and theological historians can testify that in the study of objective data, *patterns* often emerge among the data. As not everyone seems able to see every pattern, it is difficult to make the case that the pattern is actually *in* the data. Consequently, the pattern is probably best understood as a function of the interplay between the data and the observer.[122] I suggest that the term "natural evil" names just this sort of

---

[120] I owe this anecdote to Nancey Murphy.

[121] Tilley, *History, Theology, and Faith.*

[122] See reception theory of Hans Robert Jauss with respect to philosophical her-

pattern, the sort of uptake that requires a high degree of involvement by the observer with the data.

4.1.1 "Natural Evil" Names a Pattern

When we consider any event under the rubric of "evil," we are implying that more than objectivity is required to correctly name it.[123] For this reason, in some cases, objectivity serves to obscure rather than clarify descriptions.

Imagine I spy a little old lady being savagely mugged across the parking lot. I turn to you and say in a monotone voice, "It is 11:30 pm EST on the 11th of November 2005 and I am beholding a 92-year-old, 100-pound white female being struck 6 times within the span of 5 seconds by a 200-pound, 31-year-old male holding a 24-inch length of ¾-inch pipe his right hand." Imagine this statement is factually correct. Yet things are horribly wrong. The "objective" description of the event involves mis-description precisely for what it leaves off in the name of objectivity.[124] Vocabulary such as "mugging," "wielding," "accosting," "beating," as well as "savagely," "gruesomely," and "mercilessly" are all emotionally laden terms, but surely the most appropriate terms for completely and correctly describing this event. However, the ability to render just such a description accurately implies something about the self-involving character of truthful description. To spring forward with a cry, "Oh no! She's being mugged! Stop! Help! Police!" is a more truthful speech act and therefore an indication that the description ought not be reduced to the aspect of objectivity.

When a jury is later given eye-witness testimony, now properly deposed in an objective voice, it is not exclusively their objectivity that enables them to deliver a fair verdict. If objectivity alone delivers a just verdict, then justice would be the sort of thing that could be looked up in a table. ("Let's see . . . a frequency of striking that exceeds one strike per second is defined as "wanton and malicious, ergo . . .") Rather, judgment is rendered fairly by means of the jurors' abilities to reconstruct the event and imagine it in all its original resplendent horror. Such imaginative skill is common, though it is not automatic. A young child would be unable to reconstruct the original event, as are those who are emotionally or psychologically defective in certain ways. Such skill is not an innate talent, and achieving is means one is able to see events under aspects that necessarily include objectivity, but also crucially *exceed* it.

Lest the emotional component of witnessing assault and battery eclipse the point I am trying to make, let me summarize: the aspect under which a natural phenomenon is best observed, one which allows the observer to affix the label "natural *evil*," necessarily includes more from the witness than simply his or her acquisition of facts. Even in cases where emotions are not

---

meneutics and, more popularly, the Reader Response Criticism of Stanley Fish. Jauss, *Toward an Aesthetic of Reception*; compare Fish, *Is There a Text in This Class?*

[123] It is well known that data does not speak for itself. See, e.g., Vesilind, "How to Lie with Engineering Graphics."

[124] D. Z. Phillips writes of this in numerous essays. Those new to his work can find no better place to begin than his *Introducing Religion*.

so near the surface as the parking-lot mugging, to view through the lens of ethics requires of the observer an aesthetic-like engagement with the data. I am not saying that ethics is merely about taste. I am saying that ethical perception must be trained in ways similar to those that cultivate appreciation of art.[125] The upshot is this: when natural "evils" are described as mere natural phenomena, an incomplete story is told. In order to achieve truthful description, the *rest* of the story must be told. The telling of the rest of the story will require greater participation of the speaker than is required of an objective witness.

4.1.2 Aspect Seeing as Self-Involving

Walking through the mall, I was brought up short outside a store window. Inside a group of people stood gawking at a picture. Several kept moving towards the picture then away, now leaning this way then that, while their companions pointed over their shoulders at the painting. Curious, I asked what all the fuss was about. One of the bystanders told me that the picture was a computer-generated stereogram. As I listened, I learned that to see the 3-D image (in this case a dinosaur), one needed training in viewing the geodesic confusion in a certain manner.

Was he pulling my leg? Was a dinosaur *really* there in the picture? Until I had been adequately trained, I would not be able to tell for myself. Nevertheless, the fact that there was an image there could be deduced from the reactions of the other admirers. One by one they exclaimed, "Oh! I see! It's a dinosaur!" or "There's a volcano in the background." Their cries were spontaneous and uncoerced. I suppose that some may have been carried along by the enthusiasm of the others and simply pretended to see. But their skills could have been tested by asking them what they saw in the next stereogram (it was a flying saucer). As a (nonparticipating) bystander, my belief in the "reality" of the dinosaur at this moment was limited to deductions drawn from my observations of the spontaneous reactions of those whose skills were more up to the task than mine. But if I wanted to see it for myself, reason dictated that I follow the suggested protocol ("Stand three feet in front of the picture and stare at your reflection in the glass.")

For any given experience, human participants can be divided into two camps: those who see it and those who don't. Often enough, *everyone* sees. In fact, human insight is so uniform that we often slide unwittingly from the belief "The pattern is self-evident to me" to the conviction "*No one* can possibly deny that the pattern is really there." Yet not everyone *sees* an object in a technical drawing, *sees* beauty in a printed music score, *sees* numbers on a display as data, *sees* the function in a mechanism, *sees* a whole world in a book.[126] Of course, those who cannot see are *not* warranted in denying the reality of the thing in question any more than those who do indeed see the pattern are warranted in asserting the "undeniable reality" of

---

[125] Kallenberg, "Ethics as Aesthetics" in *Ethics as Grammar,* 49–82.

[126] Ferguson, *Engineering and the Mind's Eye*; Frei, *Eclipse of Biblical Narrative.* See also Barth, "Strange New World within the Bible."

that which is, after all, a pattern the perception of which depends as much upon training as upon the data.[127]

Communication *within* a linguistic group is enabled by agreement in reactions and judgments in response to the seeing of a particularly patterned world. Communication *between* two different groups, where what goes for shared judgment within one group may be only barely imaginable to the other, must begin with each group noting the similarities and differences between the two respective sets of spontaneous reactions. Even before the outsiders have had a chance to undertake insider's training in reactions and judgment, the reality of a thing or event or pattern may be charitably supposed by the outsiders on the basis of the visible behavior of members of the community fluent in such things.[128]

The word "evil" does not belong to the vocabulary of objectivity. That we instinctively employ the term "evil" with respect to some natural phenomena indicates that the door has been opened for seeing these phenomena through moral and theological lenses as well as through the lens of "objectivity." To the extent that one lacks (or refuses) training appropriate to ethics and theology, such a one remains as a deaf person who may provisionally conclude that music is real but must content themselves with someone else's description of a lively concert. Until one undertakes the training necessary for seeing the world under these additional aspects and engages the world in the correlative manner, one is left with a merely objective, which is to say impoverished, view of the world.

A second way natural evil receives attenuated descriptions is when too much of the background is left out of the description.

4.2 Overly Restricting the Narrative Context of Description

The Solvag Conference in Brussels in the fall of 1927 signaled a definitive victory for the Copenhagen school of quantum physics as Niels Bohr painstakingly solved each daily puzzle Einstein proposed. The "duel" between Einstein and Bohr (as Heisenberg later dubbed it[129]) epitomized what Thomas Kuhn would later call the "proliferation of paradigms" that follows the demise of the received account while rival schools of thought struggle for supremacy over each other.[130] One of the sticking points between Bohr and Einstein was whether mathematics referred to anything real.[131]

---

[127] Nor does the spontaneous reaction of trained observers tip the balance toward the ontological realists. For the *lack* of response among the group that cannot see is itself a primitive reaction from which the third-party conclusion might be deduced with equal validity: there's nothing there. On the role of observer skills in perception see Wykstra, "Humean Obstacle to Evidential Arguments from Suffering."

[128] Wittgenstein thought that all language worked this way. The reality of the runaway bus is not in question until it runs me over. It is already shown by the fact that everyone jumps out of the way—including, importantly, the blind person who speaks English and jumps in response to the cry: "Watch out for the bus!"

[129] "Commentary of Heisenberg (1967)."

[130] Kuhn, *Structure of Scientific Revolutions.*

[131] Quantum mechanics constituted a revolution in thought. "It is not a new chapter in the ontological tradition [of Western philosophy] but rather a phase of another evolution. The source of this evolution is mathematics, not philosophy. Its

Einstein believed that there were "perfect laws in the world of things existing as real objects."[132] For his part, Bohr could not deny this. But neither could he affirm Einstein's simple realism. Bohr saw that human language is itself the culprit that prevents us from settling the question: "we are suspended in language in such a way that we cannot say what is up and what is down."[133]

When the reality of the quantum world was at stake, Bohr observed that like all other English words, the word "'reality' is a word in our language and that this word is no different from other words in that we must *learn to use* it correctly. . . ."[134] But being the sorts of critters that human beings are, there are inescapable restrictions on how broadly we can properly use the term "reality." On Bohr's view, measurements in quantum experiments can only be framed in terms of classical physics. It is this problem that raises the issue of how much context is enough in order for a description to be a truthful one. As Bohr's problem resembles my present discussion, his solution may prove useful.

Every atomic phenomenon is closed in the sense that observation of it is based on data obtained *not* by direct detection of the quantum particle but by recording the irreversible registrations displayed by suitable amplification devices.[135] In other words, a measurement amplifies things up to the classical level, where "that" indeterminate electron becomes "this" discrete and unmistakable mark on the photographic film. Only with these amplified results can human beings, who are neither quark-sized nor photon-fast, "describe" quantum events. These marks are observed, recorded, and subsequently discussed. But by the act of amplification a necessarily *distorted* picture is rendered.[136] Consequently, rather than *settling* the age-old problem about the existence of objective reality, quantum physics is the paradigmatic case of human *inability* to say anything one way or another. "The limit," Bohr writes, "which nature herself has thus imposed upon us, of the possibility of speaking about phenomena as existing objectively finds its expression, as far as we can judge, just in the formulation of quantum mechanics."[137]

The problematic status of mathematical descriptions of the quantum world (are they depictions of an objective world or of the patterns humans happen to see?) did not lead Bohr to conclude that they weren't worth

---

trend is a search not for ultimate reality but for rigorous use of language." Petersen, *Quantum Physics and the Philosophical Tradition,* 129.

[132] "Bohr-Einstein Dialogue," vi. Einstein was also reported to have said "If, without in any way disturbing a system we can predict with certainty . . . the value of a physical quantity, then there exists an element of physical reality corresponding to this physical quantity." Petersen, *Quantum Physics and the Philosophical Tradition,* 166.

[133] Cited in Petersen, *Quantum Physics and the Philosophical Tradition,* 188.

[134] Ibid., 172. Emphasis added. Here Bohr seems to borrow from Wittgenstein (*Philosophical Investigations,* §43).

[135] Bohr, cited in "Bohr-Einstein Dialogue," 7.

[136] If electrons are non-local entities, we would have no way of detecting the property of "non-locality" since amplification devices necessarily register the effect of electrons as *localized* entities.

[137] Ibid., 3.

troubling about. On the contrary, even if mathematical descriptions were human constructs, they were constructs that challenged the tidy way human beings conceived of their world. In order to compensate for the untidiness, Bohr urged for *expanding* the context when describing quantum events in hopes that the quantum events might be fixed by a background we *do* understand.

> In electron interference, the physical "process" starting at the electron's emergence from the gun and ending at its impact on the plate has no definable course. It cannot be broken up into physically well-defined steps. Unlike classical phenomena, a quantum phenomenon is not a sequence of physical events, but a new kind of individual entity.[138]

For Bohr the "inscrutability of the quantum event's interior" is simply the result of the indivisibility that characterizes the quantum world.[139] "Indivisibility" was Bohr's term for a quantum event's resistance to dissection. There are no tinier parts which can be treated in isolation and added to make up to the larger quantum event. The "movement" of the electron from gun to plate is a seamless whole. Rather rue human inability to zero in on the electron-as-particle-in-flight, quantum indivisibility led Bohr to insist that quantum description was incomplete until it was *expanded.* In other words, instead of doubling his efforts to break down the quantum event into tinier pieces, Bohr's strategy was to expand the horizon to include not only the end conditions (most notably the markings on the photographic plate), but also other classical phenomena such as the initial conditions of the gun and even *the table top.*[140]

On Bohr's view, a perspicuous understanding of the quantum world cannot be achieved by the old-school strategy of divvying up each event into smaller and smaller events. To the extent that Bohr considered the issue of indivisibility and closure "of fundamental significance, not only in quantum physics but in the whole description of nature,"[141] when we are faced with inscrutable mysteries in nature, we ought to resist the temptation to divvy the event into smaller and smaller steps but rather follow Borh's protocol by viewing the event in a broader and broader context.

It doesn't take much imagination to expand Bohr's protocol to enfold not just the table top but also the *human observer* as part of the initial and final conditions. Aage Petersen, one of Bohr's last students, explains:

> In exploring the quantum world we are no longer detached observers, but we mold that which we describe. Thus quantum mechanics is not a description of nature as such but it has an *observer-dependent aspect.* It is a description of nature as exposed to man's method of questions.[142]

Bohr realized he was faced with two radically different domains of cause and effect. Both domains belong to the natural world. The entities of

---

138 Petersen, cited in Ibid., 6.

139 Petersen, *Quantum Physics and the Philosophical Tradition,* 173.

140 Ibid., 171.

141 Ibid., 177.

142 Ibid., 141–42. Emphasis added.

the one are entirely unlike the "new kind of individual entities" of the other. The grammatical rules for speaking about the other constitute a terribly misleading way for speaking about the one. Under these conditions, Bohr concluded that the best strategy for understanding the quantum world was to expand descriptions of it to include *more* (rather than less) of the world in which we live. Bohr's is a reasonable strategy for reconciling two dissimilar domains of cause and effect. Since religious believers are bent on claiming divine cause and effect (whatever that is) is displayed in the natural world, conversations about "natural evil" likewise straddle two dissimilar domains of cause and effect. In the fourth section of this paper, I will apply Bohr's protocol of expanding the scope of description in conversations about inscrutable evils. But first I will add to the list one more tempting way to attenuate descriptions of so-called natural evils.

4.3 Overly Restricting the Range of Acceptable Causes

Third, contemporary descriptions of natural evil tend to unnaturally restrict the sorts of causes that are acceptable answers to the question Why? From Aristotle up through the Baroque period in the West, investigators of natural phenomena framed their inquiry along four lines. They sought accounts of efficient, material, formal and final causes. These four were taken as conspiring together to precipitate any natural event.

After the Renaissance, emerging sciences tended to divvy up causal explanation in a slightly different fashion. After describing the shape of contemporary causal explanations, I will suggest why the modern taxonomy of cause and effect is insufficient for conversations about "natural evil," as it fails to account for all the natural causes in the mix.

4.3.1 Science Restricts Itself to Uncovering Material and Efficient Causes

By and large, contemporary science has restricted itself to the discovery of efficient and material causes.[143] When considering what was previously known by the names "formal" and "final" causes, science has redivided the natural world into the domains of the *technological* and the *natural*.

As regards *technology*, formal causes are lumped in with efficient causes. A given technology—say a bridge—succeeds because the human agents exert control over the materials by mastery of the formal properties that relate the materials to each other. Knowledge of the formal principles of trusses guides the engineer to construct a framework of overlapping triangles rather than to duct-tape I-beams end to end. Final causes of technologies are less often explicitly considered. In the case of a bridge, the final cause is a no-brainer: it's for getting to the other side. In the more dicey

---

[143] Strictly speaking, scientists do not seek material "causes" in the same way that medieval thinkers did. Clearly knowledge of material properties is determinative for understanding how efficient causation proceeds. But the ancients understood material causes as playing an additional, almost agent-like role in the event. Thus the ancients would raise questions about the wisdom of creating new alloys, new viruses, new nano-products precisely because one can never predict what havoc these "agents" might wreak down the road.

case of nuclear warheads, scientists tend to defer to politicians and ethicists in matters of final causes.

Unfortunately, this division of labor is not altogether satisfying for responding to cases in which technologies fail. If a bridge collapses in a dramatic way, an account of the material and efficient causes does not head off the urge of those touched by the tragedy from asking "But why?" Nor will observation of formal causes quell the question of Why? For example, just because we know that bridges tend to fail on a thirty-year cycle doesn't remove the urge for us to ask "Why this bridge, and why today?"[144] Human beings seem predisposed to react to the event as if the bridge, or perhaps technology in general, were itself an agent of evil, since the Why? question expects an answer in the shape of a description of *final* causes.

I suspect that human beings ask the Why? question in the face of natural disasters for the same reason they ask it when technology fails: they are requesting a description of final causes. If human beings do reflexively ask Why? then Aristotle's fourfold account of causation may not be an outdated mode of reasoning after all, but one that corresponds to the very shape of human existence. We simply cannot stop ourselves from requesting explanations of final causes whether or not the question itself makes sense.[145]

A similar urge to ask Why? crops up in the face of nontechnological, which is to say *natural,* disaster. Of course, technological failures differ from natural disasters *only* in the sense that one sort of efficient cause (namely human beings) is taken as *the* classifying mark for the former but not the latter. To say a disaster is "natural" cannot mean that it differs in kind from a technological one, because human beings are clearly a part of the natural world. Rather, the class "natural disasters" is what remains of the set of all disasters once human beings are disqualified from the list of efficient causes. My complaint is that we too quickly exclude human agents from descriptions of natural events. This exclusion breeds confusion and prevents us from crafting satisfactory responses to the Why? question.

Setting aside, for a moment, those disasters in which human beings are obviously complicit, we can ask whether it is intelligible to ask for an account of *final* causes as well as material, efficient, and formal. Clearly not every askable question is intelligible. So, it may turn out that our reflexive request for an account of final causes implied by the Why? question says more about human beings and our Feuerbachian projection of an Agent above, below, or behind the natural world, than about the way things really are. In response, I maintain that we ought not be overly hasty in ruling out requests for final causation in cases of natural evil. Our difficulty in seeing final (and sometimes formal) causes often tricks us into describing events in ways that exclude, without warrant, the possibility of giving an account of final causation. We describe natural events only in terms of those causes that are most readily apparent to us. But attention to

---

[144] Petroski, "Past and Future Failures."

[145] Martin Heidegger, for one, thought that this question made eminent sense and our failure to realize this only makes us slaves to our technology. In his essay, "The Question of Technology," Heidegger describes Technology, with a capital T, as something that takes on a life of its own. As such, Technology ought to be classed with other social structures, power relations, and emergent properties capable of exerting top-down causal force in our world. See Heidegger, "The Question Concerning Technology"; for a theological interpretation see also Berkhof, *Christ and the Powers.*

the grammar of the word "cause" will show that the intelligibility of requests for causes of any sort turn upon how human speakers conduct their
lives. Consequently, it may be that our inability to properly hear, much less
respond to requests for fourfold causal explanation of natural evils is more
an indication of a defective way of living more than it is a lack of data.

4.3.2 Grammar of "Cause" Demands Skillful Participation of an Observer

In this section, I argue that (contra Hume) assignment of "cause" depends
more on a skill of seeing than on a configuration of data.

In his essay "The Limits of Empiricism," Bertrand Russell argues that
causation is not always arrived at by deduction, but in its simplest forms,
causation is *intuitively perceived*: "We have reason to believe: that if any
verbal knowledge can be known in any sense derived from sense experience, we must be able, sometimes, to 'see' a relation, analogous to causation
between two parts of one specious present."[146] The simplest case occurs
when we stub our toe on a chair leg and we immediately fault the chair.
How did we *know* the chair is the cause of pain in our toe? We know this
not by deduction but by a form of perception. Russell reasons, "when I am
hurt and cry out, I can perceive not only the hurt and the cry, but the fact
that the one 'produces' the other."[147] This runs against the general grain of
the Humean insistence that "causes" are unperceivable fictions that we
posit only after observing a regular series of events. ("The chair is stationary, my foot swings forward, there is a sound, my toe hurts, the chair is
still stationary, therefore the striking of the chair *causes* the pain in my
toe. Let's try that again to make sure.") It only takes one exception to dis-
prove the empiricist's rule. Russell concludes therefore, "If we can sometimes perceive relations which are analogous to causation, we do not depend wholly upon enumerations of instances in the proof of causal
laws. . . ."[148]

Russell criticized Humean empiricism for barring claims to certainty
with respect to knowing a causal relation (short of an infinite string of data
pairs). Russell was surely correct to say that the empiricist's demand for
"proof" of causation must come to an end somewhere. On my view, he was
also right to say that it ends with a way of seeing. But in sharp contrast to
Russell, his one-time student Wittgenstein showed that "seeing" causation
did rest upon regularity, but not the kind of regularity that Hume sought,
regularity of data (B_1 follows A_1, B_2 follows A_2, . . . , B_n follows A_n, therefore
A causes B). Rather, "seeing" causal links is a function of regularities in
ways of acting for the speakers of a language. As mentioned above,
Wittgenstein sometimes called this regularity "primitive reactions." Some
of our primitive reactions are biological: we squint at bright lights, we
pucker at lemons, we sneeze at pepper. We also "instinctively look from
what has been hit to what has hit it."[149] Yet others of our primitive
reactions have been *trained* into us. For example, while social creatures

---

[146] Cited in Wittgenstein, "Cause and Effect," 370.

[147] Ibid., 371. Russell's logic is distributed across the chain: injury $\rightarrow$ sensa $\rightarrow$ cry.

[148] Cited in Ibid.

[149] Ibid., 373.

(humans, chimpanzees, etc.) instinctively follow the gaze of tribe member, we must be *trained* to respond to signs, such as this arrow "→", in this case by looking to the right. Regular ways of responding to our environment, both *natural* (following another's gaze) and *trained* (turning right at the arrow), constitute the *unquestioning certainty* from which language—in the present case, the language of causality—grows. Wittgenstein concludes, "I want to say: it is characteristic of our language that the foundation on which it grows consists in steady ways of living, regular ways of acting. Its function is determined *above all* by action, which it accompanies."[150]

Wittgenstein goes on to say that our ordinary language of cause and effect is fuzzy at both edges. On the one side, we have to admit "what a powerful urge we have to see everything in terms of cause and effect."[151] Consequently, we are overly quick to assign causes where none may exist. We have all had this experience: "Sometimes we think we are causing a sound by making a movement and then realize it is quite independent of us."[152] Yet not all nominations for an instance of cause and effect can be suspect. So on the other side, it is not the case that absolutely everything and anything can always and at any time be doubted. Those who sprinkle doubt on the most basic perceptions are confused about what it takes for language to be working properly.[153] "The basic form of our game," Wittgenstein writes in reference to the language game of cause and effect, "must be one in which there is no such things as doubt. . . . Doubting . . . has to come to an end somewhere. At some point we have to say—without doubting: *that* happens because of *this* cause."[154] In other words, there is a de facto limit both to gullibility and to skeptical demands for proof.

Imagine following a moving string down the hall and around a corner to see who is tugging it. Suppose we catch him or her red-handed. Are we able to doubt whether the culprit is the primary cause of the string's movement? So naturally do we say, "Aha! *You* did it!" that we cannot even imagine how someone could maintain the position of doubt while admitting "Yes, yes. I saw her pulling the string. But how am I to *know* that this is an instance of causation?" If we conjure up in our imaginations such a hyperskeptic, could we also imagine that the matter could be settled by a series of experiments?[155] Clearly not; if someone cannot recognize causation in its simplest form, he or she will be unable to grasp additional empirical demonstrations of it. For proofs get their force from the fact that the paradigm cases of causality cannot be questioned.

It seems then, that the concept of causation in every instance rests on human ability to "perceive" or "see" *something* as a cause. Mastery of everyday life and language governs the range of our primitive reactions with respect to causes so that we search for causes from the presence of attention-getting effects (e.g., a loud voice in a room makes us look for a microphone). Likewise someone's fluency both in *specialized* experience and in attending *specialized* language-use trains him or her to see causes within highly

---

[150] Ibid., 397.

[151] Ibid., 375.

[152] Ibid., 373.

[153] Language that isn't working properly is gibberish or "idle" or "on holiday."

[154] Wittgenstein, "Cause and Effect," 377.

[155] Ibid., 387.

specific situations. Thus familiarity with physics laboratories and fluency in the mathematical language of quantum mechanics trains the physicist to extend adapted notions of cause and effect to lab experiments. If the scope of our application of the word *cause* is related to the specificity of our training, we can conclude that the *sort* of cause we look for is related to the conceptual lenses through which we have been trained to see.

In this light, it makes sense to ask whether our regular ways of acting are sufficient for seeing broadly enough for handling cases of natural evil. It may be that our mute inability to give an "answer" to the problem of evil indicates a deficient way of living rather than insufficient data. The solution therefore, may not be to multiply words but to change the way we live. Space will not permit me to pursue this topic here.[156] Sufficient for my present purpose is an explication of how descriptions of "natural evils" might be satisfactorily expanded.

5 Broadening the Descriptive Context

5.1 Narrative Space

In one of his classically provocative statements, Wittgenstein once observed that if his sofa were moved onto the front lawn, it would be tempting to say that it had become a quite different object. In some sense, the identity of a thing includes its connections to its surroundings. For example, while visiting the Hermitage Museum I stood for many minutes in front of Rembrandt's paintings. Suppose I become quite taken by the color of one character's kind eyes and endeavor to paint my house this color because of the warmth it kindles in me. Were I to succeed in matching this hue (say, by some form of spectroscopy), I'd be terribly disappointed with the results, not because the house paint was the wrong spectroscopic hue, but because it will have become quite a different *color*. A color is what it is in a particular configuration with its surroundings.

The same phenomenon holds for descriptions of objects and events. Consider the sentence, "Cadavers are frequently in the prone position." Devoid of context, this sentence is trivially true. But the very absence of context bewitches us into thinking that what is primarily at stake is the truth or falsity of the sentence.

Imagine now a different conversation. At a casual gathering, I begin to brag how nimble I am as a forty-seven-year-old distance runner. I recount how just this afternoon while running through the forest I nimbly leaped over a boulder, a cadaver, and a fallen tree.[157] "Wait just minute! Did you say 'Cadaver'?" "What of it?" I reply, "Cadavers are frequently in the prone position." If I calmly proceed to inform you how far I ran and at what pace, you'd think something was horribly wrong with me. Part and parcel of a proper description (in this case of cadavers) is the inclusion of a broad enough context so that what is really going on is made apparent. To leave out such details is substandard, even demented, by the canons of ordinary conversation.

---

[156] See Kallenberg, "Some Things Are Worth Dying For."

[157] I must credit my friend Charles Pinches for this fanciful story. See Pinches, *Theology and Action.*

Relevant context includes narrative space. It is important to the conversation about my nimbleness that an unnamed cadaver was encountered on a trail run. Had I instead been visiting the morgue in my capacity as a forensic pathologist and reported at a dinner party (with a twinkle), "Cadavers are frequently in the prone position," you might think me humorous, but not because any of the bodies were out of place. A cadaver becomes suspicious when it is discovered in surroundings in which we normally do not encounter cadavers. It does not become suspicious *simply* in virtue of being a cadaver in the prone position.

As in the case of cadavers, important relevant details must be included from the surrounding context to render a description truthful. An example of the significance of expanding spatial context for describing natural disasters can be seen in reports that surrounded the spate of tropical storms that have recently plagued the Atlantic. To cite but one example, tropical storm *Jeanne* ravaged the tiny country of Haiti in 2004. In the city of Gonaives alone, nearly 2,900 cadavers were recovered. Such a tragedy might easily prompt the question, "Where was God in September?" But this disaster is not yet properly described. If we expand the narrative space to include not just Haiti but the entire Atlantic seaboard, we learn that *Jeanne's* strength grew after it passed Haiti, elevating it to hurricane status. Yet when hurricane *Jeanne* pummeled the Atlantic coast, only four persons died.[158] The "tragedy" is evidently understood not by viewing the storm itself, but by viewing it in light of the enormous disparity of wealth between Haitian poor and rich North Americans.

A similarly expanded re-description ought to be conducted for the Armenian earthquake. Fact #1: On December 7, 1988 an earthquake of 6.9 magnitude struck and killed 25,000 people leaving some 400,000 homeless.[159] Fact #2: When a similar-sized earthquake struck California two days before Christmas of the same year, only three persons died.[160] Fact #3: Civil engineers attest to the fact that the difference between the earthquake-safe construction and its alternative is an additional construction cost of a mere 15%. Taken together, these three facts shift the description Fact#1 from a theodical problem to an economic, political, and moral problem.

Not only must the spatial context of so-called natural evils be expanded to achieve a truthful description, the temporal context must also be expanded.

5.2 Narrative Time

5.2.1 Expanding the Narrative Context Prior to the Event

In the first place, the temporal context must be described so as to include enough of what came *before* the event so as to render perspicuous the relevant connections between this event with events that precede it. In *Zettel*, Wittgenstein observed "Only in the stream of thought and life do words

---

[158] Lawrence and Cobb, "Tropical Cyclone Report: Hurricane Jeanne."

[159] See US Geological Survey, "Earthquakes Facts and Lists."

[160] See Branigan and Whitaker, "Dangerous Buildings."

have meaning."[161] So, for example, we may misunderstand a command issued to us unless we have an adequate grasp of what came before the command. Once again, a concrete example can help show what I mean.

Diseases such as tuberculosis are designated as "natural evils" since people, especially children, contract it through no fault of their own. It is my claim that not every case of TB implicates God as deeply as every other. In the opening years of the new millennium, the *increase* of incidence of TB on U.S. soil has epidemiologists worried. But this trend is made all the more troubling by what events preceded it. First, as Harvard cultural anthropologist and physician Paul Farmer has demonstrated, TB can have an astonishing 100% cure rate at minimal cost . . . provided the money is spent in the right places. In the protocol introduced to Haiti, Farmer was able to attain 100% cure rate for only $150–200 for persons treated in their homes compared to $15,000 to $20,000 for persons treated in a U.S. hospital setting. The difference? Money was spent not simply in the treatment of the disease (the same in both cases) but in providing a small but continual food allowance that reversed the malnourished state of the poverty-stricken Haitians.[162] Second, failure to eliminate TB in the West is made morally more serious by the fact that in the past 70 years the gap between the rich and the poor has dramatically increased. The rich-poor gap in the United States is the worst it has been since 1929; the top 5% of money makers owns 58% of the nation's wealth. Only the top 10%—the *very* rich—have made economic gains over the past 20 years. Ninety percent of American wage-earners have actually lost ground. Yet the top 0.001% (some 13,400 families) have had a wealth gain of 558.3%.[163] These economic facts radically alter the identity of the evil of reemergent tuberculosis. Reemergent TB is not a "natural evil," but most correctly described as a *moral* evil.

It will surely be objected that not every instance of natural evil can be so readily dispatched. My response is simply this: until we expand the description enough in every case, we dare not say one way or the other.

5.2.2 Expanding Narrative Context Subsequent to the Event

In addition to expanding the temporal context *prior* to an event, a truthful description also requires skillfully expanding the context to include what *follows* the event. Recall Wittgenstein's observation that we are able to understand a command, say one issued by an authority, to the extent we understand the context that precedes the command. Leave it to Wittgenstein to continue: "if the meaning-connection can be set up before the order [is issued], then it can also be set up *afterwards*."[164] It doesn't take much imagination to see the possibility that the last line of a play or story may convey information that casts everything that happened previously in a new light, one that forces a re-reading of the entire work. My point is this: If history is itself narratively shaped, its "ending" may one day compel a reexamination of the entire story. If the ending is in some

---

[161] Wittgenstein, *Zettel*, 173.

[162] See Kidder, *Mountains beyond Mountains*; see also Satchell, "Wiping Out TB and Aids."

[163] Morris, "Economic Injustice for Most," 12–17.

[164] Wittgenstein, *Zettel*, §289, emphasis added.

sense uncertain (i.e., contingent rather than necessary), then knowledge of the ending cannot be logically compelled or scientifically predicted (though some theologians claim that the end of history is proleptically present in the resurrection of Jesus Christ[165]). In the meantime, we need not sit on our hands waiting for history to end before we begin writing a nonreductive description of disasters. But we do need to expand our description of the events in question in order to include both what came before and what happens after.

A relevant part of what comes after tragedy is the immediate response of human animals to these events.[166] Inclusion of human responses to natural disasters renders a truthful description of the way things are.

5.3 The Way Things Are

When descriptive context is taken broadly in the ways I am advocating, conversations are apt to become more lively. As descriptions widen in narrative time and space, they become more contestable and hence more contested. To the extent that such descriptions fall short of universal verifiability, the descriptions will fail to garner unanimous acceptance. As disagreements get heated, it is tempting to strip down descriptions to something less controversial in order to gain wider acceptance. Unfortunately, acceptability comes at a price: stripped down descriptions no longer have what it takes to do the necessary conceptual work. Still, *within* a linguistic community (defined as those who rally around a particular thick description and share a determinate form of life) conversation is possible, because some statements stand firm and serve as the collective hinge upon which the rest of the community's speech turns. These statements embody the community's deepest notions about "the way things are." Examples may be as obvious as "I am now wearing clothes" or as esoteric as "God is love."

A surprising possibility emerges when the descriptive context is expanded to include statements which may be definitive for a given community yet importantly lie *outside* the set of statements acceded by *every* speaker ("The sun rises in the east.")[167] When a rich enough description of the way things are is juxtaposed with a practical response to a given occasion for suffering an arrangement properly deemed as *satisfactory* is achieved. Sometimes the proper action is to weep with those who weep. Sometimes it is most fitting to solve the presenting problem. Other times it is most appropriate to rush to relieve physical suffering. But when an appropriate enough action is taken in the presence of a rich enough description, the arrangement of the two has the possibility of striking us as satisfactory. There may be other satisfactory arrangements, and there is no way of telling in advance of talking about it which arrangement will turn out to be the most satisfactory. Yet clearly some arrangements are *un*satisfactory.

---

[165] See Robert John Russell's essay in this volume. See also Pannenberg, *Jesus, God and Man.*

[166] For a compelling account of the need for practical theodicy, see Don Howard's essay in this volume.

[167] It is important to note that statements such as "The sun rises in the east" are not more certain for being *shared* per se; they simply are shared by many rather than shared by few. On Wittgenstein's view, there is no private language.

In sum, I explained above that attenuated descriptions will yield an unsatisfactory arrangements just as surely as will the absence of an appropriate practical response. But when both pieces are present, both a broad enough description of the way things are and a proper practical response, there exists the possibility of achieving a harmonic resonance as description and response become tuned to each other. Sadly, I cannot demonstrate this in any quantitative or logically compelling fashion; for this harmony is perceivable only by the one with ears to hear and eyes to see.

5.4 "Explanation" as that which Supervenes upon the Juxtaposition of Theology and Action

In his poem "Which," R. S. Thomas writes of the religious believers' response to suffering:

> And in the book I read:
> God is love. But lifting
> my head, I do not find it
> so. Shall I return
> to my book and, between
> print, wander an air
> heavy with the scent
> of this one word? Or not trust
> language, only the blows
> that life gives me, wearing them
> like those read tokens with which
> an agreement is sealed?[168]

The poet's ultimate conclusion about whether there is a God of love, as Christians claim, turns upon the sort of data one is impelled to exclude. If one considers *only* "the blows that life gives," one is hard pressed to keep reading a Book that speaks of a divine being that is both loving and omnipotent. But, if language is to be trusted, there is more to consider than simply the blows that life deals. What is it about language that make it a reliable compass for navigating the choppy waters of natural evils? It is simply this: verbal behaviors and nonverbal behaviors are inextricably bound together in the successful functioning of a complicated linguistic form of life that a community of human beings inhabit.[169] Consequently, both actions and words conspire to give a satisfactory response to natural disasters.

Perhaps somewhat surprisingly, words do not, and cannot, succeed in isolation. In other words, evil is not the sort of thing that can be adequately responded to by an explanation offered by a single person, or by a string of persons for that matter. If it were, the uptake of the explanation(s) would have long since been perceived as satisfactory and the problematic retired.

I claim that in contrast to the stand-alone explanation, a "satisfactory" response to the enigma of suffering involves the perception of a pattern that emerges when properly rich descriptions are juxtaposed to concrete

---

[168] Thomas, "Which."

[169] See Kallenberg, *Ethics as Grammar.*

behaviors taken in response to a given occasion of suffering. The descriptions I have in mind are broad in space and time, at least broader than typical descriptions of natural evils.

I suspect that it is the *constellation* of adequate verbal and nonverbal responses taken together that strikes the believer as satisfactory. Satisfactory, yes, yet neither necessary nor sufficient. A satisfactory constellation is never the only one possible. Portions of either element (i.e., verbal or nonverbal) may be wrong and surely may be improved upon, for there is always more to be said. Moreover, an adequate description must also be "theological" in the sense that it is not only a description of an event, but more properly a description of a pattern in the data taken by religious believers to be undeniably the way things are. In sum, I have argued for the possibility of achieving something more realistic than the ever-elusive explanation to the Why? question we pose to natural disasters. This possibility is the emergence of a satisfactory constellation between description and action when these two are so tuned to each other than harmonic resonance is achieved. For persons who are capable of perceiving, the constellation not only is satisfactory, it has the force of a speech act. But who is the speaker? If the perceiver is a religious believer, he or she quite instinctively attributes his or her perception of the pattern as if it were a speech-act spoken without words by God.

What follows is a twelfth-century example of the juxtaposition of theological claims about the way things are with a seemingly isolated work of mercy. Taken together, these disclose a satisfactory response to a single natural evil in southern France.

6 The Theological World According to Hugh of St. Victor

To recap, as biological critters, human beings respond to their environment in physical ways, by making noise and moving their bodies. Often noise-making and movement are intentionally coordinated to achieve an effect that is larger than any one human animal can accomplish alone. Sometimes, though not always, a kind of synergy is attained even when noise-making and bodily movement are not intentionally coordinated. I take as my example a book and a bridge. In the late 1120s, a monk in Paris wrote a book called *Didascalicon* for his students. In it, Hugh of St. Victor reflected on the ways that the presence of evil (a.k.a., the absence of good) serves to orient human beings in their corporate quest for the God who alone is Wisdom. Several hundred kilometers to the southeast and five decades later, a teenage shepherd boy quit his day job to build a bridge over the Rhone River. Neither Bénezet nor Hugh were aware of the other. But their respective actions taken in response to the evils they saw achieve resonance.

Hugh's theology takes the brokenness of creation *as a given*. (He is, after all, an Augustinian.) But Hugh does not display any compulsion to reconcile the brokenness of creation with the goodness of God, for God is known to us as much as a source of *abundant redemption* as eternal goodness (Ps 130:7). Hugh sees the brokenness of human beings and of their world as gesturing to what human life is for. Human beings alone are creatures capable of imitating divine redemption by their appropriate response to the world's brokenness. Doing this is itself sacramental: the picture that the redemptive human community displays is the very reality of what is being depicted. (A generation later, Aquinas would describe this sacramental action as simultaneously human and divine under his doctrine of "double agency." Human beings who imitate divine redemption by taking redemptive action in response to particular cases, participate (a technical term for Thomas) not only in God's being, as do all existent things, but in both the very *agency* and *redemptive goodness* of God.[170])

For Hugh, the brokenness of the world that makes natural evil a terrifying prospect is not as much a cause for alarm as it is the impetus for a quest. As fallen creatures living in an unpredictable and threatening world, Hugh maintains that human beings "are restored through instruction, so that we may recognize our nature."[171] God in redemptive grace has intended the very condition of human fallenness and nature's brokenness as the impetus for our corporate pursuit of Wisdom, a quest which is the "highest curative in life."[172] All dimensions of human relations are directed by this quest.

> And so arose the pursuit of that Wisdom we are required to seek—a pursuit called "philosophy"—so that knowledge of truth might enlighten our ignorance, so that love of virtue might do away with wicked desire, and so that the quest for necessary conveniences might alleviate our weaknesses. These three pursuits first comprised philosophy. The one which sought truth was called theoretical; the one which furthered virtue men were pleased to call ethics; the one devised to seek conveniences custom called mechanical.[173]

In this passage, Hugh asserts that redemption is assisted by the practice of "arts" that correspond with *all* the powers of the soul. Corresponding to the understanding *(intelligentia)* are both the theoretical arts (i.e., the contemplation of necessary truths; here Hugh intends theology, physics, and mathematics) and the practical arts (namely, the practice of morality and the cultivation of virtue). Corresponding to knowledge *(scientia)* are all the mechanical arts. These latter have to do with feeding, fortifying the body against harm, and the contrivance of "remedies" for alleviating physical weakness.[174]

Unlike his predecessors, Hugh's theological vision encompasses all aspects of human knowing, including the perennially maligned "mechanical reasoning." By his day, "mechanical arts" had evolved into a very broad category. To be specific, mechanical arts was comprised of seven classes of practices: fabric-making, armament, commerce, agriculture, hunting, medi-

---

[170] These points were made clear to me by Elizabeth A. Johnson's wonderful essay, "Does God Play Dice? Divine Providence and Chance."

[171] Hugh of St. Victor, *Didascalicon* 1.1, p. 47.

[172] Here Hugh shows similarity to Irenaeus's take on the problem of evil. For a comparison of Augustinian and Irenaean theodicies see Hick, *Evil and the God of Love.* Though the term *wisdom* resonates with Platonic and Stoic philosophy, Hugh clearly equates divine Wisdom with the second person of the Trinity who is revealed as the Logos of creation and the Christ of the Gospels. See Taylor's introduction to *Didascalicon,* 14 n. 39.

[173] From Hugh's *Epitome Dindimi in philosophiam,* cited in Taylor, introduction, 12.

[174] Hugh of St. Victor, *Didascalicon* 1.8, p. 55.

cine, and theatrics. (Granted, "theatrics" seems like a stretch, but Hugh purposed to make the list seven in number so that it matched the perfection of the seven liberal arts. Besides, under theatrics Hugh envisioned any coordinated activity of a *group* of people. Not just drama, but marching bands and gymnastics would fit under this heading. Had Hugh lived to see Ford's assembly line, he surely would have treated it as a type of theatrics.) Each mechanical art names a *family* of practices. For example, "hunting . . . includes all the duties of bakers, butcher, cooks, and tavern keepers" as well as those who actually do the gaming, fowling and fishing.[175] And "armament" included material science, even metallurgy: "To this science belong all such materials as stones, woods, metals, sands, and clays."[176] With this last move Hugh has managed to bestow honor even upon the grimy-faced smithy[177] so consistently maligned for sixteen centuries.

On Hugh's account, all modes of rationality are mutually supportive. The ends of mechanical arts are displayed by the physical things contrived by the artificer. The "ends" of a device may be final with respect to the device but only provisional with respect to the ultimate end of human life revealed in Christ: reconciliation of God and humans. In this sense, theology can benefit mechanical arts by providing a benchmark for assessing the aptness of its aims. But the benefit also works the other way around: mechanical arts benefits theology by rendering visible invisible things. Mechanical arts yields artifacts (and processes) that are inherently sacramental because they render visible the final ends of mechanical reasoning, and in fact the final end of all human reasoning. Not only is the provisional or natural end of mechanical reasoning (namely, the alleviation of physical weakness) embodied in mechanical artifacts and processes. Mechanical reasoning also contributes to the incarnation of the ultimate or final end of human life (namely, our journeying toward reconciliation with all peoples and with God).

Hugh is quick to emphasize the difference between *worldly theology* (a theology that moves from human knowledge to God) and *graced theology* (a theology that moves from God to human knowledge). Grace figures prominently in Hugh's account because he has no other way of explaining how human beings come to "see" more clearly. In his *Exposition of the Heavenly Hierarchy,* Hugh writes:

> Invisible things can only be made known by visible things, and therefore the whole of theology must use visible demonstrations. But worldly theology adopted the works of creation and the elements of this world that it might make its demonstration in these. . . . And for this reason, namely, because it used a demonstration which revealed little, it lacked ability to bring forth the incomprehensible truth without stain of error. . . . In this were the wise men of this world fools, namely, that proceeding by natural evidences alone and fol-

---

[175] Ibid., 2.25, pp. 77–78.

[176] Ibid., 2.22, p. 76.

[177] Just as persons who were skilled at philosophy were called "wise," so too were persons skilled in mechanical reasoning *(techne)* called "technicians." Unfortunately, the term *technai* was used pejoratively rather than complimentarily of those who worked metal in front of the large furnaces. Such activity was regarded as beneath the dignity of free persons from Socrates' day until the twelfth century.

lowing the elements and appearances of the world, they lacked the
lessons of grace.[178]

---

[178] Cited in Taylor, introduction, 35.

What are these lessons of grace? For Hugh grace is not something added on top of nature, but the very Spirit of God that already permeates the created world and with which human beings may keep step *(gratia cooperans)*. Although the mechanism of grace's operation in the natural sphere is elusive, if not ineffable, theologians who write of grace do not conceive grace as a cause that stands in competition with other material or efficient causes. More commonly, theologians speak of the efficacy of grace in terms of illumination. "Grace," writes Hugh, is the powerful medicine that "was fittingly given both to illuminate the blind and to cure the weak; to illuminate ignorance, to cool concupiscence; to illuminate unto knowledge of truth; to inflame unto love of virtue."[179] How do we account for a person coming to perceive for the first time a pattern that has been "in" the data all along? Hugh asserts that it is here we see God at work; the moment of illumination is called "grace."

7 A Premodern Practical Theodicy: St. Bénezet the Bridgemaker

Hugh's vision was not without practical corollary. During the solar eclipse of 1177, a teenage shepherd in southern France reputedly heard the voice of Jesus Christ telling him to build a bridge at Avignon. Because he was young and untrained, Bénezet was scoffed by the bishop. The river was surprisingly swift at Avignon, not to mention over a half-mile across. Even good swimmers died in the Rhone. But after the boy erected into place the first stone, one larger than could be lifted by 30 men, ecclesiastical support and financial backing followed. Bénezet died shortly before the bridge's completion some eight years later.

We have no way of knowing how well Bénezet understood the significance of his project. We do know that in the twelfth century rivers posed a number of dangers. The Rhone river was capable of devastating floods. (For sake of comparison, in December of 2003, the floodwaters of the lower Rhone claimed the lives of seven people.) The river was also cold, swift, and wide.[180] Crossing the Rhone was very hazardous, especially for the region's meagerly equipped peasants. In view of these dangers, the Middle Ages looked on bridge-building as a charitable act, a work of mercy.

But rivers contributed to more "natural" dangers than flooding and drowning. Human beings are part of the natural world, and for them rivers serve as political boundaries and as a form of protection against one's enemies. (The fact that rivers typically divided property owners meant that bridges frequently fell into disrepair as neither side of the river felt that upkeep fell on their shoulders.[181])

In 1177, the Rhone divided the County of Toulouse from the County of Provence and the County of Maurienne in the old Kingdom of Arles (the northern neighbor of Provence, identified with lower Burgundy). Turmoil in the area was exacerbated by two features. First, political boundaries and allegiances were constantly shifting.

---

[179] Cited in Rydstrom-Poulsen, *The Gracious God,* 206.

[180] Students of history will recall the trepidation and care with which Hannibal moved his troops across the Rhone just a few miles north of *Pont d'Avignon.*

[181] Not surprisingly, St. Bénezet's bridge today is in total disrepair. Of the original 22 arches, only four remain standing.

Capable of magnificent feats in order to achieve expansion, the dukes [of Aquitaine, just to the north of Toulouse] had poor control over a vaguely defined area, subject to anarchic forces. The collapse of the Carolingian structures had given way to a whole system of relations, more or less binding, based on a temporary *convenientiae*. Ducal suzerainty was inconsistent, many-layered and unstable, castellanies virtually independent. All this was further aggravated by ecclesiastical privileges and a rapid decline in the public peace.[182]

To cite but one example of the region's political instability, land-hungry nobility north of Toulouse took Pope Innocent III's call to a crusade (1204) as pretext for invading Toulouse en route to Jerusalem![183]

This political picture is complicated, second, by the fact that a count or viscount's political clout was sometimes shared with a bishop (such was the case for the Archbishop of Narbonne in C. Toulouse) or even simply transferred to a bishop (such as at Albi in C. Toulouse).[184] That churchmen were major players on the political scene is made more significant by the fact that the doctrinal stakes were extremely high in this region. Toulouse had been originally settled by the Visigoths who held to a heresy resembling that of the Arians.[185] Although the Visigoths were conquered in the sixth century, the Arian heresy of the Albigenses had such a stronghold in Toulouse that the entire region "constituted a cancer in the body of European civilization that had to be rooted out at all costs."[186] So many of the nobility of Toulouse (perhaps even the count himself) and Provence were allied with the Arian sect that eventually a crusade (1208) was to be declared directly against Toulouse and Provence (the two counties were to become united by marriage).[187] But even as early as 1163, the Council of Tours had called upon the secular powers to dispossess the "heretics" of their land.[188] In the meantime, the heretics "were to be subject to a social and economic boycott so that they may be forced through the loss of human comfort to repent of the error of their way of life."[189] Sadly, as is often the case, it was the region's poor who suffered most. For the least members of the planet, there is no difference between starvation by boycott and starvation by famine.

It was in this tumultuous context that Bénezet set out to build a bridge over the treacherous Rhone at the geographical corner of fierce political

---

[182] Bur, "Kingdom of the Franks," 543.

[183] In the ninth and tenth centuries, the Duchy of Aquitaine had contained the County of Toulouse within its borders. Ever since, northern noblemen had vied, though unsuccessfully, for these lands. Bur, "Kingdom of the Franks," 542; Cantor, *Civilization of the Middle Ages,* 300.

[184] Bur, "Kingdom of the Franks," 545.

[185] Cantor, *Civilization of the Middle Ages,* 113. Arians denied the full divinity of God the Son and were defined as heterodox at Nicea in 325 CE.

[186] Ibid., 389.

[187] In 1208, a papal legate had been murdered in Toulouse. The Count of Toulouse was himself implicated in the crime. Cantor, *Civilization of the Middle Ages,* 424.

[188] Robinson, "The Papacy, 1122–1198," 337.

[189] Ibid., 336.

and doctrinal adversaries.[190] Whether Bénezet, the mechanical genius, was even vaguely aware of the multiple layers of evil represented by the Rhone is immaterial to my point. Bénezet's exercise of practical reasoning, born of need, is at once a redemptive and sacramental response to evil in the natural world. In the words of Hugh of St. Victor, the right use of human reason, even mechanical reason, "reconciles nations, calms wars, strengthens peace, and commutes the private good of individuals into the common benefit of all."[191]

8 Conclusion

A three-dimensional jigsaw puzzle cannot be completed on a table top. Some of the pieces simply do not fit together on a flat surface. Any attempt to force the fit in two dimensions will be not only frustrating, it likely ruins the pieces. Even the best puzzlers may be stumped until they are given the right tip. "Well, that explains everything!" is the likely exclamation when they learn the puzzle is 3D rather than 2D. In this scenario, the key to understanding is not another puzzle piece, but a *protocol for proceeding.*

In this essay I have argued that one way human beings respond to evil is by talking. Yet "natural evil" frustrates our attempts to speak directly about it. Natural evil is like a puzzle whose pieces do not easily fit together in a satisfactory way. My *protocol* suggestion is that what functions as a satisfactory "explanation" does not take the form of a "missing piece" nor of a bird's-eye view of the "boxtop" (a 2D picture of what's inside). In contrast, I suggest that a satisfactory defense supervenes upon the constellation of two components: (1) a rich enough description and (2) a set of practical actions. My essay has spent the greatest energy on the first element of the protocol. In particular, I've argued for the *expansion* of the descriptions that are given to natural evils. I've advocated increasing the participation of the observer in the speech-act of describing, expanding the range of acceptable causes, and broadening the narrative context of the description. What results is precisely the sort of conversation that constitutes theology. When theological dialogue serves as the backdrop to practical action at the same time practical action grounds the theological conversation, conditions are ripe for the emergence of an aspect, the dawning of which may strike some with the force of a satisfactory explanation.[192]

---

[190] In 1309 Pope Clement V took up residence at Avignon. The fact that "the papacy eventually bought it outright from its ruler the Countess of Provence" implies that Avignon in Bénezet's day had been a part of the County of Provence and thus more likely than not tainted by its heretical associations with Toulouse. MacCulloch, *The Reformation,* 35. See also "Map 12: The Angevin Empire."

[191] Hugh of St. Victor, *Didascalicon* 2.23, p. 77.

[192] Where would I be without friends? I am extremely thankful to Michael Barnes, Aaron James, Ethan Smith, Terry Tilley, Nancey Murphy and my other compatriots at the CTNS for pointing out problems with an earlier version of my essay. Several of these problems turned out to be insuperable and forced me to write a (hopefully) better paper. Any difficulties that remain are, I'm afraid, completely mine.

PHYSICS AS THEODICY

Don Howard

1 Introduction: "Gravity: Our Enemy Number One"

On Saturday, August 26, 1893, thirteen-year-old Edith Low Babson was swimming in her favorite swimming hole on the Annisquam river in her home town of Gloucester, Massachusetts. Though she was a strong swimmer, something went wrong, and she drowned. A tragedy like all such. But this drowning had unusual consequences. Edith's older brother was Roger W. Babson, who grew up to become one of America's most prominent businessmen of the early twentieth century. A statistician, prolific author, philanthropist, founder of Babson College, in Wellesley, Massachusetts, and the Prohibition Party's Presidential candidate in 1940, Roger Babson was deeply affected by his sister's death, as he was again many years later, in 1947, by the death of his grandson, Michael, who drowned while saving the life of a companion who had been knocked off of a sailboat in Lake Sunapee, New Hampshire. But Roger Babson was a man of action, not one quietly to acquiesce when confronted by suffering inflicted by a seemingly impersonal and uncaring nature. One year after his grandson's death, Babson dedicated a significant part of his vast personal wealth to the establishment of the Gravity Research Foundation in New Boston, New Hampshire, which thereafter awarded an annual prize for theoretical research on gravitation, a prize whose winners include the likes of Stephen Hawking. Why? As Babson explained in a pamphlet published by the new foundation, "Gravity: Our Enemy Number One,"[193] the goal was to alleviate the suffering for which gravity was responsible, the gravity that seized his sister "like a dragon and brought her to the bottom," by developing a partial insulator against the force of gravity, and to reap a host of other benefits for health and human welfare by taming gravity.[194] Needless to say, Babson's obsession was a bit nutty, and the insulator was never found. Nevertheless, the foundation's support for gravitation research in the late 1940s and 1950s, a lean time in the history of theoretical work on general relativity, helped importantly to maintain and grow a community of younger theorists who were responsible for the tremendous flowering of theoretical general relativity that started in the 1960s and continues to this day.[195]

---

[193] Babson, "Gravity: Our Enemy."

[194] The range of expected benefits can be inferred from the following list of other publications from the Gravity Research Foundation, which can be found at the Foundation's home page, http://www.gravityresearchfoundation.org/: Roger W. Babson, "Gravity and Sitting"; Roger W. Babson, "Gravity and Ventilation"; Grace K. Babson, "Gravity Aids for Weak Hearts"; W. Stewart Whittemore, MD, "Gravity and Health"; Mary E. Moore, "Gravity and Posture"; Raymond H. Wheeler, "Gravity and Psychology"; Raymond H. Wheeler, "Gravity and the Weather"; Arthur D. Baldwin, MD, "Gravity Effect in Relation to Health"; William Drake, MD, "Gravity and Your Feet"; William R. Esson, "Possibility of Free Heat"; George M. Rideout, "Is Free Power Possible?"

[195] For details on the drownings and the history of the Foundation, see Babson, *Actions and Reactions,* 15–16, and "Gravity: Our Enemy," as well as Kaiser, "Roger

We all have our ways of coping with grief and guilt. One might think that a professional grief counselor, had such been available, would have done Babson more good than proselytizing against the evils of gravity. But to think thus would be to miss the importance of the model Babson offers us. For while he names gravity as the "dragon," the natural root of the evil that befell his sister, his doing this is not the end point of his moral reflections, nor does he then just yield to an implacable nomic gravitational necessity. He is not content just to blame gravity. His naming gravity as the cause is not for the purpose of absolving himself and others of responsibility. No, his naming the evil that is gravity is but the first step in taking responsibility. For Babson, it is not gravity that is, ultimately, the problem. The problem, ultimately, is our ignorance of how gravity works and our consequent inability to control it or, at least, to mitigate the sometimes unfortunate consequences of its operation. On Babson's view, it is, in the end, we who are to blame.

Sadly, we have recently been afforded other demonstrations of the lesson Babson would have us learn. Hurricane Katrina's devastation of New Orleans and the Gulf coast will be remembered as our Lisbon earthquake, even if scale, alone, should have earned the December 2004 Indian Ocean tsunami that title. In both instances, overwhelming natural forces wrought unprecedented destruction. It was nature's wrath, nature's fury. Humanity was humbled before the power of impersonal natural forces. Or was it? Only years from now will we be capable of the critical detachment needed for a sober analysis. Even the quick-witted Voltaire needed four years after the Lisbon earthquake to write *Candide* (1759). Still, some things are clear. Had there been an Indian Ocean tsunami warning system, as there was such a warning system for the Pacific basin, many lives would have been saved. Why was there no warning system for the Indian Ocean? No blaming nature there. In the case of Katrina, why were the New Orleans levees built only to withstand a category 3 hurricane, not a category 4 or 5? Why, in recent years, did the Bush administration and the congress repeatedly make drastic cuts in the funds requested by the Army Corps of Engineers for upgrading the New Orleans levees?[196] Forces of nature? No.

The big question about Katrina, however, the question we shall be debating for years, the question most relevant to the purposes of this chapter, is whether a global warming trend caused by human action is partly responsible for the storm's unprecedented size and savagery. That we are to expect global warming to produce an increase in the severity, if not also the frequency, of hurricanes is something that climatologists have been saying for some time.[197] Precisely how large a role is played by global warming alone is unknown. Katrina is, after all, part of a long recognized decadal

---

Babson and the Rediscovery of General Relativity." I thank David Kaiser for first bringing this fascinating story to my attention. Additional details on the drownings and the history of the Babson family are to be found in Finney-McDougal, *Babson Genealogy*. On the rebirth of work on general relativity, see Will, *Was Einstein Right?* 3–18; and Goldberg, "US Air Force Support of General Relativity."

[196] See Bunch, "Did New Orleans Catastrophe Have to Happen?"

[197] See, for example, Knutson and Tuleya, "Impact of CO_2-Induced Warming," and Trenberth, "Uncertainty in Hurricanes and Global Warming."

hurricane cycle. Much research has been done; much more is needed. For a scientific understanding of this aspect of global climate change is a crucial prerequisite for planned actions to counteract the trend, just as modeling the fluctuations in the Earth's ozone layer was crucial to the success that has just been reported in halting that frightening trend.[198] Action to prevent or alleviate suffering requires knowledge. Willful ignorance, or a collective failure to promote and fund the relevant research is a moral failing, not a defect in the natural order of things. A big part of the blame lies with those of us who did not ask and seek to answer the question scientifically.[199]

Writing in 1759, Voltaire mocked the simple-minded optimism of Leibniz's *Théodicée* (1710). Could the Lisbon earthquake, the flames of the Inquisition, pestilence, war, greed, and human savagery all be part of God's plan for the "best of all possible worlds"? One shudders to think what the poorer worlds must be like. But Candide, Pangloss, Martin, and Cunegonde are also portrayed as more or less helpless victims of a fate that doles out happiness and misery in roughly equal measure, in accord with no discernible plan. There is no mastering of the overwhelming forces that destroyed one of Europe's great cities with, in sequence, an earthquake, a tidal wave, and an all-consuming firestorm. Hope for no more than to be able to cultivate one small corner of the earthly garden. Perceptive students always note with puzzlement the irony. How is it that one of Newton's greatest French champions and one of the leading theorists of the Enlightenment could have been so shaken by the Lisbon earthquake as to be left with so little confidence in our ability to shape our fate by reason-guided human action? However much he mocks Leibniz's road to resignation, Voltaire walks us down a different road to the same end. Nature will have her way with us. Resistance is futile. Resistance, however, is not futile. For all that we are finite beings of limited capacity, we can know more, and we can do better. The Dutch are already sending us their experts to show us how to build a better sea wall. Our not trying to do better is the moral imperfection upon which we should focus.

2 What's Wrong with Blaming Nature? What Are We Blaming When We Blame Nature?

Is this the best of all possible worlds? I don't know. Being a finite being, I humbly admit that I do not know the bounds of the possible and so do not know even the meaning of "possibility" in this cosmic setting. My finitude debars me also from a knowledge of the only metric whereby the "better" and the "best" are to be judged. Even if, therefore, I assume an omnipotent, omniscient, and benevolent God, I am in no position to draw Leibniz's conclusion. Lacking the comfort of that conclusion, I still must think about the problem of evil. Let us start with so called "natural" evil.

---

[198] See "Study: Ozone Layer."

[199] Of course, the example is not a perfect one, for, in the case of global climate change, we already have more than enough science to serve as a basis for some actions, at least. We are now at a point where the call to do more science, especially the call to do more "sound science," is the favorite refuge of procrastination masking greed (see Mooney, "Beware 'Sound Science'").

To speak of "natural" evil is to speak of a genus of evil for which "nature" is mainly responsible. But what is this "nature" on which we thus pin the blame? For the theologian, it is the divinely created natural order. Many philosophers of science use expressions like "order of nature" too, usually meaning thus to highlight a nomic structure, a structure of laws, whether divinely created or not. Both the theologian and the philosopher speak from within a long tradition of regarding nature, whether harsh or benign, not as a realm of randomness but as an arena of events governed by rules. Within this tradition, to blame nature for evil is thus to blame mostly the laws of nature. Does that make sense?

Here is a place where the philosopher must ask many questions. The first is why we so readily model the order of nature as encoded in laws. One might think it more helpful to model the relevant notion of the natural order as an ontic structure, not a nomic structure. Think of the force of gravity, not the law of gravity. A structure of forces is a problematic alternative to a structure of laws, however, because the notion of force is an even more recent invention than that of laws, arguably just an artifact of the way we happened to do physics for about three hundred years. Newton's physics was a physics of forces. So was the physics of Michael Faraday and James Clerk Maxwell, but already in the 1890s Heinrich Hertz made a famous attempt to rid mechanics of forces altogether. And while we still speak today of seeking a grand unification of the four fundamental forces—gravity, electromagnetism, and the strong and weak nuclear forces—the basic notion of "force" in this setting is a highly abstract one, little more than a quirk of the mathematics that does the explanatory heavy lifting.

The theist likes to model the order of nature as a structure of laws in part because laws are plausible objects of knowledge in a way that forces or other ontologies are not. I can feel the force of gravity pulling me down (at least when something like the surface of the Earth impedes the resulting motion), but I cannot know the force of gravity in the way in which I can know Newton's law of universal gravitation. So, if one is enjoined to know God through his works, better to think of his created natural order as being encoded in laws. Of course some like to think of nature in the guise of laws of nature because that comports well with the image of the legislator God, but I shall leave as a homework exercise for you to figure out why this is not a good argument.

Still, the view of nature as an order of laws prevails. Let us assume that and ask a second question, this one with theological consequences in other settings. Ask whether, in all domains of nature, there exist exceptionless laws. Noteworthy is the fact that this question grows more acute as one goes from fundamental sciences, like physics and chemistry, to sciences dealing with more complex phenomena at higher levels of structure or organization, like biology. That there should be exceptionless laws in all domains is a view toward which intuition schooled by tradition strongly inclines many of us. Yet even the careful physics student learns to say, "*other things being equal,* $F = mc^2$." The problem is that other things are never quite equal. There are no frictionless surfaces, no perfect vacuua, no perfectly adiabatic processes, no completely isolated systems. The physics student learns to regard this circumstance as but a distraction. With more complex systems, however, one has to add so many *caeteris paribus* clauses that the operation of law alone might be difficult to discern. Those who want to find an underlying exceptionless nomic order will explain the

seeming exceptions to laws as but consequences of the operations of other laws, as when van der Waals forces are invoked in explaining the friction that hides inertial motion. Those drawn to a more anarchic metaphysics will see in this strategy more evasion than explanation, especially, again, in the biological arena. I don't know who is right, and, frankly, I don't have a pony in this race.

Let us make our task easier by assuming, for the sake argument, that exceptionless law lurks everywhere behind nature's heterogeneity. The next question the philosopher might ask is whether laws reflect an ontic order–as with the Scholastic's real universals, the contemporary anti-Humean's causal powers, or the modal logician's nomic necessitation–or are, instead, just a shorthand for constant conjunctions of events, as the Humean would have it, or mere perfect functional correlations, as Ernst Mach said. Here, too, theological questions are implicated. God's creating a nomic order in the form of real universals has long made sense to many, whereas many have seen a threat to theism in the Humean view of laws. Some would also say that it is harder to put the blame for natural evil on mere correlations, however perfect they be, than on, say, causal powers. But however much causal powers might seem like potentially scary things, the last point is lost on me, the point that it is easier to blame laws realisti-cally construed than laws regarded as constant conjunctions. The stone that crushes my big toe hurts just as much whether such stones merely *happen* always to fall when dislodged from an embankment or *must* do so.

We assume, then, that there are exceptionless laws and that, in thinking about natural evil, it makes little difference whether we view laws as did David Hume and Mach or as did Aquinas and Fred Dretske. What is, for our purposes, a more important question concerns the role of laws in producing natural phenomena. Here too a distinction or two helps. Physi-cists, at least, tend to think about events as explained, perhaps even pro-duced by laws, boundary conditions, and initial conditions acting in collu-sion with one another. How would Newton explain the motion of a meteoroid? First, he would need his three laws of motion plus his law of universal gravitation. Second, he would need to know how many massive objects are interacting with the meteoroid gravitationally and whether other forces do or do not perturb the system. Lastly, for all of the masses including the meteoroid, he needs to know for some initial time their pre-cise positions and momenta. Suppose, now, that our meteoroid is the one that struck the Yucatan peninsula at the end of the Cretaceous period causing the extinction of the dinosaurs. What do we blame? Is there any good reason for apportioning more of the blame to the laws than to the boundary conditions and the initial conditions? I see none. Yes, gravity played a role, but just as crucial to understanding the deaths of the dino-saurs is the accident, if you will, of exactly where and when the meteoroid was formed millions of years earlier.

Philosophers of science have their own reasons for arguing about the distinction between laws and boundary conditions. Of late, debates about inflationary cosmology have made this an important problem, because a powerful argument for the inflation model has been that it reduces cosmol-ogy's dependence on otherwise arbitrary seeming boundary conditions by such things as its demonstrating that cosmic flatness (zero global curva

ture) would be the expected post-inflation state of affairs regardless of how, within limits, we fine tune some otherwise crucial boundary conditions. The idea seems to be that a better theory is one that makes more of the structure in the domain of description nomic. But serious people have questioned the cogency of the laws–boundary conditions distinction in the cosmological setting (and others). One extreme view holds that it makes no sense to talk of laws in cosmology at all, since talk of laws assumes a generality hard to reconcile with the uniqueness of *the* universe. Another criticism of the distinction notes that what are boundary conditions from the point of view of one set of laws might well be straightforward consequences of the operation of other laws—or even those same laws—in other domains. Thus, it is conceivable that heretofore unknown laws of planetary genesis require there to be exactly nine planets orbiting a star like our Sun, just as we now know that there are laws determining the numbers and properties of the elements, whereas these were arbitrary externalities from the point of view of the mid-nineteenth-century chemist.

However interesting the philosophers' worry about the laws–boundary conditions distinction, those debating natural evil might find the question less than compelling. After all, are not both the laws and the boundary conditions part of a created natural order? I am no theologian, and I surely do not pretend to understand all that is written about divine action, but surely there is here an issue of theological importance. We might agree about God's role in laying down the laws of nature. But we might just as well disagree about his role in fixing boundary conditions and initial conditions. Need we here rehearse seventeenth-century debates about occasionalism? Did God once set the cosmic wheel spinning and then just step back for ever after to enjoy the show? Or does God find it necessary to steady the wheel and give it push from time to time? Or is God content to let it wobble, thus allowing in nature a contingency that is not a violation of the laws but a consequence of what are, from the point of view of the laws, externalities? Note that the very formulation of these alternatives, and, hence, theories of divine action generally, make, thus, a strong assumption about the cogency of the physicists' distinction between laws and boundary conditions. That is why it is probably not an historical accident that debates about occasionalism only came sharply into focus after the rise of the mechanical world view.

Consider now the third option mentioned above, the God who lets the cosmic wheel wobble. If God does not bear a full measure of responsibility for the externalities, and if those externalities are deeply implicated in suffering like the Cretaceous extinction, then do we blame the natural nomic order or do we blame just plain bad luck? Is that why some moral theologians add to natural and moral evil a third genus, existential evil? One skier is swept away by an avalanche. One skiing a few yards away escapes harm. Do we blame the first skier's demise on gravity, statics, and the physics of snow melt, or on the fact that a brief stiff breeze happened to push him ahead of his companion, putting him in harm's way? Both laws and boundary conditions play an explanatory and causal role in all such situations. But to the extent that there is something accidental in the setting of the boundary conditions, then blaming the boundary conditions makes no more sense than would a finding of criminal liability in a case of unforeseeable accidental injury.

What if the boundary conditions are not accidental? How can they fail to be? One way would be if God contrived them or reset them in line with a plan. Another would be if it were we humans who set or reset them. Such is, arguably, the case in many situations where suffering is the issue. What if the unlucky or unfortunate skier had been pushed ahead by his companion rather than a stiff breeze? What if the companion knew of the avalanche risk and failed to advise his friend to avoid the danger area? What if the ski patrol could have prevented this avalanche by triggering a smaller, controlled avalanche with a percussion shell? What if more research on avalanche risk could have identified more reliably the areas of greatest danger? With questions such as these, we enter a new arena.

3 What's Wrong with Blaming Nature? Perhaps It's Not Nature's Fault Alone

No one could have prevented the Lisbon earthquake. Further research on global climate change might prove otherwise in the case of hurricane Katrina. Our best current modeling shows a robust link between the concentration of atmospheric CO_2 and the intensity of Atlantic hurricanes. The exact mechanism is not understood. It is, after all, a very complex system. But the picture is growing clearer with each new study. What does this tell us about the problem of natural evil?

What it illustrates is that, with the progress of science, ever more of the blame for much of the suffering previously deemed a consequence of natural evil will have to be accorded to human action or the lack thereof. The growing human role in and responsibility for such suffering is to be seen in at least three different ways.

3.1 We Control the Boundary Conditions

Human action has changed the global environment since humankind first emerged. Slash and burn agriculture expanded deserts and altered rain patterns over vast areas, as did deforestation. Lebanon was once a forest. Sicily was once a wet island. Carthage was the breadbasket of Rome. As long as we did not know what we were doing to the environment, our actions were, from a moral point of view, but another part of nature, like in kind to buffalo overgrazing their feeding grounds. Now that we are coming to know more, the scales are gradually tipping. By the early 1960s, we knew that burning fossil fuels polluted the air in the Los Angeles basin. We realized that our failure to act on that knowledge made us responsible for the death and disease that ensued. We did something about it, even if we have still not done enough. Fewer people now get sick. Fewer people now die.

The laws governing the Earth's atmosphere are the same today as they were 500,000 years ago. The laws did not change. But by our human actions, we changed the boundary conditions. Photosynthesis still follows the same laws, but phytoplankton populations have crashed in some areas because of toxic chemical waste, while algae flourishes in places where it was formerly rare because of agricultural run-off. It is hard to blame nature when it is our actions that fix so many of the crucial inputs to the system.

3.2 We Know the Laws

Voltaire did not know the real causes of earthquakes. We do, at least in part. But with knowledge comes responsibility. Knowing why things happen and how things happen, we can sometimes prevent their happening, and sometimes, when we cannot, we can at least mitigate the effects. We cannot stop an earthquake anymore than Voltaire could, but we know enough now to inform the public about possible earthquakes, enabling those who will to evacuate and preparing others to seek safe shelter at the first hint of a tremor. When it comes to floods, some of those we can stop, thanks to what we know now about why they happen. We know that the devastation from the 1993 Mississippi River Valley floods would have been vastly less had we more aggressively restored the lost wetlands that once absorbed excess water.

3.3 New Technologies Make Intervention and Mitigation Possible

We cannot stop an earthquake, but now we know how to construct buildings, bridges, and highways that better withstand earthquakes. In the case of meteoroid impact, as even popular cinema makes clear, we now possess technology sufficient to prevent all but the most massive such bodies from striking the Earth. When you think about it, that is pretty amazing. We are almost certainly not yet well enough organized actually to deploy those technologies in a timely and effective fashion. But at least we now routinely scan the skies for evidence of such threats, and there is a system in place for issuing warnings.

4 What's Wrong with Blaming Nature? We've Got to Take the Good with the Bad

Ignorant of the causes of "natural" catastrophes, it is easy to convince oneself that some of nature's ways are mainly malign. What good can come of earthquakes and floods? Would anyone welcome a hurricane or a tornado? Whatever, then, the underlying forces or laws, do they not bring only misery? Should this be another homework exercise? Only a little knowledge or even just a little careful thought is enough to demonstrate that the same laws that give rise to disasters always produce also good. Gravity pulls a meteoroid down upon us. But Gravity also structures the solar system so as to make possible life on Earth. Tectonic shifts cause earthquakes and tsunamis, but they also build up the land masses that we call home.

Nothing in nature produces only ill (except, perhaps, the appendix). If not exactly morally neutral, natural laws operate in such a way as to yield good and ill in roughly equal measure. How then blame a nature of laws for suffering? If we blame nature for the bad, give credit for the good as well. Better still, stop playing this particular blame game altogether. In nature, things happen. That's it. If that nature is God's creation and God foresees all of the consequences of the operation of natural law, then I suppose that he intended both the suffering and the prospering. But what do we gain with that insight? Not much, really. The important question is not whether nature acting alone produces evil as well as good. Surely it does. The important question is: "What are we going to do about it?" The answer to that question is: "Do more physics."

5 What's Wrong with Blaming Nature? Blaming Nature Is an Evasion

As we survey the shattered Gulf Coast or the shattered coast of the Indian basin, as we count the shattered bodies and shattered dreams, as we see ourselves caught in the grip of forces so much larger than ourselves, it would be so easy to assign all of the blame to nature. We do it unthinkingly every time we utter the phrases "natural catastrophe" and "natural disaster." No anaesthetic for bodily pain, it is an anodyne for an injured conscience. It gets us off the hook. But however much blaming nature soothes the soul, it is really just an evasion of responsibility.

We are not alone responsible. Savage hurricanes lashed the Gulf Coast long before the human habitation of North America, and tidal waves have reshaped the coast of Sri Lanka for millions of years. But Katrina was probably made more intense by the consequences of our petroleum gluttony, and a bit of hydrology, a bit of estuarine ecology, and a bit of engineering would have spared many lives. Of the evil done in Alabama, Mississippi, and Louisiana, more of it than we might care to admit is the result of our acts and omissions. This is moral evil.

6 Hubris and Humility

No. We cannot know everything. No. We cannot prevent all or even most natural disasters. Yes. Intellectual arrogance and intellectual ambition have often produced more suffering than good. In our age look no farther than to nuclear weapons for proof of the consequences of unchecked intellectual arrogance. That is a case in which our doing more physics brought upon the world more suffering than Einstein could have imagined when first he wrote $E = mc^2$, and we have surely not yet taken the full measure of the suffering that will ensue. I would never argue that all knowledge is inherently good. One cannot foresee all of the consequences of new discoveries. Perhaps fusion will turn out to be the ultimate solution to the problem of filling our energy needs. Still, prudence dictates that some lines of research are better left unexplored. Would that we had never built the H-bomb. We must be humble. And yet we cannot and should not stop inquiring. For inquiry is the only way to grow the understanding of nature that gives us our only hope for fixing our own mistakes and for lessening the harm that might be done by a nature unopposed by human will.

7 Conclusion

In a different volume I would want to press other questions. For example, I think that I don't buy the basic moral evil/natural evil distinction because I dissent from assumptions about human "nature" that typically underlie the distinction. Without a robust soul-body distinction, the moral evil/natural evil distinction is hard to float. One does not have to be a crude materialist reductionist to think that we have not yet worked through all of the problems dragged in with such dualisms. But, again, that is a topic for another time.

The main point that I wish to emphasize in this volume is that knowledge brings with it responsibility, and that we should not shirk that responsibility but embrace it, along with the obligation to learn still more. Of course no one really disagrees with this point. Well, perhaps corporate apologists for the hydrocarbon fuels economy disagree. They seem to prefer ignorance to knowledge as once did the corporate apologists for smoking. But no reasonable person disagrees. We will not stop the funding for research on global climate change. We will not stop the quest to unite a theory of gravitation to a theory of the other three fundamental forces. I doubt that we'll find Babson's hoped-for gravity insulator. But we might find another way to the stars.

Do you want a better world? Do physics.[200]

---

[200] My sincere thanks to my colleague, Jerry McKenny, in the Department of Theology at Notre Dame, for suggesting the title for this paper and for much other very helpful advice about the issues addressed. I wish to thank also David Burrell, Niels Christian Hvidt, Ernan McMullan, David Oldroyd, and Peter van Inwagen for help of various kinds, including conversations through which I came to realize that I had, in fact, been thinking about the topic of this paper for a long time.

LIST OF RESOURCES

Abernethy, Alexis D., H. Theresa Chang, Larry Seidlitz, James S. Evinger, and Paul R. Duberstein. "Religious Coping and Depression among Spouses of People with Lung Cancer." *Psychosomatics* 43, no. 6 (2002): 456–63.

Adams, Marilyn McCord. "Horrendous Evils and the Goodness of God." In *The Problem of Evil,* edited by Marilyn McCord Adams and Robert Merrihew Adams, 209–21. Oxford: Oxford University Press, 1990.

Adams, Marilyn McCord, and Robert Adams, eds. *The Problem of Evil.* Oxford: Oxford University Press, 1990.

Albert, David Z. *Quantum Mechanics and Experience.* Cambridge: Harvard University Press, 1992.

Allen, Diogenes. "Natural Evil and the Love of God." In *The Problem of Evil,* edited by Marilyn McCord Adams and Robert Merrihew Adams, 189–207. Oxford: Oxford University Press. 1990.

———. *The Traces of God in a Frequently Hostile World.* Cambridge, MA: Cowley, 1981.

Alston, William P. "Religion, History of Philosophy of." In *Routledge Encyclopedia of Philosophy,* edited by E. Craig, 8:238–48. London; New York: Routledge, 1998. Available at http://www.rep.routledge.com/article/K067SECT7 (accessed December 13, 2005).

Ano, Gene G., and Erin B. Vasconcelles. "Religious Coping and Psychological Adjustment to Stress: A Meta-Analysis." *Journal of Clinical Psychology* 61, no. 4 (2005): 461–80.

Aquinas, Thomas. *Summa theologica.* Translated by the Fathers of the English Dominican Province. London: Burns, Oates, and Washbourne, 1912–1936; New York, Benziger Brothers, 1947–1948; New York: Christian Classics, 1981. Available online from InteLex Corporation, http://www.nlx.com.

———. *On Evil.* Translated by Jean T. Oesterle. Notre Dame, IN: University of Notre Dame Press, 1995.

Aspect, Alain, Jean Dalibard, and Gérard Roger. "Experimental Test of Bell's Inequalities Using Time-Varying Analyzers." *Physical Review Letters* 49, no. 25 (1982): 1804–7.

Augustine [of Hippo]. *Aurelius Augustinus: Enchiridion de fide spe et caritate; Handbüchlein über Glaube Hoffnung und Liebe.* Testimonia: Schriften de altchristlichen Zeit, Band 1. Translated by Joseph Barbel. Dusseldorf: Patmos-Verlag, 1960.

———. *The City of God against the Pagans.* Translated by G. E. McCracken. Loeb Classical Library. Cambridge: Harvard University Press, 2001.

———. *Concerning the City of God against the Pagans.* Translated by Henry Bettenson. New York: Penguin Books, 1984.

———. *The Confessions.* Translated by Philip Burton. New York: Knopf, 2001.

———. *The Enchiridion on Faith, Hope and Love.* Edited by Thomas S. Hibbs. Washington, DC; Lanham, MD: Regnery, 1996.

———. *On Free Choice of the Will.* Translated by Thomas Williams. Indianapolis: Hackett, 1993.

Babson, Roger W. *Actions and Reactions: An Autobiography of Roger W. Babson.* [1935]. 2d rev. ed. New York: Harper & Brothers, 1950.

———. "Gravity: Our Enemy Number One." New Boston, NH: Gravity Research Foundation, no date (ca. 1948). Original in the files of the Gravity Research Foundation, Box 2, Folder 3.

Barrow, John D. *The Constants of Nature: From Alpha to Omega—the Numbers That Encode the Deepest Secrets of the Universe.* New York: Vintage Books/Random House, 2002.

———. *Impossibility: The Limits of Science and the Science of Limits.* Oxford: Oxford University Press, 1998.

Barrow, John D., and Frank J. Tipler. *The Anthropic Cosmological Principle.* Oxford, New York: Oxford University Press, 1986.

Barrow, John D., and John K. Webb, "Inconstant Constants: Do the Inner Workings of Nature Change with Time?" *Scientific American,* May 23, 2005.

Barth, Karl. *Church Dogmatics.* Edited by G. W. Bromiley and T. F. Torrance. Vol. 1/1: *The Doctrine of the Word of God.* Translated by G. T. Thomson and Harold Knight. 2d ed. Vol. 3/1: *The Doctrine of Creation.* Translated by Harold Knight, G. W. Bromiley, J. K. Reid, and R. H. Fuller. Edinburgh: T. & T. Clark, 1975, 1958.

———. "The Strange New World within the Bible." In *The Word of God and the Word of Man,* edited by Douglas Horton, 28–50. New York: Harper & Row, 1957.

Beaudoin, John. "Skepticism and the Skeptical Theist." *Faith and Philosophy* 22 (2005): 42–56.

Beker, Johan Christiaan. *Suffering and Hope: The Biblical Vision and the Human Predicament.* Philadelphia: Fortress, 1987.

Bell, John S. "Introduction to the Hidden-Variable Question." In *Speakable and Unspeakable in Quantum Mechanics: Collected Papers in Quantum Mechanics,* 29–39. Cambridge: Cambridge University Press, 1987.

———. "On the Einstein Podolsky Rosen Paradox." In *Quantum Theory and Measurement,* edited by J. A. Wheeler and W. H. Zurek, 403–8. Princeton, NJ: Princeton University Press, 1983.

Berkhof, Hendrikus. *Christ and the Powers.* Translated by John H. Yoder. Scottdale, PA: Herald Press, 1977.

Berkovitz, Joseph. "The Nature of Causality in Quantum Phenomena." *Theoria* 15, no. 37 (2000): 87–122.

Bernhardt, Reinhold. *Was heisst "Handeln Gottes"? Eine Rekonstruktion der Lehre von der Vorsehung.* Gütersloh: Kaiser Gütersloher, 1999.

Berry, Michael. "Chaos and the Semiclassical Limit of Quantum Mechanics (Is the Moon There When Somebody Looks?)." In *QM.*

Berthold, Fred. *God, Evil, and Human Learning: A Critique and Revision of the Free Will Defense in Theodicy.* Albany: State University of New York Press, 2004.

Boesch, Christophe, and Michael Tomasello, "Chimpanzee and Human Cultures." *Current Anthropology* 39, no. 5 (December 1998): 591–; available on CogWeb's Evolutionary Psychology webpage, http://cogweb.ucla.edu/Abstracts/Boesch_Tomasello_98.html, dated December 1998 (accessed August 5, 2005).

Boethius. *The Consolation of Philosophy.* Translated by P. G. Walsh. Oxford; New York: Clarendon, 1999.

Boff, Leonardo. *Trinity and Society.* Translated by Paul Burns. Maryknoll, NY: Orbis, 1988.

Bohm, David. *Quantum Theory.* New York: Dover, 1989.

———. "A Suggested Interpretation of the Quantum Theory in Terms of 'Hidden Variables' I." In *Quantum Theory and Measurement,* edited by John Archibald Wheeler and Wojciech Hubert Zurek, 369–82. Princeton, NJ: Princeton University Press, 1983.

"Bohr-Einstein Dialogue." In *Quantum Theory and Measurement,* edited by John Archibald Wheeler and Wojciech Hubert Zurek. Princeton, NJ: Princeton University Press, 1983.

Bosworth, H. B., K. S. Park, D. R. McQuoid, J. C. Hays, and D. C. Steffens. "The Impact of Religious Practice and Religious Coping on Geriatric Depression." *International Journal of Geriatric Psychiatry* 18, no. 10 (2003): 905–14.

Bowker, John Westerdale. *Problems of Suffering in Religions of the World.* Cambridge: Cambridge University Press, 1970.

Branigan, Tania, and Brian Whitaker. "Dangerous Buildings, Lax Rules: Why Bam Death Toll Was So High." *Guardian Unlimited,* December 27, 2003, http:// www.guardian.co.uk/iran/story/0,12858,1112938,00.html (accessed November 15, 2005).

Broad, William J. "Deadly and Yet Necessary, Earthquakes Renew the Planet." *New York Times,* January 11, 2005, Science Section, 1, 4.

Brown, Julian. *The Quest for the Quantum Computer.* New York: Simon & Schuster, 2000.

Bunch, Will. "Did New Orleans Catastrophe Have to Happen? 'Times-Picayune' Had Repeatedly Raised Federal Spending Issues." *Editor & Publisher,* August 31, 2005, http://www.editorandpublisher.com.

Bur, Michel. "The Kingdom of the Franks from Louis VI to Philip II: The Signeuries." In *The New Cambridge Medieval History,* vol. IV C. 1024-C. 1198, part II, edited by David Luscombe and Jonathan Riley-Smith, 530–48. Cambridge: Cambridge University Press, 1995.

Bussing, A., T. Ostermann, and P. F. Matthiessen. "Role of Religion and Spirituality in Medical Patients: Confirmatory Results with the SpREUK Questionnaire." *Health and Quality of Life Outcomes* 3, no. 1 (2005): 10.

Butter, Eric M. "Development of a Model for Clinical Assessment of Religious Coping: Initial Validation of the Process Evaluation Model." *Mental Health, Religion & Culture* 6, no. 2 (2003): 175.

Cabello, Adán. "Ladder Proof of Nonlocality without Inequalities and without Probabilities." *Physical Review A* 58, no. 3 (1998): 1687–93.

Callicott, J. Baird. *Companion to a Sand County Almanac.* Madison: University of Wisconsin Press, 1987.

Cantor, Norman F. *The Civilization of the Middle Ages: A Completely Revised and Expanded Edition of Medieval History, the Life and Death of a Civilization.* San Francisco: HarperCollins, 1993.

Carter, Brandon. "Large Number Coincidences and the Anthropic Principle in Cosmology." In *Confrontation of Cosmological Theories with Observational Data,* edited by M. S. Longair. Dordrecht: Reidel, 1974.

Caygill, Howard. *A Kant Dictionary.* Blackwell Philosopher Dictionaries. Oxford, UK; Cambridge, MA: Blackwell Reference, 1995.

Chadwick, Douglas H. "Investigating a Killer." *National Geographic,* April 2005, pp. 86–105.

Chang, Hasok, and Nancy Cartwright. "Causality and Realism in the EPR Experiment." *Erkenntnis* 38, no. 2 (1993): 169–90.

Chappell, T. D. J. *Aristotle and Augustine on Freedom.* New York: St. Martin's Press, 1995.

Chidester, David. *Christianity: A Global History.* London; New York: Allen Lane, 2000.

Clayton, Philip, and Paul Davies, eds. *The Re-emergence of Emergence: The Emergentist Hypothesis from Science to Religion.* Oxford: Oxford University Press, 2006.

"Commentary of Heisenberg (1967)." In *Quantum Theory and Measurement,* edited by John Archibald Wheeler and Wojciech Hubert Zurek. Princeton, NJ: Princeton University Press, 1983.

Corry, L., J. Renn, and J. Stachel. "Belated Decision in the Hilbert-Einstein Priority Dispute." *Science,* no. 14 (November 1997).

Cottingham, John. *Western Philosophy: An Anthology.* Blackwell Philosophy Anthologies 1. Oxford, UK; Cambridge, MA: Blackwell, 1996.

Cramer, John G. "The Transactional Interpretation of Quantum Mechanics." *Reviews of Modern Physics* 58 (1986): 647–87.

Cunningham, David S. *These Three Are One: The Practice of Trinitarian Theology.* Oxford: Blackwell, 1998.

Cushing, James T. "A Background Essay." In *Philosophical Consequences of Quantum Theory: Reflections on Bell's Theorem,* edited by James T. Cushing and Ernan McMullin, 1–24. Notre Dame, IN: University of Notre Dame Press, 1989.

Cushing, James T., and Ernan McMullin, eds. *Philosophical Consequences of Quantum Theory: Reflections on Bell's Theorem.* Studies in Science and the Humanities from the Reilly Center for Science, Technology, and Values 2. Notre Dame, IN: University of Notre Dame Press, 1989.

Datson, Lorraine. "Marvelous Facts and Miraculous Evidence in Early Modern Europe." In *Questions of Evidence: Proof, Practice and Persuasion across the Disciplines,* edited by James Chandler, Arnold I. Davidson, and Harry Harootunian, 243–89. Chicago: University of Chicago Press, 1994.

Davidson, Donald. "On the Very Idea of a Conceptual Scheme." In *Inquiries into Truth and Interpretation,* 183–98. Oxford: Clarendon, 1984.

de Waal, Frans. *Good Natured: The Origins of Right and Wrong in Humans and Other Animals.* Cambridge: Harvard University Press, 1996.

Deacon, Terrence. *The Symbolic Species: The Co-evolution of Language and the Brain.* New York: Norton, 1997.

———. "Three Levels of Emergent Phenomena." In *Evolution and Emergence: Systems, Organisms, Persons,* edited by Nancey Murphy and William R. Stoeger, SJ, 88–110. Oxford: Oxford University Press, 2007.

Dickson, William Michael. *Quantum Chance and Non-Locality: Probability and Non-Locality in the Interpretations of Quantum Mechanics.* Cambridge: Cambridge University Press, 1998.

Dillistone, F. W. *The Christian Understanding of Atonement.* London: SCM Press, 1984.

Dirac, Paul A. M. *The Principles of Quantum Mechanics.* 4th ed. Oxford: Clarendon, 1958.

Donald Davidson. "The Irreducibility of Psychological and Physiological

Description, and of Social to Physical Sciences." In *The Study of Human Nature,* edited by Leslie Stevenson. Oxford: Oxford University Press, 1981.

———. "Mental Events." In *Essays on Actions and Events,* 207–24. Oxford: Clarendon, 1980.

Dostoevski, Fyodor. *The Brothers Karamazov.* Translated by David Magarshack. Baltimore: Penguin Books, 1958.

Dransart, Philippe. *La maladie cherche à me guérir.* Vol. 2. Grenoble: Le Mercure Dauphinois, 1999.

Dyson, Freeman. "Time without End: Physics and Biology in an Open Universe." *Review of Modern Physics* 51 (1979): 447–460.

Echeverria, Eduardo J. "The Gospel of Redemptive Suffering: Reflections on John Paul II's *Salvifici Doloris.*" In *Christian Faith and the Problem of Evil,* edited by Peter Van Inwagen, 111–47. Grand Rapids, MI and Cambridge, UK: Eerdmans, 2004.

Edwards, Denis. "Every Sparrow that Falls to the Ground: The Cost of Evolution and the Christ-Event." *Ecotheology* 11, no. 1 (March 2006): 103–23.

Einstein, Albert. "Zur Elektrodynamik bewegter Körper." *Annalen der Physik* 17 (1905): 891–921.

Einstein, Albert, Max Born, and Hedwig Born. *The Born-Einstein Letters: Correspondence between Albert Einstein and Max and Hedwig Born from 1916 to 1955 with Commentaries by Max Born.* Translated by Irene Born. New York: Walker and Co., 1971.

Einstein, Albert, Boris Podolsky, and Nathan Rosen. "Can Quantum-Mechanical Description of Physical Reality Be Considered Complete?" *Physical Review* 47 (1935): 777–80.

Ekstrom, Laura Waddell. "Suffering as Religious Experience." In *Christian Faith and the Problem of Evil,* edited by Peter van Inwagen, 95–110. Grand Rapids, MI and Cambridge, UK: Eerdmans, 2004.

Eliot, George. *Middlemarch.* [1871–1872.] London: Penguin Classics, 1994.

Eliot, T. S. *The Complete Poems and Plays of T. S. Eliot.* [1969.] London: Faber & Faber, 1978.

Ellis, George F. R. "On the Nature of Emergent Reality." In *Evolution and Emergence: Systems, Organisms, Persons,* edited by Nancey Murphy and William R. Stoeger, SJ, 113–40. Oxford: Oxford University Press, 2007.

———. "The Theology of the Anthropic Principle." In *QCLN.*

Ellis, George F. R., U. Kirchner, and W. R. Stoeger. "Multiverses and Physical Cosmology." *Mon. Not. R. astr. Soc.* 347 (2003): 921–36 [http://arXiv.org/astro-ph/0305292v3].

Evans, Gillian R. *Augustine on Evil.* Cambridge; New York: Cambridge University Press, 1982.

Fabricatore, Anthony N. "Stress, Religion, and Mental Health: The Role of Religious Coping." *Dissertation Abstracts International, Section B: The Sciences and Engineering* 63, no. 4-B (2002): 2053.

Fabricatore, Anthony N., Paul J. Handal, Doris M. Rubio, and Frank H. Gilner. "Stress, Religion, and Mental Health: Religious Coping in Mediating and Moderating Roles." *The International Journal for the Psychology of Religion* 14, no. 2 (2004): 97–98.

Farrer, Austin, *Love Almighty and Ills Unlimited.* Garden City, NY:

Doubleday, 1961; London: Collins, 1962.

Feinberg, John S. *The Many Faces of Evil: Theological Systems and the Problem of Evil.* Rev. ed. Grand Rapids, MI: Zondervan, 1994.

Ferguson, Eugene S. *Engineering and the Mind's Eye.* Cambridge, MA and London, UK: MIT Press, 1993.

Fiddes, Paul S. *The Creative Suffering of God.* Oxford: Clarendon, 1988.

Finney-McDougal, Catherine. *The Babson Genealogy, 1637–1977.* Watertown, MA: Eaton Press, 1978.

Fish, Stanley. *Is There a Text in This Class?* Cambridge: Harvard University Press, 1980.

Fitchett, G. "Religious Struggle: Prevalence, Correlates and Mental Health Risks in Diabetic, Congestive Heart Failure, and Oncology Patients." *The International Journal of Psychiatry in Medicine* 34, no. 2 (2004): 179.

Fitzgerald, Allan, and John C. Cavadini. *Augustine through the Ages: An Encyclopedia.* Grand Rapids, MI: Eerdmans, 1999.

Frankl, Viktor Emil. *Man's Search for Meaning.* Rev. and updated ed. New York: Washington Square Press, Pocket Books, 1985.

Frei, Hans. *The Eclipse of Biblical Narrative: A Study in Eighteenth- and Nineteenth-Century Hermeneutics.* New Haven, CT: Yale University Press, 1974.

Fuller, Reginald Horace. *Preaching the Lectionary: The Word of God for the Church Today.* Rev. ed. Collegeville, MN: Liturgical Press, 1984.

Gall, Terry Lynn, and Mark W. Cornblat. "Breast Cancer Survivors Give Voice: A Qualitative Analysis of Spiritual Factors in Long-Term Adjustment." *Psycho-Oncology* 11, no. 6 (2002): 524.

Gall, Terry Lynn, and Karen Grant. "Spiritual Disposition and Understanding Illness." *Pastoral Psychology* 53, no. 6 (2005): 515.

Garber, Daniel. "Leibniz, Gottfried Wilhelm." In *Routledge Encyclopedia of Philosophy,* edited by E. Craig, 5:541–62. London; New York: Routledge, 1998. Available at http://www.rep.routledge.com/article/DA052SECT3 (accessed December 13, 2005).

Gerlitz, Peter. "Theodizee I – Religionsgeschichtlich." In *Theologische Realenzyklopädie,* edited by Gerhard Möller and Gerhard Krause, 22:210–15. Berlin: W. de Gruyter, 1997.

Geyer, Carl-Friedrich. "Theodizee VI – Philosophisch." In *Theologische Realenzyklopädie,* edited by Gerhard Möller and Gerhard Krause, 22:231–37. Berlin: W. de Gruyter, 1997.

———. "Das Theodizeeproblem, ein historischer und systematischer Überblick." In *Theodizee, Gott vor Gericht?* edited by Willi Oelmüller and Carl-Friedrich Geyer, 9–32. Munich: Wilhelm Fink, 1990.

Goldberg, Joshua. "US Air Force Support of General Relativity: 1956–1972." In *Studies in the History of General Relativity,* edited by Jean Eisenstaedt and A. J. Kox, 89–102. Boston: Birkhäuser, 1992.

Greene, William Chase. "Fate, Good, and Evil, in Early Greek Poetry." *Harvard Studies in Classical Philology* 46 (1935): 1–36.

Greenstein, George, and Arthur G. Zajonc. *The Quantum Challenge: Modern Research on the Foundations of Quantum Mechanics.* Boston: Jones & Bartlett, 1997.

Griffin, David Ray. "Creation Out of Nothing, Creation Out of Chaos, and the Problem of Evil." In *Encountering Evil: Live Options in Theodicy,* new ed., edited by Stephen T. Davis, 108–44. Louisville, KY: Westminster John Knox, 2001.

———. *God, Power, and Evil: A Process Theodicy.* Philadelphia: Westminster Press, 1976; 2d ed., Lanham, MD: University Press of America, 1991.

Griffin, David Ray, ed. *Sacred Interconnections: Postmodern Spirituality, Political Economy, and Art.* Albany, NY: SUNY Press, 1990.

Gunton, Colin E. *The Triune Creator: A Historical and Systematic Study.* Grand Rapids, MI: Eerdmans, 1998.

Gwynne, Paul. *Special Divine Action: Key Issues in the Contemporary Debate, 1965–1995.* Tesi Gregoriana, Serie Teologia 12. Roma: Pontificia Universitá Gregoriana, 1996.

Haga, Tsutomu. *Theodizee und Geschichtstheologie: Ein Versuch der Überwindung der Problematik des deutschen Idealismus bei Karl Barth.* Forschungen zur systematischen und Ökumenischen Theologie 59. Göttingen: Vandenhocek & Ruprecht, 1991.

Hasker, William. *The Emergent Self.* Ithaca, NY: Cornell University Press, 1999.

Haught, John F. "Darwin and the Cardinal." *Commonweal* 132, no. 14 (August 12, 2005): 39.

Hay, David. "Colossians, Letter to the." In *Eerdmans Dictionary of the Bible,* edited by David Noel Freedman, Allen C. Myers, and Astrid B. Beck, 270–71. Grand Rapids, MI: Eerdmans, 2000.

Healey, Richard A. "Holism and Nonseparability in Physics." *Stanford University's Encyclopedia of Philosophy,* 1999, http://plato.stanford.edu/entries/physics-holism (accessed 12 October, 2002).

Hebblethwaite, Brian. "God and the World as Known to Science." In *The Human Person in God's World: Studies to Commemorate the Austin Farrer Centenary,* edited by Brian Hebblethwaite and Douglas Hedley. London: SCM Press, 2006.

Hegel, Georg Wilhelm Friedrich. *The Philosophy of History.* Translated by John Sibree. Rev. ed. New York: Willey, 1944.

———. *Vorlesungen über die Geschichte der Philosophie.* Edited by Pierre Garniron and Walter Jaeschke. Vol. 3. Hamburg: Felix Meiner, 1986.

Heidegger, Martin. "The Question Concerning Technology." In *The Question Concerning Technology and Other Essays,* 3–36. New York: Harper & Row, 1977.

Hick, John. *The Center of Christianity.* New and enlarged ed. San Francisco: Harper & Row, 1978.

———. *Dialogues in the Philosophy of Religion.* Houndmills, Basingstoke, Hampshire; New York: Palgrave, 2001.

———. *Evil and the God of Love.* London: Macmillan, 1966; rev. ed., San Francisco: Harper & Row, 1977.

———. "An Irenaean Theodicy." In *Encountering Evil: Live Options in Theodicy,* edited by John B. Cobb and Stephen T. Davis, 39–52. Atlanta: John Knox Press, 1981. Pages 38–72 in new ed., edited by Stephen T. Davis. Louisville, KY: Westminster John Knox, 2001.

———. *The Myth of God Incarnate.* Philadelphia: Westminster Press, 1977.

———. "Soul-Making and Suffering." In *The Problem of Evil,* edited by Marilyn McCord Adams and Robert Merrihew Adams, 168–88. Oxford:

Oxford University Press, 1990. Reprinted from his 1978 edition of *Evil and the God of Love,* first published 1966.

Hoekema, Anthony A. *Saved by Grace.* Grand Rapids, MI: Eerdmans/ Exeter, UK: Paternoster, 1989.

Hugh of St. Victor. *Didascalicon.* Translated with an introduction and notes by J. Taylor. New York and London: Columbia University Press, 1961.

Hume, David. *Dialogues Concerning Natural Religion.* Edited with an introduction by Norman Kemp Smith. Indianapolis: Bobbs-Merrill, n.d.

Hütter, Reinhard. "Bound to Be Free: Liberated for What?" *Christian Century,* August 20, 2004, http://www.findarticles.com/p/articles/ mi_m1058/is_16_121/ai_n8702385 (accessed March 30, 2005).

Inbody, Tyron. *The Faith of the Christian Church: An Introduction to Theology.* Grand Rapids, MI: Eerdmans, 2005.

James, William. *Varieties of Religious Experience: A Study in Human Nature.* London; New York: Routledge, 2002.

Janssen, Hans-Gerd. *Gott, Freiheit, Leid: Das Theodizeeproblem in der Philosophie der Neuzeit.* Darmstadt: Wissenschaftliche Buchgesellschaft, 1989.

Jarrett, Jon P. "Bell's Theorem: A Guide to the Implications." In *Philosophical Consequences of Quantum Theory: Reflections on Bell's Theorem,* edited by James T. Cushing and Ernan McMullin, 60–79. Notre Dame, IN: University of Notre Dame Press, 1989.

―――. "On the Physical Significance of the Locality Conditions in the Bell Arguments." *Noûs* 18 (1984): 569–89.

Jauss, Hans Robert. *Toward an Aesthetic of Reception.* Translated by Timothy Bahti. Introduction by Paul de Man. Minneapolis: University of Minnesota Press, 1982.

Johnson, Elizabeth A. "Does God Play Dice? Divine Providence and Chance." *Theological Studies* 57, no. 1 (March 1996): 3–18.

―――. *She Who Is: The Mystery of God in Feminist Theological Discourse.* New York: Crossroad, 1995.

Journet, Charles. *The Meaning of Evil.* Translated by Michael Barry. New York: P. J. Kenedy & Sons, 1963.

Jungerman, John A. *World in Process: Creativity and Interconnection in the New Physics.* Albany, NY: SUNY Press, 2000.

Kahn, Sholom J. "The Problem of Evil in Literature." *The Journal of Aesthetics and Art Criticism* 12, no. 1 (1953): 98.

Kaiser, David. "Roger Babson and the Rediscovery of General Relativity." Chapter 10 of "Making Theory: Producing Physics and Physicists in Postwar America." PhD dissertation, Harvard University, 2000.

Kallenberg, Brad J. *Ethics as Grammar: Changing the Postmodern Subject.* Notre Dame, IN: University of Notre Dame Press, 2001.

―――. "Some Things Are Worth Dying For." *New Blackfriars* 87, no. 1007 (2006): 50–71.

Kane, R. "Principles of Reason." *Erkenntnis* 24, no. 2 (1986): 115–36.

Kant, Immanuel. *Kant on History and Religion, with a Translation of Kant's "On the Failure of All Attempted Philosophical Theodicies."* Translated by Michel Despland. Montreal: McGill-Queen's University Press, 1973.

———. *Religion within the Boundaries of Mere Reason, and Other Writings.* Translated by Allen W. Wood and George Di Giovanni. Cambridge Texts in the History of Philosophy. Cambridge; New York: Cambridge University Press, 1998.

Karavites, Peter. *Evil—Freedom—and the Road to Perfection in Clement of Alexandria.* Supplements to Vigiliae Christianae 43. Leiden; Boston: Brill, 1999.

Kasper, Walter. *The God of Jesus Christ.* London: SCM Press, 1983.

Kauffman, Stuart A. *Investigations.* Oxford: Oxford University Press, 2000.

Kearns, Cleo McNelly. "Suffering in Theory." In *Suffering Religion,* edited by Elliot R. Wolfson and Robert Gibbs, 56–72. London; New York: Routledge, 2002.

Kelly, Joseph F. *The Problem of Evil in the Western Tradition: From the Book of Job to Modern Genetics.* Collegeville, MN: Liturgical Press, 2002.

Kenny, Anthony. "Practical Reasoning and Rational Appetite." In *Will, Freedom and Power,* 70–96. New York: Barnes and Noble, 1976.

Kidder, Tracy. *Mountains beyond Mountains: The Quest of Dr. Paul Farmer, a Man Who Would Cure the World.* New York: Random House, 2004.

Kierkegaard, Søren. *Fear and Trembling.* Translated by Alastair Hannay. Harmondsworth, Middlesex, England: Penguin, 1987.

Kim, Jaegwon. "Making Sense of Emergence." *Philosophical Studies* 95 (1999): 3–36.

King, William. *An Essay on the Origin of Evil.* Translated by Edmund Law. London: W. Thurlbourn, 1731.

Kleinknecht, Karl Theodor. *Der leidende Gerechtfertigte: Die alttestamentlich-jüdische Tradition vom "leidenden Gerechten" und ihre Rezeption bei Paulus.* Tübingen: Mohr, 1984.

Kropf, Richard W. *Evil and Evolution: A Theodicy.* 2d ed. Eugene: OR: Wipf and Stock, 2004.

Knutson, Thomas R., and Tuleya, Robert E. "Impact of CO_2-Induced Warming on Simulated Hurricane Intensity and Precipitation: Sensitivity to the Choice of Climate Model and Convective Parameterization." *Journal of Climate* 17 (2004): 3477–95.

Koenig, H. G., H. J. Cohen, D. G. Blazer, C. Pieper, K. G. Meador, F. Shelp, V. Goli, and B. DiPasquale. "Religious Coping and Depression among Elderly, Hospitalized Medically Ill Men." *American Journal of Psychiatry* 149, no. 12 (1992): 1693–1700.

Koenig, H. G., D. B. Larson, and S. S. Larson. "Religion and Coping with Serious Medical Illness." *The Annals of Pharmacotherapy* 35, no. 3 (2001): 352–59.

Köhlmoos, Melanie. "Theodizee II – Altes Testament." In *Theologische Realenzyklopädie,* edited by Gerhard Möller and Gerhard Krause, 22:215–18. Berlin: W. de Gruyter, 1997.

Kozak, Jan T., and Charles D. James. "Historical Depictions of the 1755 Lisbon Earthquake." National Information Service for Earthquake Engineering, University of California, 1998. Available at http://nisee.berkeley.edu/lisbon/ (accessed December 4, 2005).

Kuehn, Manfred. *Kant: A Biography.* New York: Cambridge University Press, 2001.

Kuhn, Thomas S. *The Structure of Scientific Revolutions.* 2d enlarged ed. Chicago: University of Chicago Press, 1970.

Küppers, Bernd-Olaf. *Information and the Origin of Life.* Cambridge, MA: MIT Press, 1990.

———. *The Molecular Theory of Evolution.* New York: Springer-Verlag, 1983.

LaCugna, Catherine Mowry. *God for Us: The Trinity and Christian Life.* San Francisco: HarperSanFrancisco, 1991.

Lane, Dermot A. *Keeping Hope Alive: Stirrings in Christian Theology.* New York: Paulist Press, 1996.

Lawrence, Miles B., and Hugh B. Cobb. "Tropical Cyclone Report: Hurricane Jeanne, 18–26 September 2004" (revised January 7, 2005). *National Weather Service,* http://www.nhc.noaa.gov/2004jeanne.shtml (accessed November 11, 2005).

Leibniz, Gottfried Wilhelm. *Discourse on Metaphysics and Related Writings.* Translated by R. N. D. Martin and Stuart C. Brown. Manchester: Manchester University Press; distributed exclusively in the USA and Canada by St. Martin's Press, 1988.

———. *Essais de Théodicée sur la bonté de Dieu, la liberté de l'homme et l'origine du mal.* Amsterdam: Isaac Troyel, 1710. English translation by Austin Marsden Farrer: *Theodicy: Essays on the Goodness of God, the Freedom of Man, and the Origin of Evil.* La Salle, IL: Open Court, 1985.

Leslie, John. "How to Draw Conclusions from a Fine-Tuned Universe." In *PPT.*

———. *Universes.* London: Routledge, 1989.

Lewis, C. S. *A Grief Observed.* New York: HarperCollins, 1996.

———. *The Problem of Pain.* New York: Macmillan, 1943.

Loder, James Edwin, and Jim W. Neidhardt. "Barth, Bohr, and Dialectic." In *Religion and Science: History, Method, Dialogue,* edited by W. Mark Richardson and Wesley J. Wildman, 271–90. New York: Routledge, 1996.

Luther, Martin. *Bondage of the Will.* Translated and edited by Ernst F. Winter. New York: Ungar, 1961.

MacCulloch, Diarmaid. *The Reformation: A History.* New York: Penguin, 2003.

Mackie, J. L. "Evil and Omnipotence." [1955.] In *The Problem of Evil,* edited by Marilyn McCord Adams and Robert Merrihew Adams, 25–37. Oxford: Oxford University Press, 1990.

"Mama Dolphins Teach Their Babies." *Animals in Translation* website, dated June 8, 2005, http://animalsintranslation.blogspot.com/2005/06/mama-dolphins-teach-their-babies.html (accessed August 5, 2005).

"Map 12: The Angevin Empire." In *The New Cambridge Medieval History,* vol. IV C. 1024-C. 1198, part II, edited by David Luscombe and Jonathan Riley-Smith, 551. Cambridge: Cambridge University Press, 1995.

Marquard, Odo. *Abschied vom prinzipiellen: Philosophische Studien.* Stuttgart: Reclam, 1981.

Mason, Richard. *Before Logic.* Albany, NY: SUNY Press, 2000.

Maturana, Humberto, and Francisco Varela. *The Tree of Knowledge: The Biological Roots of Human Understanding.* Translated by Robert Paolucci. Rev. ed. New York: Random House, 1992.

Maudlin, Tim. *Quantum Non-Locality and Relativity: Metaphysical Intimations of Modern Physics.* 2d ed. Oxford: Blackwell, 2002.

McDaniel, Jay. *Of God and Pelicans: A Theology of Reverence for Life.* Louisville, KY: Westminster John Knox, 1989.

McDermott, John M. *The Bible on Human Suffering.* Middlegreen, Slough, UK: St Paul Publications, 1990.

McFague, Sallie. *The Body of God: An Ecological Theology.* Minneapolis: Fortress, 1993.

————. *Life Abundant: Rethinking Theology and Economy for a Planet in Peril.* Minneapolis: Fortress, 2000.

————. *Metaphorical Theology: Models of God in Religious Language.* Philadelphia: Fortress, 1982.

McGrath, Alister E. *Christian Theology: An Introduction.* 3d ed. Oxford; Malden, MA: Blackwell, 2001.

Mercer, Christia. *Leibniz's Metaphysics: Its Origins and Development.* Cambridge; New York: Cambridge University Press, 2001.

Metz, Johann Baptist. *Faith in History and Society: Towards a Practical Fundamental Theology.* London: Burns and Oates, 1980.

Milton, John. *Paradise Lost and Paradise Regained.* Edited by Christopher Ricks. New York: New American Library, 1982.

Mitchell, Stephen, trans. *The Book of Job.* San Francisco: North Point Press, 1987.

Moltmann, Jürgen. *The Trinity and the Kingdom: The Doctrine of God.* Translated by Margaret Kohl. New York: Harper & Row, 1981.

Mooney, Chris. "Beware 'Sound Science' It's Doublespeak for Trouble." *Washington Post,* February 29, 2004, B02.

Morris, Brian. *Anthropological Studies of Religion: An Introductory Text.* Cambridge; New York: Cambridge University Press, 1987.

Morris, Charles R. "Economic Injustice for Most: From the New Deal to the Raw Deal." *Commonweal* (2004): 12–17.

Mueller, Gustav E. "The Hegel Legend of 'Thesis-Antithesis-Synthesis.'" *Journal of the History of Ideas* 19, no. 3 (1958): 411.

Murdoch, Iris. "Vision and Choice in Morality." In *Christian Ethics and Contemporary Philosophy,* edited by Ian T. Ramsey, 195–218. New York: Macmillan, 1966.

Murphree, Wallace A. "Can Theism Survive Without the Devil?" *Religious Studies* 21 (June 1985): 231–44.

Murphy, Nancey. *Anglo-American Postmodernity: Philosophical Perspectives on Science, Religion, and Ethics.* Boulder, CO: Westview Press, 1997.

————. "Divine Action in the Natural Order: Buridan's Ass and Schrödinger's Cat." In *CC.*

————. "Emergence and Mental Causation." In *The Re-Emergence of Emergence,* edited by Philip Clayton and Paul Davies. Oxford: Oxford University Press, 2006.

Murphy, Nancey, and Warren S. Brown. *Did My Neurons Make Me Do It? Philosophical and Neurobiological Perspectives on Moral Responsibility and Free Will.* Oxford: Oxford University Press, 2007.

Murphy, Nancey, and George F. R. Ellis. *On the Moral Nature of the Universe: Theology, Cosmology, and Ethics.* Theology and the Sciences Series. Minneapolis: Fortress, 1996.

Neiman, Susan. *Evil in Modern Thought: An Alternative History of Philosophy.* Princeton, NJ: Princeton University Press, 2002.

Neuhaus, Gerd. *Theodizee Abbruch oder Anstoss des Glaubens.* Freiburg: Herder, 1993.

"The NIST Reference on Constants, Units, and Uncertainty: Fundamental Physical Constants." http://physics.nist.gov/cuu/Constants/introduction. html, reproduced with permission from *Encyclopaedia Britannica,* 15th ed., Encyclopaedia Britannica, Inc., 1974.

O'Flaherty, Wendy Doniger. *The Origins of Evil in Hindu Mythology.* Hermeneutics, Studies in the History of Religions 6. Berkeley: University of California Press, 1976.

O'Murchu, Diarmuid. *Quantum Theology: Spiritual Implications of the New Physics.* New York: Crossroad, 1997.

Oelmüller, Willi. *Die unbefriedigte Aufklärung: Beiträge zu einer Theorie der Moderne von Lessing, Kant und Hegel.* Frankfurt am Main: Suhrkamp, 1969.

"One Man's Pain May Be Another Woman's Agony." *The Dominion,* March 8, 2000, http://flatrock.org.nz/topics/science/dont_wilt_have_a_pill.htm (accessed August 5, 2005).

Page, Ruth. *God and the Web of Creation.* London: SCM Press, 1996.

Pannenberg, Wolfhart. *Jesus—God and Man.* Translated by Lewis Wilkins and Duane Priebe. 2d ed. Philadelphia: Westminster Press, 1977.

———. *Systematic Theology.* Vols. 1–2. Translated by G. W. Bromiley. Grand Rapids, MI: Eerdmans, 1991, 1994.

Pargament, Kenneth I., et al. "God Help Me (I): Religious Coping Efforts as Predictors of the Outcomes to Significant Negative Life Events." *American Journal of Community Psychology* 18, no. 6 (1990): 793–824.

Pargament, Kenneth I. "God Help Me (II): The Relationship of Religious Orientations to Religious Coping with Negative Life Events." *Journal for the Scientific Study of Religion* 31, no. 4 (1992): 504–13.

Pargament, Kenneth I., et al. "Red Flags and Religious Coping: Identifying Some Religious Warning Signs among People in Crisis." *Journal of Clinical Psychology* 59, no. 12 (2003): 1335–48.

Pargament, Kenneth I., et al. "Religion and HIV: A Review of the Literature and Clinical Implications." *Southern Medical Journal* 97, no. 12 (2004): 1201.

Pargament, Kenneth I., Harold G. Koenig, Nalini Tarakeshwar, and June Hahn. "Religious Coping Methods as Predictors of Psychological, Physical and Spiritual Outcomes among Medically Ill Elderly Patients: A Two-year Longitudinal Study." *Journal of Health Psychology* 9, no. 6 (2004): 713–30.

———. "Religious Struggle as a Predictor of Mortality among Medically Ill Elderly Patients: A 2-Year Longitudinal Study." *Archives of Internal Medicine* 161, no. 15 (2001): 1881–85.

Peacocke, Arthur R. "Biological Evolution: A Positive 'Theological Appraisal.'" In *EMB,* 357–76.

———. *Creation and the World of Science: The Bampton Lectures, 1978.* Oxford: Clarendon, 1979.

———. "God's Interaction with the World: The Implications of Deterministic 'Chaos' and of Interconnected and Interdependent Complexity." In *CC,* 263–87.

————. *An Introduction to the Physical Chemistry of Biological Organization.* Oxford: Oxford University Press, 1983.

————. *Theology for a Scientific Age.* Revised and expanded ed. London: SCM Press, 1993.

Peirce, Charles S. *The Collected Papers of Charles Sanders Peirce.* Edited by Charles Hartshorne and Paul Weiss. 8 vols. Cambridge: Harvard University Press, 1931–1958.

————. "The Fixation of Belief." First published in *Popular Science Monthly,* November 12, 1877, 1–15. Reprinted in *The Essential Peirce: Selected Philosophical Writings.* Vol. 1: *1867–1893,* edited by Nathan Houser and Christian Kloesel, 109–23. Bloomington: Indiana University Press, 1992.

Penrose, Roger. *The Emperor's New Mind.* Oxford: Oxford University Press, 1989.

————. *The Road to Reality: A Complete Guide to the Laws of the Universe.* New York: Alfred A. Knopf, 2005.

Peters, Ted. *God as Trinity: Relationality and Temporality in the Divine Life.* Louisville, KY: Westminster John Knox, 1993.

Peters, Ted, and Martinez Hewlett. *Evolution from Creation to New Creation: Conflict, Conversation and Convergence.* Nashville, TN: Abingdon, 2003.

Petersen, Aage. *Quantum Physics and the Philosophical Tradition.* New York: Yeshiva University and Belfer Graduate School of Science; Cambridge, MA: MIT Press, 1968.

Petroski, Henry. "Past and Future Failures." *American Scientist* 92, no. 6 (2004): 500–504.

Peukert, Helmut. *Wissenschaftstheorie, Handlungstheorie, fundamentale Theologie: Analysen zu Ansatz und Status theologischer Theoriebildung.* Frankfurt am Main: Suhrkamp, 1978.

Phillips, D. Z. *Introducing Philosophy: The Challenge of Scepticism.* Oxford, UK and Cambridge, MA: Blackwell, 1996.

————. "The Problem of Evil." In *Reason and Religion,* edited by S. C. Brown. Ithaca, NY: Cornell University Press, 1977.

————. *The Problem of God and the Problem of Evil.* London: SCM Press, 2004; Minneapolis: Fortress, 2005.

————. *Wittgenstein and Religion.* New York: St Martin's Press, 1993.

Phillips, Russell E., III, Kenneth I. Pargament, Quinten K. Lynn, and Craig D. Crossley. "Self-Directing Religious Coping: A Deistic God, Abandoning God, or No God at All?" *Journal for the Scientific Study of Religion* 43, no. 3 (2004): 409–18.

Pike, Nelson. "Hume on Evil." [1963.] In *The Problem of Evil,* edited by Marilyn McCord Adams and Robert Merrihew Adams, 38–52. Oxford: Oxford University Press, 1990.

Pinches, Charles. *Theology and Action: After Theory in Christian Ethics.* Grand Rapids, MI: Eerdmans, 2002.

Plantinga, Alvin. "The Free Will Defense." In *Philosophy of Religion: Selected Readings,* edited by William L. Rowe and William J. Wainwright, 3d ed., 259–83. New York: Oxford, 1998.

————. "The Free Will Defense." In *Readings in the Philosophy of Religion: An Analytic Approach,* 2d ed., edited by Baruch A. Brody, 292–304. Engelwood Cliffs, NJ: Prentice-Hall, 1992.

————. "God, Evil and the Metaphysics of Freedom." In *The Problem of Evil,* edited by Marilyn McCord Adams and Robert Merrihew Adams, 83–109. Oxford: Oxford University Press, 1990.

————. *God, Freedom and Evil.* New York: Harper & Row, 1974.

————. *The Nature of Necessity.* Clarendon Library of Logic and Necessity. Oxford: Clarendon, 1982.

————. "The Probabilistic Argument from Evil." *Philosophical Studies* 26 (1979): 1–53.

————. *Warranted Christian Belief.* New York: Oxford University Press, 2000.

Polkinghorne, John. *Belief in God in an Age of Science.* New Haven, CT: Yale University Press, 1998.

————. "Eschatology: Some Questions and Some Insights from Science." In *The End of the World and the Ends of God: Science and Theology on Eschatology,* edited by John Polkinghorne and Michael Welker, 29–41. Harrisburg, PA: Trinity Press International, 2000.

————. "Eschatological Credibility: Emergent and Teleological Processes." In *Resurrection: Theological and Scientific Assessments,* edited by Ted Peters, Robert John Russell, and Michael Welker, 43–55. Grand Rapids, MI: Eerdmans, 2002.

————. *The Faith of a Physicist: Reflections of a Bottom-Up Thinker.* Theology and the Sciences Series. Minneapolis: Fortress, 1994. Published in the UK as *Science and Christian Belief: Reflections of a Bottom-Up Thinker.* London: SPCK, 1994.

————. *Reason and Reality: The Relationship Between Science and Theology.* London: SPCK, 1991.

————. *Science and Providence.* London: SPCK, 1989.

————. *Science and Theology: An Introduction.* London: SPCK, 1998.

Polkinghorne, John, ed. *The Work of Love: Creation as Kenosis.* London: SPCK; Grand Rapids, MI and Cambridge, UK: Eerdmans, 2001.

Power, Will. "The Splendor of Divine Perfection." Paper delivered at the Society for Philosophy of Religion, Hilton Head, South Carolina, February 2001.

Powers, Daniel G. *Salvation through Participation: An Examination of the Notion of the Believers Corporate Unity with Christ in Early Christian Soteriology.* Contributions to Biblical Exegesis and Theology 29. Leuven; Stirling, VA: Peeters, 2001.

Price, Robert M. "Illness Theodicies in the New Testament." *Journal of Religion and Health* 25, no. 4 (1986): 309–15.

Quirk, Michael J. Review of *Theology and the Problem of Evil* by Kenneth Surin. *Theology Today* 44, no. 3 (1987): 405–8.

Rahner, Karl. "Book of God—Book of Human Beings." In *Theological Investigations.* 23 vols. (London: Darton, Longman & Todd/New York: Seabury Press, 1974–1992 [imprint varies for some volumes]), 22:214–24. (*Theological Investigations* hereafter *TI.*)

————. "Christology in the Setting of Modern Man's Understanding of Himself and of His World." *TI* 11:215–29.

————. "Christology within an Evolutionary View of the World." *TI* 5:157–92.

————. *The Content of Faith: The Best of Karl Rahner's Theological Writings.* Edited by Karl Lehmann, Albert Raffelt, and Harvey D. Egan. New York: Crossroad, 1993.

————. "Dogmatic Questions on Easter." *TI* 4:121–33.

————. *Foundations of Christian Faith: An Introduction to the Idea of Christianity.* New York: Seabury Press, 1978.

————. "A Fragmentary Aspect of a Theological Evaluation of the Concept of the Future." *TI* 10:235–42.

————. "The Hermeneutics of Eschatological Assertions." *TI* 4:323–46.

————. "Hidden Victory." *TI* 7:151–58.

————. *Hominisation: The Evolutionary Origin of Man as a Theological Problem.* London: Burns and Oates, 1965.

————. "Immanent and Transcendent Consummation of the World." *TI* 10:273–89.

————. "Marxist Utopia and the Christian Future of Man." *TI* 6:59–88.

————. "Natural Science and Christian Faith." *TI* 21:16–55.

————. "The Question of the Future." *TI* 12:181–201.

————. "Resurrection: D. Theology." In *Encyclopedia of Theology: A Concise Sacramentum Mundi,* edited by Karl Rahner, 1440–42. London: Burns and Oates, 1975.

————. "The Specific Character of the Christian Concept of God." *TI* 21:185–95.

————. "The Theological Problems Entailed in the Idea of the 'New Earth.'" *TI* 10:260–72.

Rahner, Karl, and Karl-Heinz Weger. *Our Christian Faith.* London: Burns and Oates, 1980.

Rahula, Walpola. *What the Buddha Taught.* Rev. ed. New York: Grove Press, 1974.

Redhead, Michael L. G. *Incompleteness, Nonlocality, and Realism: A Prolegomenon to the Philosophy of Quantum Mechanics.* Oxford: Clarendon, 1987.

Rees, Martin. *Before the Beginning: Our Universe and Others.* London: Touchstone, 1997.

————. *Our Cosmic Habitat.* London: Phoenix, 2002.

Regan, Tom. *The Case for Animal Rights.* Berkeley: University of California Press, 1983.

Rhees, Rush. *Without Answers.* London: Routledge & Kegan Paul, 1969.

Ricœur, Paul. *Le mal: Un défi à la philosophie et à la théologie.* Autres temps 5. Genève: Labor et Fides, 1996.

Robinson, I. S. "The Papacy, 1122–1198." In *The New Cambridge Medieval History,* vol. IV C. 1024-C. 1198, part II, edited by David Luscombe and Jonathan Riley-Smith, 317–83. Cambridge: Cambridge University Press, 1995.

Roche, Mark William. *Tragedy and Comedy: A Systematic Study and a Critique of Hegel.* SUNY Series in Hegelian Studies. Albany, NY: SUNY Press, 1998.

Rodin, R. Scott. *Evil and Theodicy in the Theology of Karl Barth.* Edited by Paul G. Molnar. Issues in Systematic Theology 3. New York: P. Lang, 1997.

Rolston, Holmes, III. *Environmental Ethics: Duties to and Values in the Natural World.* Philadelphia: Temple University Press, 1988.

————. "Kenosis and Nature." In *The Work of Love,* edited by J. Polkinghorne, 43–65. London: SPCK; Grand Rapids, MI and Cambridge, UK: Eerdmans, 2001.

————. "Naturalizing and Systematizing Evil." In *Is Nature Ever Wrong, Evil, or Ugly?* edited by Willem B. Drees, 67–86. London and New York: Routledge, 2003.

————. *Science and Religion: A Critical Survey.* New York: Random House, 1987.

Rosenau, Hartmut. "Theodizee IV – Dogmatisch." In *Theologische Realenzyklopädie,* edited by Gerhard Möller and Gerhard Krause, 22:222–29. Berlin: W. de Gruyter, 1997.

Rossi, Philip. "Kant's Philosophical Development." In *Stanford Encyclopedia of Philosophy,* edited by Edward N. Zalta. The Metaphysics Research Lab, 2005. Available at http://plato.stanford.edu/entries/kant-religion/#3.6 (accessed January 2, 2006).

Rozental, I. L. "Physical Laws and the Numerical Values of Fundamental Constants." *Soviet Physics: Uspekhi* 23, no. 6 (June 1980): 293–305; reprinted 1981 American Institute of Physics.

Ruse, Michael. *Can a Darwinian Be a Christian? The Relationship Between Science and Religion.* Cambridge: Cambridge University Press, 2001.

Russell, Robert John. "Bodily Resurrection, Eschatology and Scientific Cosmology: The Mutual Interaction of Christian Theology and Science." In *Resurrection: Theological and Scientific Assessments,* edited by Ted Peters, Robert John Russell, and Michael Welker, 3–30. Grand Rapids, MI: Eerdmans, 2002.

————. "Cosmology and Eschatology: The Implications of Tipler's 'Omega-Point' Theory for Pannenberg's Theological Program." In *Beginning with the End: God, Science, and Wolfhart Pannenberg.* Edited by Carol Rausch Albright and Joel Haugen. Chicago: Open Court, 1997.

————. "Cosmology, Creation, and Contingency." In *Cosmos as Creation: Theology and Science in Consonance,* edited by Ted Peters, 177–210. Nashville, TN: Abingdon, 1989.

————. "Divine Action and Quantum Mechanics: A Fresh Assessment." in *QM.*

————. "Entropy and Evil." *Zygon* 19, no. 4 (December 1984): 449–68.

————. "Eschatology and Physical Cosmology: A Preliminary Reflection." In *The Far Future: Eschatology from a Cosmic Perspective,* edited by George F. R. Ellis, 266–315. Philadelphia: Templeton Foundation Press, 2002.

————. "Finite Creation without a Beginning: The Doctrine of Creation in Relation to Big Bang and Quantum Cosmologies." In *QCLN.*

————. Introduction to *QM.*

————. "Natural Theodicy in an Evolutionary Context: The Need for an Eschatology of New Creation." In *Theology and Eschatology,* edited by David Neville and Bruce Barber, Task of Theology Today 5, 121–52. Hindmarsh, Australia: ATF Press, 2005.

————. "Religion and Peace in the Nuclear Age." In *ORC Conference Proceedings,* edited by Ryusei Takeda. Kyoto, Japan: Ryokoku University, 2005.

————. "Special Providence and Genetic Mutation: A New Defense of Theistic Evolution." In *EMB.*

————. "The Theological Consequences of the Thermodynamics of a Moral Universe." *CTNS Bulletin* 18, no. 4 (Fall 1998): 19–24.

Russell, Robert John, Nancey Murphy, and William R. Stoeger, SJ., eds. *Scientific Perspectives on Divine Action: Twenty Years of Problems and Progress.* Vatican City State: Vatican Observatory; Berkeley, CA: Center for Theology and the Natural Sciences, 2007.

Rydstrom-Poulsen, Aage. *The Gracious God: Gratia in Augustine and the Twelfth Century.* Copenhagen: Akademisk Forlag, 2002.

Sakurai, J. J., and San Fu Tuan. *Modern Quantum Mechanics.* Rev. ed. Reading, MA: Addison-Wesley, 1995.

Sanders, E. P. *Paul and Palestinian Judaism: A Comparison of Patterns of Religion.* Philadelphia: Fortress, 1977.

Satchell, Michael. "Wiping Out TB and Aids." *U.S. News & World Report,* October 31, 2005. http://www.usnews.com/usnews/news/articles/051031/31farmer.htm (accessed 15 Nov, 2005).

Schillebeeckx, Edward. *Christ: The Christian Experience in the Modern World.* London: SCM Press, 1980.

———. *Church: The Human Story of God.* New York: Crossroad, 1990.

Schneider, John. "Seeing God Where the Wild Things Are: An Essay on the Defeat of Horrendous Evil." In *Christian Faith and the Problem of Evil,* edited by Peter van Inwagen, 226–62. Grand Rapids, MI and Cambridge, UK: Eerdmans, 2004.

Schönfeld, Martin. "Kant's Philosophical Development." In *Stanford Encyclopedia of Philosophy,* edited by Edward N. Zalta. The Metaphysics Research Lab, 2003. Available at http://plato.stanford.edu/entries/kant-development/ (accessed January 2, 2006).

———. *The Philosophy of the Young Kant: The Precritical Project.* Oxford; New York: Oxford University Press, 2000.

Schrödinger, Erwin. "Discussion of Probability Relations between Separated Systems." *Proceedings of the Cambridge Philosophical Society* 31 (1935): 555–63.

Schumacher, Thomas. *Theodizee: Bedeutung und Anspruch eines Begriffs.* Frankfurt am Main; New York: P. Lang, 1994.

Schuon, Frithjof. *Understanding Islam.* Bloomington, IN: World Wisdom Books, 1998.

Sideris, Lisa. *Environmental Ethics, Ecological Theology and Natural Selection.* New York: Columbia University Press, 2003.

Silberstein, Michael. "Reduction, Emergence, and Explanation." In *The Blackwell Guide to the Philosophy of Science,* edited by Peter Machamer and Michael Silberstein, 80–107. Oxford: Blackwell, 2002.

Silcox, Mark. "Mind and Anomalous Monism." In *The Internet Encyclopedia of Philosophy,* http://www.iep.utm.edu/m/anom-mon.htm. Verified August 12, 2005.

Singer, Peter. *Animal Liberation: A New Ethics for Our Treatment of Animals.* New York: Random House, 1975.

Southgate, Christopher. "God and Evolutionary Evil: Theodicy in the Light of Darwinism." *Zygon* 37, no. 4 (2002): 803–21.

Sparn, Walter. *Leiden, Erfahrung und Denken: Materialien zum Theodizeeproblem.* Theologische Bücherei 67. Munich: Kaiser, 1980.

Stoeger, William R., SJ. "Conceiving Divine Action in a Dynamic Universe." In *Scientific Perspectives on Divine Action: Twenty Years of Problems and Progress,* edited by Robert John Russell, Nancey Murphy, and

William R. Stoeger, SJ. Vatican City State: Vatican Observatory; Berkeley, CA: Center for Theology and the Natural Sciences, 2007.

————. "Divine Action in a Broken World." In *Science and Religion . . . and Culture in the Jesuit Tradition: Perspectives from East Asia,* edited by Jose Mario C. Francisco, SJ, and Roman Miguel G. de Jesus, 7–22. Adelaide, Australia: ATF Press, 2006.

————. "The Emergence of Novelty in the Universe and Divine Action." *Omega* 2, no. 2 (December 2003): 25–50.

————. "Epistemological and Ontological Issues Arising from Quantum Theory." In *QM,* 81–98.

————. "The Mind-Brain Problem, the Laws of Nature, and Constitutive Relationships." In *NP,* 129–46.

————. "The Problem of Evil: The Context of a Resolution." In *Can Nature Be Evil or Evil Natural? A Science-and-Religion View on Suffering and Evil,* Proceedings of the 12th Seminar of the South African Science and Religion Forum (SASRF), edited by Cornel W. Toit, 1–15. Pretoria: Research Institute for Theology and Religion, University of South Africa, 2006.

Stoeger, William R., George F. R. Ellis, and Ulrich Kirchner. "Multiverses and Cosmology: Philosophical Issues." arXiv.org, Cornell University Library's e-prints in Astrophysics, July 2004, http://arxiv.org/abs/astro-ph/0407329 (accessed May 16, 2006).

Stout, Jeffrey. *The Flight from Authority: Religion, Morality and the Quest for Autonomy.* Notre Dame, IN: University of Notre Dame Press, 1981.

Strobel, Lee. *The Case for a Creator: A Journalist Investigates Scientific Evidence that Points toward God.* Grand Rapids, MI: Zondervan, 2004.

"Study: Ozone Layer Has Stopped Shrinking." *CNN.com,* August 31, 2005, http://www.cnn.com.

Sullivan, Roger J. *Immanuel Kant's Moral Theory.* Cambridge; New York: Cambridge University Press, 1989.

Surin, Kenneth. "The Problem of Evil." In *The Blackwell Encyclopedia of Modern Christian Thought,* edited by Alister E. McGrath, 192–99. Oxford, UK; Cambridge, MA: Blackwell, 1993.

————. *Theology and the Problem of Evil.* Signposts in Theology. Oxford; New York: Basil Blackwell, 1986.

Swinburne, Richard. *The Existence of God.* Oxford: Clarendon, 1979.

————. "The Problem of Evil." In *Reason and Religion,* edited by S. C. Brown. Ithaca, NY: Cornell University Press, 1977.

"Table of Physical Constants." SearchSMB.com Definitions website, http://searchsmb.techtarget.com/sDefinition/0,,sid44_gci860994,00.html (updated September 9, 2003).

"Take Two Aspirin for Re-Leaf." Reuters Online, August 5, 1999, http://flatrock.org.nz/topics/science/dont_wilt_have_a_pill.htm (accessed August 5, 2005).

Taylor, Jerome. Introduction to *Didascalicon.* New York and London: Columbia University Press, 1961.

Taylor, Paul. *Respect for Nature: A Theory of Environmental Ethics.* Princeton, NJ: Princeton University Press, 1986.

Tegmark, Max. "Is 'the Theory of Everything' Merely the Ultimate Ensemble Theory?" *Annals of Physics* 270 (1998).

Thomas, R. S. "Which." In *Laboratories of the Spirit,* 54. London: Macmillan, 1975.

Tilley, Terence W. *The Evils of Theodicy.* Washington, DC: Georgetown University Press, 1991; reprint, Spokane, WA: Wipf and Stock, 2000.

——. *History, Theology, and Faith: Dissolving the Modern Problematic.* Maryknoll, NY: Orbis, 2004.

——. *Inventing Catholic Tradition.* Maryknoll, NY: Orbis Books, 2000.

——. "The Philosophy of Religion and the Concept of Religion: D. Z. Phillips on Religion and Superstition." *Journal of the American Academy of Religion* 68, no. 2 (June 2000): 345–56.

——. "'Superstition' as a Philosopher's Gloss on Practice: A Rejoinder to D. Z. Phillips." *Journal of the American Academy of Religion* 68, no. 2 (June 2000): 363–65.

——. "The Use and Abuse of Theodicy." *Horizons* 11, no. 2 (Fall 1984): 304–19.

——. *The Wisdom of Religious Commitment.* Washington, DC: Georgetown University Press, 1995.

Tillich, Paul. *Systematic Theology.* Vol. 2. [1957.] London: James Nisbet, 1964.

Tipler, Frank. "The Omega Point Theory: A Model of an Evolving God." In *PPT.*

Tittel, Wolfgang, Jürgen Brendel, Bernard Gisin, Thomas Herzog, Hugo Zbinden, and Nicolas Gisin. "Experimental Demonstration of Quantum Correlations over More Than 10 Km." *Physical Review A* 57, no. 5 (1998): 3229–32.

Torgerson, Justin Roald, David A. Branning, Carlos H. Monken, and Leonard Mandel. "Experimental Demonstration of the Violation of Local Realism without Inequalities." *Physics Letters A* 205 (1995): 323–28.

Tracy, David. "The Hidden God: The Divine Other of Liberation." *Cross Currents* 46, no. 1 (1996): 5–16.

Tracy, Thomas F. "Evil, Human Freedom, and Divine Grace." In *Human and Divine Agency,* edited by F. Michael McLain and W. Mark Richardson. Lanham, MD: University Press of America, 1999.

——. "Evolution, Divine Action and the Problem of Evil." In *EMB,* 511–30.

——. "Victimization and the Problem of Evil." *Faith and Philosophy* 9 (1992): 301–19.

Tracy, Thomas F., ed. *The God Who Acts: Philosophical and Theological Explorations.* University Park: Pennsylvania State University Press, 1994.

Trenberth, Kevin. "Uncertainty in Hurricanes and Global Warming." *Science* 308 (2005): 1753–54.

US Geological Survey. "Earthquakes Facts and Lists: Earthquakes with 1,000 or More Deaths since 1900," updated June 7, 2005, http://neic.usgs.gov/neis/eqlists/ eqsmajr.html (accessed November 15, 2005).

Vardy, Peter. *The Puzzle of Evil.* London: Fount, 1992.

Vaught, Carl G. *The Quest for Wholeness.* Albany, NY: SUNY Press, 1982.

Vermeer, Paul. *Learning Theodicy: The Problem of Evil and the Praxis of*

Religious Education. Empirical Studies in Theology 3. Leiden; Boston: Brill, 1999.

Vesilind, P. Aarne. "How to Lie with Engineering Graphics." *Chemical Engineering Education* 33, no. 4 (1999): 304–9.

Voltaire [François Marie Arouet de]. *Candide, ou l'Optimisme.* 1759.

———. *Philosophical Dictionary.* Translated by Theodore Besterman. Harmondsworth: Penguin Books, 1984.

Wainwright, William J. *Philosophy of Religion.* Belmont, CA: Wadsworth, 1988.

Ward, Keith. *God, Chance and Necessity.* Oxford: Oneworld, 1996.

Warren, Henry Clarke. *Buddhism in Translations.* Cambridge: Harvard University Press, 1979.

Watt, W. Montgomery. *Free Will and Predestination in Early Islam.* London: Luzac, 1948.

Watt, W. Montgomery, and Michael Marmura. *Politische Entwicklungen und theologische Konzepte.* Vol. 2 of *Der Islam.* Edited by W. Montgomery Watt. Stuttgart: Kohlhammer, 1985.

Webb, J. K., et al. "Further Evidence for Cosmological Evolution of the Fine Structure Constant." *Phys. Rev. Lett.* 87, 091301 (2001).

Weber, Bruce H., et al. *Entropy, Information, and Evolution.* Cambridge, MA: MIT Press, 1990 [2d printing].

Weber, Max. *Economy and Society: An Outline of Interpretive Sociology.* Berkeley: University of California Press, 1978.

———. *The Sociology of Religion.* Translated by Ephraim Fischoff. Boston: Beacon Press, 1993.

Wegter-McNelly, Kirk. "Atoms May Be Small, But They're Everywhere: Robert Russell's Theological Engagement with the Quantum Revolution." In *God's Action in Nature's World: Essays in Honour of Robert John Russell,* edited by Ted Peters and Nathan Hallanger, 93–112. Burlington, VT: Ashgate, 2006.

———. "The World, Entanglement, and God: Quantum Theory and the Christian Doctrine of Creation." PhD dissertation, Graduate Theological Union, 2003.

Weihs, Gregor, Thomas Jennewein, Christoph Simon, Harald Weinfurter, and Anton Zeilinger. "Violation of Bell's Inequality under Strict Einstein Locality Conditions." *Physical Review Letters* 81, no. 23 (1998): 5039–43.

Weil, Simone. *Gravity and Grace.* [1952.] London: Routledge, 1992.

Westerholm, Stephen. *Perspectives Old and New on Paul: The "Lutheran" Paul and His Critics.* Grand Rapids, MI: Eerdmans, 2004.

Whitehead, Alfred North. *Process and Reality: An Essay in Cosmology.* New York: Macmillan, 1929. Corrected ed., edited by David Ray Griffin and Donald W. Sherburne. New York and London: The Free Press, 1978.

Wildman, Wesley. "A Review and Critique of the 'Divine Action Project': A Dialogue among Scientists and Theologians, Sponsored by Pope John Paul II." In *Philosophy and Theology in the New Millennium,* edited by Fr. Val. A. McInnes, OP. Tulane: Tulane University Press, 2006.

Will, Clifford. *Was Einstein Right? Putting General Relativity to the Test.* New York: Basic Books, 1986.

Williams, Stephen N. *Revelation and Reconciliation: A Window on Modernity.* Cambridge; New York: Cambridge University Press, 1995.

Wilson, Walter T. *The Hope of Glory: Education and Exhortation in the Epistle to the Colossians.* Supplements to Novum Testamentum 88. Leiden; New York: Brill, 1997.

Wittgenstein, Ludwig. "Cause and Effect: Intuitive Awareness." In *Philosophical Occasions, 1912–1952,* edited by James C. Klagge and Alfred Nordmann, 370–426. Indianapolis; Cambridge, UK: Hackett, 1993.

———. *Culture and Value.* Translated by Peter Winch. Edited by G. H. von Wright and Heikki Nyman. English translation with the amended 2d ed. Oxford: Basil Blackwell, 1980.

———. *Philosophical Investigations.* Translated by G. E. M. Anscombe. Edited by G. E. M. Anscombe and Rush Rhees. New York: Macmillan, 1953.

———. *Remarks on the Philosophy of Psychology.* 2 Vols. Translated by C. G Luckhardt and M. A. E. Aue. Edited G. H. von Wright and Heikki Nyman. Chicago: University of Chicago Press, 1980.

———. *Tractatus Logico-Philosophicus.* Translated by D. F. Pears and B. F. McGuinness. London: Routledge & Kegan Paul, 1961; first English edition 1922.

———. *Zettel.* Translated by G. E. M. Anscombe. Edited by G. E. M. Anscombe and G. H. von Wright. Berkeley and Los Angeles: University of California Press, 1970.

Woodward, Kenneth L., with Sudip Mazumdar. "Countless Souls Cry Out to God." *Newsweek,* January 10, 2005, 37.

Wykstra, Stephen J. "The Humean Obstacle to Evidential Arguments from Suffering: On Avoiding the Evils of 'Appearance.'" [1984.] In *The Problem of Evil,* edited by Marilyn McCord Adams and Robert Merrihew Adams, 138–60. Oxford: Oxford University Press, 1990.

Yinger, Kent L. *Paul, Judaism, and Judgment according to Deeds.* Cambridge; New York: Cambridge University Press, 1999.

Zbinden, Hugo, J. Brendel, Nicolas Gisin, and Wolfgang Tittel. "Experimental Test of Nonlocal Quantum Correlation in Relativistic Configurations." *Physical Review A* 63, no. 2 (2001): 1–10.

CONTRIBUTORS

Philip Clayton, Ingraham Professor, Claremont School of Theology, and Professor of Philosophy and of Religion, Claremont Graduate University, Claremont, California, USA

Denis Edwards, Senior Lecturer in Theology, Flinders University, Adelaide College of Divinity, and Catholic Theological College, Adelaide, South Australia

Don Howard, Director, Program in History and Philosophy of Science; Reilly Fellow, John J. Reilly Center for Science, Technology, and Values; Professor of Philosophy, University of Notre Dame, Notre Dame, Indiana, USA

Niels Christian Hvidt, Associate Professor, Institute of Public Health, Research Unit of Health, Man and Society, University of Southern Denmark

Brad J. Kallenberg, Associate Professor of Theology, University of Dayton, Dayton, Ohio, USA

Steven Knapp, Professor of English and President, the George Washington University, Washington, DC, USA

Nancey Murphy, Professor of Christian Philosophy, Fuller Theological Seminary, Pasadena, California, USA

Andrew Robinson, Honorary University Fellow in Theology, University of Exeter, Exeter, Devon, UK

Robert John Russell, Founder and Director, The Center for Theology and the Natural Sciences and Ian G. Barbour Professor of Theology and Science in Residence, The Graduate Theological Union, Berkeley, California, USA

Christopher Southgate, Research Fellow in Theology, University of Exeter, Exeter, Devon, UK

William R. Stoeger, SJ, Staff Astrophysicist and Adjunct Associate Professor of Astronomy, Vatican Observatory, Vatican Observatory Research Group, Steward Observatory, University of Arizona, Tucson, Arizona, USA

Terrence W. Tilley, Professor and Chair of the Department of Theology, Fordham University, Bronx, New York, USA

Thomas F. Tracy, Phillips Professor of Religion, Department of Philosophy and Religion, Bates College, Lewiston, Maine, USA

Kirk Wegter-McNelly, Assistant Professor of Theology, Boston University School of Theology, Boston, Massachusetts, USA

Wesley J. Wildman, Associate Professor of Theology and Ethics, Boston University, Boston, Massachusetts, USA

INDEX OF TERMS AND NAMES